Springer Series in
ADVANCED MICROELECTRONICS 1

Springer
*Berlin
Heidelberg
New York
Barcelona
Hong Kong
London
Milan
Paris
Singapore
Tokyo*

Springer Series in
ADVANCED MICROELECTRONICS

The Springer Series in Advanced Microelectronics provides systematic information on all the topics relevant for the design, processing, and manufacturing of microelectronic devices. The books, each prepared by leading researchers or engineers in their fields, cover the basic and advanced aspects of topics such as wafer processing, materials, device design, device technologies, circuit design, VLSI implementation, and subsystem technology. The series forms a bridge between physics and engineering and the volumes will appeal to practicing engineers as well as research scientists.

1 **Cellular Neural Networks**
   Chaos, Complexity and VLSI Processing
   By G. Manganaro, P. Arena, and L. Fortuna
2 **Noise Simulation of Semiconductor**
   Devices in Linear and Nonlinear RF Applications
   By G. Ghione and F. Bonani

G. Manganaro   P. Arena   L. Fortuna

Cellular Neural Networks

Chaos, Complexity and VLSI Processing

With 164 Figures

 Springer

Dr. Gabriele Manganaro
Texas Instruments Inc.
Advanced Analog Products
Data Converter Design Group
P.O. Box 660199, M/S 8755
Dallas TX 75266, USA
e-mail: g-manganaro@ti.com

Dr. Paolo Arena
Professor Luigi Fortuna
Dip. Elettrico, Elettronico e Sist.
V. le A. Doria 6
I-95125 Catania, Italy
e-mail: parena@dees.unict.it
e-mail: lfortuna@dees.unict.it

Series Editors:

Dr. Kiyoo Itoh
Hitachi Ltd., Central Research Laboratory
1-280 Higashi-Koigakubo
Kokubunji-shi
Tokyo 185-8601
Japan

Professor Takayasu Sakurai
Center for Collaborative Research
University of Tokyo
7-22-1 Roppongi, Minato-ku,
Tokyo 106-8558
Japan

ISSN 1437-0387
ISBN 3-540-65202-7 Springer-Verlag Berlin Heidelberg New York

Library of Congress Cataloging-in-Publication Data.

Cellular neural networks: chaos, complexity and VLSI processing / Gabriele Manganaro, Paolo Arena, Luigi Fortuna. p. cm. – (Advanced microelectronics ; v. 1). Includes bibliography references and index. ISBN 3-540-65202-7 (hardcover). 1. Neural networks (Computer science) 2. Chaotic behavior in systems. I. Manganaro, Gabriele, 1969– . II. Arena, Paolo, 1966– . III. Fortuna, L. (Luigi), 1953– . IV. Series. QA76.87.C43 1999 006.3'2–dc21 98-45460

This work is subject to copyright. All rights are reserved, whether the whole or part of the material is concerned, specifically the rights of translation, reprinting, reuse of illustrations, recitation, broadcasting, reproduction on microfilm or in any other way, and storage in data banks. Duplication of this publication or parts thereof is permitted only under the provisions of the German Copyright Law of September 9, 1965, in its current version, and permission for use must always be obtained from Springer-Verlag. Violations are liable for prosecution under the German Copyright Law.

© Springer-Verlag Berlin Heidelberg 1999
Printed in Germany

The use of general descriptive names, registered names, trademarks, etc. in this publication does not imply, even in the absence of a specific statement, that such names are exempt from the relevant protective laws and regulations and therefore free for general use.

Typesetting: Data conversion by Satztechnik Katharina Steingraeber, Heidelberg
Cover concept by eStudio Calmar Steinen using a background picture from Photo Studio "SONO". Courtesy of Mr. Yukio Sono, 3-18-4 Uchi-Kanda, Chiyoda-ku, Tokyo
Cover design: *design & production* GmbH, Heidelberg
Computer-to-plate and printing: Mercedesdruck, Berlin
Binding: Buchbinderei Lüderitz & Bauer, Berlin

SPIN: 10698635   57/3144 - 5 4 3 2 1 0 – Printed on acid-free paper

Foreword

The field of *cellular neural networks* (CNNs) is of growing importance in nonlinear circuits and systems and it is maturing to the point of becoming a new area of study in general nonlinear theory. CNNs emerged through two seminal papers co-authored by Professor Leon O. Chua back in 1988. Since then, the attention that CNNs have attracted in the scientific community has been vast. For instance, there are international workshops dedicated to CNNs and their applications, special issues published in both the *International Journal of Circuit Theory* and in the *IEEE Transactions on Circuits and Systems*, and there are also Associate Editors appointed in the latter journal especially for the CNN field. All of this bears witness the importance that CNNs are gaining within the scientific community.

Without doubt this book is a primer in the field. Its extensive coverage provides the reader with a very comprehensive view of aspects involved in the theory and applications of cellular neural networks. The authors have done an excellent job merging basic CNN theory, synchronization, spatio-temporal phenomena and hardware implementation into eight exquisitely written chapters. Each chapter is thoroughly illustrated with examples and case studies. The result is a book that is not only excellent as a professional reference but also very appealing as a textbook. My view is that students as well professional engineers will find this volume extremely useful.

I find the sequence of chapters to be an excellent and didactical model that ranges from basic theory and applications founded on the Chua and Yang model, through sophisticated analyses of complex chaotic phenomena, to the actual monolithic implementation of a one-dimensional CNN structure. All along this path there is an in-depth coverage of methods to characterize nonlinear dynamics. For example, the authors introduce the reader to the unique concepts that together define *state controlled CNNs* (SC-CNNs) as well as to the synchronization of various such cells through experimental verification. Of additional importance is the treatise on the spatio-temporal behavior of coupled cells yielding Turing patterns and spiral waves. The book finishes with a novel monolithic implementation of one-dimensional discrete-time CNNs. The latter exploits an innovative multiplexing technique to minimize silicon area without speed penalties. These topics certainly reflect the current state-of-the-art in cellular neural networks.

In conclusion, original work and excellent tutorial write up are merged in this volume. The material covered is fundamental and unique and it will positively set a new pace in the field. Practitioners in the field have to understand these concepts and to update their knowledge to remain competitive in the long run. As has happened many times in the past, nonlinear theory of circuits and systems has added a new component to everyday practice: cellular neural networks.

College Station, June 1998 *Jose Pineda de Gyvez*

Preface

Cellular neural networks (CNNs) constitute a family of nonlinear circuits. This family was introduced by Prof. Leon O.Chua around ten years ago and has achieved a growing interest and success in the Circuits and Systems community. This popularity has grown to the point that, in the last two years, any monthly issue of IEEE Transactions on Circuits and Systems - Part I, one the most important journals in the electrical and electronic engineering area, has published at least one or two new papers on CNNs reporting new results, applications and implementations.

But the success of this paradigm has not been limited to electronic circuit specialists. Its infectious charm has spread through a wide range of different disciplines including robotics, system theory, physics, neurophysiology, biology, information processing, just to cite a few. As a consequence, the diversity of topics that can be gathered under this common frame is quite impressive and indeed sometimes bewildering. Dealing with all the many aspects of this subject is nowadays no mean task.

This monograph tries to bring some of these elements to a wide audience of readers, including both beginners and specialists, through the experience of the authors. The content ranges from the highest-level topics, which are treated in a rigorous analytic way, through some of the applications, to the technological challenges and the circuit design aspects. Examples, simulation and experimental results complement theoretical results throughout the book.

Mirroring the current trends, the book has been subdivided into two main parts. The first part, composed of six chapters, is mainly devoted (though not limited) to the circuit-theoretical issues and the applications of CNNs. The second part, organized in two chapters, deals with the design and implementation in silicon of a CNN architecture. Two appendices complement the eight chapters providing some of the mathematical tools and a library of commonly used templates. These appendices can be used both as a quick reference for a deeper understanding of the topics and for future work. A description of the chapters follows.

The fundamental concepts and definitions of the world of cellular neural networks are reviewed in Chap. 1. The original model introduced by L.O.Chua and L.Yang in 1988 is first discussed. Then an overview is given of the most common generalizations in the family tree of CNNs. The chapter ends with a recent broad and formal definition, given by L.O.Chua et al., that simultaneously brings order and clarity in the growing jungle of CNN models.

Chapter 2 discusses some new applications of CNNs to industrial problems. In particular, two different cases in which a CNN can be used to perform pre-filtering of two-dimensional arrays of data are considered. Furthermore some CNN-based models for environmental modeling and simulation are reported.

In Chap. 3 it is shown that a so-called *state-controlled CNN* is able to reproduce the dynamics of a wide class of nonlinear circuits including several well-known chaotic and hyperchaotic ones. The idea that such a circuit can be considered a building block for nonlinear dynamic generation is discussed.

The ability to generate many different dynamics from a single architecture finds immediate application in the field of chaos-based transceivers. Some of the advantages and limits of synchronization of chaotic circuits are examined through an experimental study reported in Chap. 4. Moreover a new method for the identification of the circuit parameters of nonlinear circuits based on the principles of synchronization is presented.

Self-organizing patterns, autowaves and other spatio-temporal phenomena have recently been observed in CNN arrays composed of coupled Chua oscillators and degenerate Chua oscillators. Chapter 5 shows how many of these phenomena can equally be obtained in a traditional and simpler two-layer Chua and Yang CNN model. A thorough analytic study of the fundamental second-order circuit used as cell sets up the basis for generating and understanding the phenomena considered.

Starting from the complex phenomena previously discussed, Chap. 6 presents experimental results obtained in a discrete-component set-up. An extremely interesting application is then presented: mechatronic devices are driven by the spatio-temporal dynamics of the CNN with the goal of mimicking some natural locomotion patterns.

Chapter 7 discusses the design and implementation of a new switched-current multiplier. This analogue block is one of the fundamental components for the realization of a programmable CNN. Specifically it is the component determining the coupling among the CNN cells. A thorough analytical study of the circuit and its non-idealities is followed by the experimental characterization of a fabricated prototype.

In Chap. 8 an analogically programmable 1-D discrete-time CNN is presented. The main blocks of the proposed design are discussed together with the implied technological limitations. One of the most original features of the design is the inherent multiplexing scheme that aims at reaching a highly efficient use of the silicon. Simulation, examples and experimental results for a fabricated CNN chip are included.

Dallas, USA, July 1998 *Gabriele Manganaro*
Catania, Italy, October 1998 *Paolo Arena*
Luigi Fortuna

Acknowledgements

The authors wish to express their appreciation to the many persons that, in different ways, helped them in carrying out the research work presented in this book. A particular acknowledgment goes to Prof. Jose Pineda de Gyvez, Texas A&M University, for his frequent advice and significant support. Without him the second part of this book would never have come into existence. We thank Prof. Leon O. Chua, University of California at Berkeley, for his continuous encouragement.

A word of thanks also goes to Salvo Baglio, University of Catania (Italy), Riccardo Caponetto, Mario Lavorgna and the R&D-Soft Computing Group of SGS-Thomson Microelectronics, and to the many students that have been involved in some of the practical realizations presented in the first part of the book.

Finally we are grateful to our families for their patience and support.

Contents

Part I. Circuit Theory and Applications of CNNs

1. **CNN Basics** 3
   1.1 The CNN of Chua and Yang 3
       1.1.1 The Cell 4
       1.1.2 The CNN Array 5
       1.1.3 More About Templates 7
       1.1.4 Multilayer CNNs 9
       1.1.5 The CNN as an Analog Processor 10
       1.1.6 Some Stability Results 11
   1.2 Main Generalizations 15
       1.2.1 Nonlinear CNNs and Delay CNNs 15
       1.2.2 Nonuniform Processor CNNs
             and Multiple Neighborhood Size CNNs 17
       1.2.3 Discrete-Time CNNs 17
       1.2.4 The CNN Universal Machine 19
   1.3 A Formal Definition 20
       1.3.1 The Cells and Their Coupling 20
       1.3.2 Boundary Conditions 22
   1.4 Summary 23

2. **Some Applications of CNNs** 25
   2.1 CNN-Based Image Pre-processing
       for the Automatic Classification of Fruits 25
       2.1.1 The Pre-filtering 26
   2.2 Processing of NMR Spectra 28
       2.2.1 Two-Dimensional NMR Spectra 29
       2.2.2 Processing of NMR Spectra with CNNs 30
       2.2.3 Description of the Dual Algorithm 32
   2.3 Air Quality Modeling 35
       2.3.1 Models 35
       2.3.2 CNNs for Air Quality Modeling 36
       2.3.3 Examples 39

| | 2.4 | Conclusions | 40 |

3. **The CNN as a Generator of Nonlinear Dynamics** 43
   3.1 The State Controlled CNN Model 44
       3.1.1 Discrete Components Realization of SC-CNN Cells ... 45
   3.2 Chua Oscillator Dynamics Generated by the SC-CNN 47
       3.2.1 Main Result 47
       3.2.2 Experimental Results 49
   3.3 Chaotic Dynamics of a Colpitts Oscillator 51
   3.4 Hysteresis Hyperchaotic Oscillator 55
   3.5 n-Double Scroll Attractors 58
       3.5.1 A New Realization of the n-Double Scroll Family 58
       3.5.2 n-Double Scrolls in SC-CNNs 65
   3.6 Nonlinear Dynamics Potpourri 66
       3.6.1 A Non-autonomous Second Order Chaotic Circuit 66
       3.6.2 A Circuit with a Nonlinear Reactive Element 68
       3.6.3 Canards and Chaos 70
       3.6.4 Multimode Chaos in Coupled Oscillators 72
       3.6.5 Coupled Circuits 74
   3.7 General Case and Conclusions.......................... 75
       3.7.1 Theoretical Implications.......................... 77
       3.7.2 Practical Implications 77

4. **Synchronization** 79
   4.1 Background... 79
       4.1.1 Pecora–Carroll Approach 80
       4.1.2 Inverse System Approach 81
   4.2 Experimental Signal Transmission
       Using Synchronized SC-CNN 81
       4.2.1 Circuit Description 82
       4.2.2 Synchronization: Results of Experiment
             and Simulation.................................. 83
       4.2.3 Non-ideal Channel Effects 86
       4.2.4 Effects of Additive Noise and Disturbances
             on the Channel 87
   4.3 Chaotic System Identification 93
       4.3.1 Description of the Algorithm...................... 93
       4.3.2 Identification of the Chua Oscillator 95
       4.3.3 Examples 96
   4.4 Summary and Conclusions.............................. 101

5. **Spatio-temporal Phenomena** 105
   5.1 Analysis of the Cell 105
       5.1.1 Fixed Points.................................... 107
       5.1.2 Limit Cycle and Bifurcations 109

	5.1.3 Slow–Fast Dynamics 110

        5.1.3 Slow–Fast Dynamics 110
        5.1.4 Some Simulation Results 112
  5.2 The Two-Layer CNN 112
  5.3 Traveling Wavefronts 115
        5.3.1 Autowaves.. 116
        5.3.2 Labyrinths 118
  5.4 Pattern Formation... 120
        5.4.1 Condition for the Existence of Turing Patterns
              in Arrays of Coupled Circuits 121
        5.4.2 Turing Patterns in the Two-Layer CNN 122
        5.4.3 Simulation Results 124
  5.5 Sensitivity to Parametric Uncertainties and Noise........... 127
        5.5.1 Spiral Wave: Parametric Uncertainty................ 127
        5.5.2 Spiral Waves: Presence of Noise
              in the Initial Conditions........................... 129
        5.5.3 Patterns: Parametric Uncertainties 130
  5.6 Summary and Conclusions................................ 132

6. Experimental CNN Setup and Applications to Motion Control 133
  6.1 The Experimental Setup 134
        6.1.1 Realization of the Cell for Autowave Generation...... 135
        6.1.2 Realization of the Cell for Pattern formation 136
        6.1.3 Realization of the Laplacian Couplings
              and Boundary Conditions 138
        6.1.4 Realization of the Main Board 139
        6.1.5 Autowave Experiments............................ 140
  6.2 Pattern Formation and Propagation 142
  6.3 CNNs for Generating and Controlling Artificial Locomotion.. 149
        6.3.1 Links to Biological Locomotion..................... 149
        6.3.2 WORMBOT: A Ring-Worm-like Walking Robot...... 152
        6.3.3 REXABOT: An Hexapode Reaction–Diffusion
              Walking Robot................................... 154
        6.3.4 READIBELT:
              Reaction Diffusion Conveyor Belt Autowave Driven ... 159
  6.4 Conclusion 160

Part II. Implementation and Design

7. A Four Quadrant S^2I Switched-Current Multiplier 167
  7.1 Detailed Analysis of the S^2I Memory Cell.................. 168
  7.2 The Multiplier Architecture................................ 173

7.3 Analysis and Design of the S^2I Multiplier 177
    7.3.1 Circuit Analysis of the Multiplier 177
    7.3.2 Circuit Design 180
7.4 Experimental Performance Evaluation..................... 182
7.5 Summary... 186

8. **A One-Dimensional Discrete-Time CNN Chip for Audio Signal Processing**............................. 189
   8.1 System Architecture 189
   8.2 The Tapped Delay Line 190
   8.3 CNN Cells... 192
      8.3.1 Multiplier and Ancillary Circuitry 193
   8.4 Cell Behavior and Hardware Multiplexing 197
   8.5 Results and Example 199
   8.6 Summary.. 203

A. **Mathematical Background** 205
   A.1 Topology 205
   A.2 Operations and Functions 206
   A.3 Matrices 207
   A.4 Dimension... 207
   A.5 Dynamical Systems: Basic Definitions 208
   A.6 Steady-State Behavior 209
      A.6.1 Classification of Asymptotic Behavior 210
   A.7 Stability.. 212
      A.7.1 Stability of equilibrium points...................... 212
      A.7.2 Stability of Limit Cycles 217
      A.7.3 Lyapunov Exponents 219
   A.8 Topological Equivalence and Conjugacy, Structural Stability and Bifurcations 220
   A.9 Šilnikov Method.. 221
   A.10 Particular Results for Two-Dimensional Flows 222

B. **Library of Templates** 225

References 259

Index 271

Part I

Circuit Theory and Applications of CNNs

1. CNN Basics

The concept of the *Cellular Neural Network* (CNN, also called *Cellular Nonlinear Network*) was introduced in 1988 with the seminal papers of Leon O.Chua and Lin Yang [1–3]. Chua proposed the idea of using an array of, essentially simple, non-linearly coupled dynamic circuits to process large amounts of information in real time [4]. The concept was inspired by the architecture of the *Cellular Automata* [5,6] and the *Neural Networks* [7,8]. Chua showed that his new architecture was able to efficiently perform time-consuming tasks, such as image processing and partial differential equation solution, and it was suitable VLSI implementation.

The new architecture rapidly attracted the attention of the scientific community, especially from the circuit theory and analog VLSI area. Indeed, a vast number of applications were proposed in just a couple of years as were the first IC implementations [9].

This tremendous interest has been witnessed by the publication of several special issues of international scientific journals (*International Journal of Circuit Theory and Applications* [10,11], *IEEE Transactions on Circuits and Systems* [12–14]) as well as the institution of a special biennial international meeting (the *IEEE International Workshop on Cellular Neural Networks and Their Applications* [15–19]) which has now been held five times.

Over the years, the original CNN model proposed by Chua and Yang, has been generalized so that successive formal definitions [2,20,21] have been given to include the many extensions under a common general framework.

In this chapter the basic concepts and definitions will be reviewed. The model of Chua and Yang is discussed first. An illustrative description of the main generalizations of this model are presented. Finally a more general and formal definition of CNN is given.

1.1 The CNN of Chua and Yang

The original model was introduced by Chua and Yang [1–3] in 1988. In spite of the various generalizations that followed, it remains the most widely used model due to the fact that it offers to a good compromise between simplicity and versatility, and is also easier to implement.

4   1. CNN Basics

In this section, formalism will be traded for clarity. Once an intuitive understanding of the concepts involved has been developed, it will be possible to discuss a more general and rigorous theory.

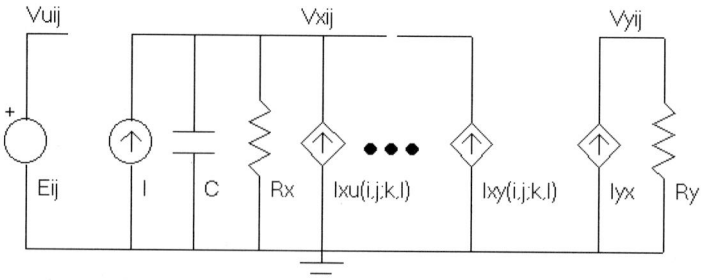

Fig. 1.1. The basic cell

1.1.1 The Cell

The fundamental building block of the CNN is the *cell*, a lumped circuit containing linear and nonlinear elements, shown in Fig. 1.1. The cell $C(i,j)$ has one input u_{ij}, one output y_{ij} and one state variable x_{ij} (represented in Fig. 1.1 by the node voltages v_{uij}, v_{yij} and v_{xij} respectively). The output y_{ij} is a memoryless nonlinear function of the state x_{ij}:

$$y_{ij} = f(x_{ij}) = \frac{1}{2}(|x_{ij} + 1| - |x_{ij} - 1|). \tag{1.1}$$

as depicted in Fig. 1.2. The elements of the cell are all linear but the *voltage controlled current source* (VCCS) $I_{yx} = (1/R_y) \cdot f(v_{xij})$. The core of the cell is constituted by a capacitor C connected in parallel to a resistor R_x, an independent current source I called *bias* and a group of VCCSs

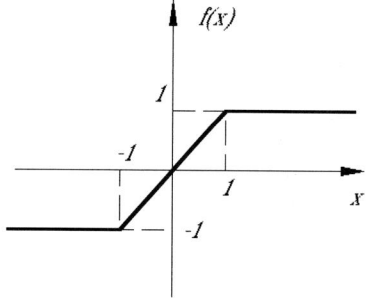

Fig. 1.2. The output nonlinearity

$I_{xu}(i,j;k,l), \ldots, I_{xy}(i,j;k,l)$ that will be discussed later. The cell's dynamics can be formally described by a single state equation (essentially the node equation for the grounded capacitor C). It is assumed that $|x_{ij}(0)| \leq 1$ (*initial condition constraint*) and that the input, obtained by the independent voltage source E_{ij}, is constant and $|u_{ij}| \leq 1$ (*input constraint*).

1.1.2 The CNN Array

A CNN is an array of cells. Each cell is coupled only to its neighboring cells. Adjacent cells can interact directly with each other while more distant cells can have influence indirectly by propagation. For instance, a two-dimensional 4×4 CNN is depicted in Fig. 1.3. The squares represent the cells while the links represent the direct coupling. It is possible to define CNNs of any dimension, but in the following we restrict ourselves to two-dimensional arrays.

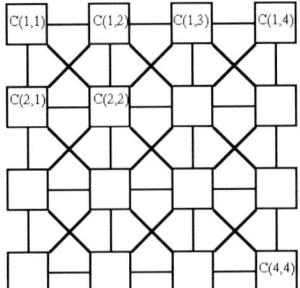

Fig. 1.3. A 4×4 CNN

Let us consider a two-dimensional array of $M \times N$ identical cells arranged in M rows and N columns. This is a $M \times N$ CNN. The generic cell placed on the ith row and jth column will be denoted by $C(i,j)$. Moreover, it is assumed that all the cells in the CNN have the same parameters (the CNN is *space invariant*).

Let us define the *neighborhood* of $C(i,j)$.

Definition 1.1.1 (r-neighborhood set). *The r-neighborhood of $C(i,j)$ is:*

$$N_r(i,j) \doteq \Big\{ C(k,l) \mid \max\{|k-i|, |l-j|\} \leq r, \quad 1 \leq k \leq M;\ 1 \leq l \leq N \Big\}, \quad (1.2)$$

where $r \in \mathbb{N} - \{0\}$ is the radius.

6   1. CNN Basics

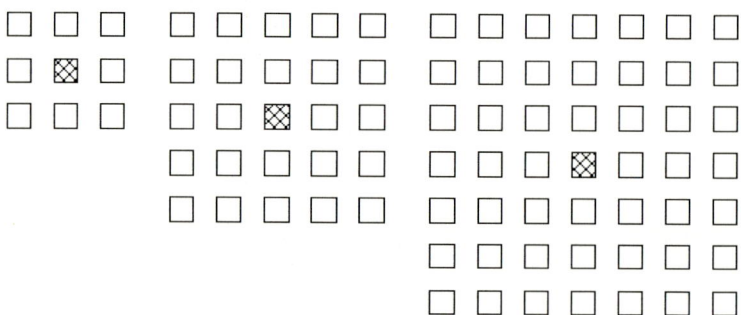

Fig. 1.4. The $N_r(i,j)$ sets for $r = 1, 2, 3$ respectively

Examples of *neighborhoods* of the same cell (highlighted in the center) for $r = 1, 2, 3$ are shown in Fig. 1.4. It is also common practice to talk about "3 × 3 neighborhood", "5 × 5 neighborhood", "7 × 7 neighborhood" and so forth.

The coupling between $C(i,j)$ and the cells belonging to its neighbor set $N_r(i,j)$ is obtained by means of the *linear* VCCSs $I_{xu}(i,j;k,l), \ldots, I_{xy}(i,j;k,l)$ mentioned above. In fact, the input and output of any cell $C(k,l) \in N_r(i,j)$ influence the state x_{ij} of $C(i,j)$ by means of two VCCSs, in $C(i,j)$, defined by the equations:

$$I_{xy}(i,j;k,l) = A(i,j;k,l) \cdot v_{ykl}, \tag{1.3a}$$
$$I_{xu}(i,j;k,l) = B(i,j;k,l) \cdot v_{ukl}, \tag{1.3b}$$

where the coupling coefficients $A(i,j;k,l), B(i,j;k,l) \in \mathbb{R}$ are the *feedback template coefficient* and the *control template coefficient* respectively.

It is important to emphasize that the coupling between the cells in the CNN is only local. This restriction is extremely important for the feasibility of implementation. As previously pointed out, however, cells that do not belong to the same neighbor set can still affect each other indirectly because of the propagation effects of the continuous-time dynamics of the network.

At this point, by performing a nodal analysis[1], it is possible to write the circuit equations of the CNN:

$$C\frac{dv_{xij}}{dt} = -\frac{1}{R_x}v_{xij}(t) + \sum_{C(k,l) \in N_r(i,j)} A(i,j;k,l)v_{ykl}(t)$$
$$+ \sum_{C(k,l) \in N_r(i,j)} B(i,j;k,l)v_{ukl} + I, \tag{1.4}$$
$$1 \leq i \leq M, \ 1 \leq j \leq N.$$

---

[1] It is quite evident that networks of this kind are suited to nodal analysis. However, for large CNN arrays, extremely sparse matrices must be expected because of the local connectivity.

It must be noted that there are essentially two classes of cells: *inner* and *boundary* cells. The inner cell is a cell which has $(2r+1)^2$ neighbor cells. All the other cells are boundary cells. For the latter class it is assumed that the missing cells in the neighbor set contribute with zero input, state and output. Other possible boundary conditions will be discussed in the following. As expected, boundary conditions are going to affect the behavior of boundary cells, and, by virtue of the indirect propagation, can also affect the whole array dynamic [22, 23].

1.1.3 More About Templates

It is now clear that the template coefficients[2] are going to completely define the behavior of the network with given input and initial condition. It is worth remembering that it has been assumed that all the cells in the CNN have equal parameters and hence equal templates (*space invariance*). The term *cloning templates* is used to emphasize this property of invariance. This means that the set of $2 \cdot (2r+1)^2 + 1$ real numbers $A(i,j;k,l), B(i,j;k,l)$ and I completely determine the behavior of an arbitrary large two-dimensional CNN (they define the *dynamic rules* of the network).

Besides, if any template set corresponds to a different behavior, then any choice for the template set corresponds to a particular processing of the inputs and initial conditions. And so it can be thought of as a primitive instruction for an analog processor (the CNN). This will be discussed in more detail later.

The templates are often expressed in a compact form by means of tables or matrices. For instance the following two square matrices are used for a CNN with $r = 1$:

$$A = \begin{pmatrix} A(i,j;i-1,j-1) & A(i,j;i-1,j) & A(i,j;i-1,j+1) \\ A(i,j;i,j-1) & A(i,j;i,j) & A(i,j;i,j+1) \\ A(i,j;i+1,j-1) & A(i,j;i+1,j) & A(i,j;i+1,j+1) \end{pmatrix},$$
$$B = \begin{pmatrix} B(i,j;i-1,j-1) & B(i,j;i-1,j) & B(i,j;i-1,j+1) \\ B(i,j;i,j-1) & B(i,j;i,j) & B(i,j;i,j+1) \\ B(i,j;i+1,j-1) & B(i,j;i+1,j) & B(i,j;i+1,j+1) \end{pmatrix}.$$
(1.5)

This form enables us to rewrite the state equations of the CNN in a more compact form by means of the *two-dimensional convolution operator* $*$.

Definition 1.1.2 (Convolution operator). *For any cloning template T:*

$$T * v_{ij} \doteq \sum_{C(k,l) \in N_r(i,j)} T(k-i, l-j) v_{kl}, \qquad (1.6)$$

where $T(m,n)$ denotes the entry in the mth row and nth column of the cloning template, $m, n = -1, 0, 1$.

---
[2] The bias I is considered as a template coefficient as well.

Therefore, the state equations (1.4) can be rewritten as follows:

$$C\frac{dv_{xij}}{dt} = -\frac{1}{R_x}v_{xij}(t) + A * v_{yij}(t) + B * v_{uij} + I,$$
$$1 \leq i \leq M, \ 1 \leq j \leq N. \tag{1.7}$$

C, R_x and R_y can be conveniently chosen by the designer. CR_x determines the rate of change[3] of the dynamics of the circuit and is usually chosen to be 10^{-8}–10^{-5} s. The circuit parameters, however, can be scaled/normalized for convenience. For instance the dynamics can be simply scaled in time by changing the value of C only. Similar changes of scales can be obtained for the currents and voltages to fit the real design specifications. In practice then, it is very often convenient to describe the dynamics of the array by using the following normalized/dimensionless representation:

$$\frac{dx_{ij}}{dt} = -x_{ij}(t) + A * y_{ij}(t) + B * u_{ij} + I,$$
$$1 \leq i \leq M, \ 1 \leq j \leq N, \tag{1.8}$$

and to adopt a suitable scaling of the circuit parameters when the network is to be implemented.

It is easily understood that templates are going to determine the stability properties of the network. Several classes of templates are defined and studied. These ones are classified according to the signs of the coefficients, the structure of the template matrices and so on [24, 25]. Here only a short mention of the so-called *reciprocal templates* will be made.

Definition 1.1.3 (Symmetric / reciprocal templates). *A template is symmetric or reciprocal if:*

$$A(i,j;k,l) = A(k,l;i,j), \ 1 \leq i,k \leq M, \ 1 \leq j,l \leq N. \tag{1.9}$$

It has been proved [2] that CNNs with reciprocal templates are *completely stable* [4].
Another important and useful result for the Chua and Yang model is:

Theorem 1.1.1 (Sufficient condition for binary outputs [2]). *Let $A(i,j;i,j)$ be the normalized self-feedback coefficient (i.e. $R_x = 1$, $C = 1$). Then:*

$$A(i,j;i,j) > 1 \Rightarrow \lim_{t \to \infty} |y_{ij}| = 1 \ \forall ij. \tag{1.10}$$

---

[3] The term time constant is not really appropriate for a nonlinear network.
[4] See Def. A.7.9 in Appendix A for a definition of completely stable system.

1.1.4 Multilayer CNNs

In the *single-layer* model considered in the previous sections, each cell contributes with one state variable only. This can be generalized if the system order (the number of the state variables) of the cell is increased. A *multilayer* CNN (or MCNN) will be composed of cells having several state variables, one for each layer. Moreover, the interaction between the state variables of the same cell can be complete while the cell-to-cell interaction remains local (restricted to neighbors). One can imagine a multilayer CNN to be composed of several single-layer arrays, stacked one above the other, in which a full layer-to-layer interaction is possible.

A formal model is easily obtained by making use of the previous definitions. Then the state equations for a MCNN can be expressed in compact vector form:

$$\frac{d\boldsymbol{x}_{ij}}{dt} = -\boldsymbol{x}_{ij}(t) + A * \boldsymbol{y}_{ij}(t) + B * \boldsymbol{u}_{ij} + \boldsymbol{I}, \qquad (1.11a)$$
$$1 \leq i \leq M,\ 1 \leq j \leq N,$$

where:

$$A = \begin{pmatrix} A_{11} & 0 & 0 & & 0 \\ & \cdot & \cdot & & 0 \\ & \cdot & & \cdot & 0 \\ & \cdot & & \cdot & 0 \\ A_{m1} & \cdots & & & A_{mm} \end{pmatrix},\quad B = \begin{pmatrix} B_{11} & 0 & 0 & & 0 \\ & \cdot & \cdot & & 0 \\ & \cdot & & \cdot & 0 \\ & \cdot & & \cdot & 0 \\ B_{m1} & \cdots & & & B_{mm} \end{pmatrix}, \qquad (1.11b)$$

$$\boldsymbol{I} = (I_1, \ldots, I_m)',\quad \boldsymbol{x}_{ij} = (x_{1ij}, \ldots, x_{mij})',$$
$$\boldsymbol{y}_{ij} = (y_{1ij}, \ldots, y_{mij})',\quad \boldsymbol{u}_{ij} = (u_{1ij}, \ldots, u_{mij})',$$

and where m denotes the number of state variables in the multilayer cell circuit, or, in other words, the number of layers. Here the convolution operator $*$ between a matrix and a vector is to be interpreted like matrix multiplication but with the operator $*$ inserted between each entry of the matrix and of the vector. Observe that A and B are block triangular matrices.

Any layer can be used to perform a different processing and, of course, any layer works in parallel to the others.

As a final remark, it can be added that the capacitors corresponding to the various layers can have different values. In this way it is possible to have layers with different time rates. As a limiting case, for some layer, the capacitors can be zero (or so small compared to the others to be considered zero for all the practical purposes), thereby yielding a set of differential and algebraic equations.

1.1.5 The CNN as an Analog Processor

A CNN is mainly used as a real-time processor for arrays of data. Different approaches are possible and here only the simplest ones are mentioned; other alternatives will be presented in the following.

Let us suppose, for example, that we want to process a gray-scale image composed of $M \times N$ pixels [26,27]. This image can be represented as an array of $M \times N$ voltages normalized into the allowed input range $[-1, 1]$. Then it can be fed in at the $M \times N$ inputs of the CNN. Provided that the network parameters are such that the CNN is *completely stable*, the state will settle to an equilibrium point. This will correspond to an array of $M \times N$ outputs in the range $[-1, 1]$. This output array/image represents the result of the processing performed by the CNN according to the dynamic rule fixed by the templates. As a matter of fact, the network works as a map that associates an input $U \in [-1, 1]^{MN}$ to an output $Y \in [-1, 1]^{MN}$.

Depending on the templates, this mapping may or may not depend on the initial condition $X(0) \in [-1, 1]^{MN}$ of the CNN.

Therefore, in some cases, it can be appropriate to take $X(0) = U$ or $X(0) = 0$. Alternatively it is possible to operate with $U = 0$ and to feed in the input image as initial condition $X(0)$.

These approaches are considered when the desired processing is applied to a single operand. When, instead, as in a *binary* operation[5], two different operands (e.g. two different $M \times N$ images) are processed to give a combined result, they can be provided as input and initial condition respectively.

Several other approaches are also possible. For instance, the designer may decide to consider the state X of the network at a fixed instant τ to be the result of the desired operation (relaxing in this way the hypothesis of complete stability).

From these simple examples it is understood how a CNN is able to process arrays of data according to a chosen template and criteria on the representation of the data involved. Hence the CNN can be thought of as an analog processor. The template would then represent the *instruction* of this processor. Besides, from an alternative point of view, in analogy with *systolic arrays* [28], the *cell* could be considered the elementary processor while the *CNN* would become an *array* of processors.

Indeed, a vast *library* of cloning templates, performing the most disparate operations has been developed by several research groups over the years.

In this framework the natural next step is to realize that, if some data are processed by a CNN with a particular template, then the corresponding results can be further processed by changing the template and so on. In this way it is possible to perform algorithms. In this context the expression *dual algorithms* is used [29,30].

Image processing is just one application chosen for the purpose of illustration. Applications of CNNs have been developed for a very wide range

---

[5] Trivial examples of *binary* operations are summation, multiplication, etc.

of disparate disciplines: image and signal processing [31], solution of partial differential equations, physical system simulations, modeling of nonlinear phenomena, generation of nonlinear dynamics, associative memories [32–35], neurophysiology, robotics, biology, to cite but a few [9]. A library of well-known templates and dual algorithms is given in Appendix B. The reader is encouraged to experiment with them to achieve a better understanding of the dynamics involved and the wide range of applications of CNNs.

1.1.6 Some Stability Results

To conclude this section it is worth mentioning that several important results concerning the stability and the dynamic behavior of CNNs have been reported in the literature [2, 3, 15–18, 36–38]. Moreover, from the above discussion, it is apparent that these theoretical results have significant impact in terms of applications and implementations. A complete discussion of all the published CNN stability results would deserve much more space and it is beyond the scope of this book. Here only some of these results are briefly mentioned without proof. For a Chua and Yang CNN the following result holds:

Theorem 1.1.2. *All state variables x_{ij} are bounded $\forall t > 0$ by:*

$$x_{\max} = 1 + |I| \\ + \max_{1 \leq i \leq M, 1 \leq j \leq N} \left[\sum_{C(k,l) \in N_r(i,j)} (|A(i,j;k,l)| + |B(i,j;k,l)|) \right] . \quad (1.12)$$

The concept of symmetric/reciprocal templates has already been introduced in Sect. 1.1.3. For these, the following theorem holds:

Theorem 1.1.3 (Sufficient condition for complete stability [2]). *Let $A(i,j;k,l) = A(k,l;i,j)$, $\forall (i,j)$ (symmetric) Then $\dot{x}_{ij}(\infty) = 0$, $\forall (i,j)$, namely the CNN is completely stable.*

Since this is a sufficient but not necessary condition, one can expect to find classes of stable non–reciprocal CNN. Let us thus introduce the following classes and definitions:

Definition 1.1.4 (Positive template). *A cloning template A is called positive if:*

$$A(i,j;k,l) \geq 0 \quad \forall C(k,l) \in N_r(i,j), \; i,j \neq k,l . \quad (1.13)$$

Definition 1.1.5 (Template graph). *We associate with each template A, a labeled (+ or −) directed graph (or digraph) Γ_A. The $(2r+1)^2$ nodes of Γ_A will be the cells in $N_r(i,j)$ for some i and j, with $C(i,j)$ fixed as the central cell. $\forall C(k,l) \in N_r(i,j)$ we have a directed edge from $C(k,l)$ to $C(i,j)$ iff $A(i,j;k,l) \neq 0$. The label associated with this edge will be the sign of $A(i,j;k,l)$. We call Γ_A the template graph.*

Similarly, for each $M \times N$ CNN we associate a labeled directed graph Γ:

Definition 1.1.6 (CNN graph). *The MN nodes in Γ will be the cells in the CNN. $\forall C(i,j), C(k,l)$ in the graph there is an edge from $C(k,l)$ to $C(i,j)$ iff $A(i,j;k,l) \neq 0$. We call Γ the CNN graph.*

In the case of spatio-invariant templates, the graph Γ can be obtained from Γ_A, considering the latter as a sub-graph of Γ and periodically replicating Γ_A on each node of Γ.

Definition 1.1.7 (Irreducible digraph). *A directed graph (digraph) is irreducible if $\forall i \neq j$ there is a directed path from node i to node j.*

Note that if Γ is a connected graph this does *not* imply that it is irreducible. The concept of connectivity does not take into account the direction of the connecting path.

Definition 1.1.8 (Symmetric graph). *A digraph is symmetric if an edge from node i to node j implies that an edge exists from node j to node i.*

Definition 1.1.9 (Acyclic graph). *A digraph is called acyclic if it has no cycles, i.e. there are not nodes $v_1, \ldots, v_j, v_{j+1}$ with $v_{j+1} = v_1$ such that there is an edge from v_{j+1}, $i = 1, \ldots, j$.*

Definition 1.1.10 (cell–linking template). *A $n \times n$ A-template is cell linking if the CNN graph of the associated $n \times n$ CNN is irreducible.*

Definition 1.1.11 (Positive cell–linking template). *The template A is called positive cell linking iff the following two conditions are satisfied:*

1. *it is a positive template*
2. *it is a cell–linking template*

Definition 1.1.12 (Strictly sign–symmetric template). *A template is called strictly sign–symmetric if:*
whenever $A(i,j;k,l) \neq 0$, $A(i,j;k,l) \cdot A(k,l;i,j) > 0$.

Definition 1.1.13 (opposite–sign templates). *The class of opposite–sign templates is defined by template values having one of the following structural and sign patterns:*

$$A = \begin{pmatrix} 0 & s & 0 \\ 0 & p & 0 \\ 0 & -s & 0 \end{pmatrix}, \quad A = \begin{pmatrix} 0 & 0 & 0 \\ s & p & -s \\ 0 & 0 & 0 \end{pmatrix}, \tag{1.14a}$$

$$A = \begin{pmatrix} s & 0 & 0 \\ 0 & p & 0 \\ 0 & 0 & -s \end{pmatrix}, \quad A = \begin{pmatrix} 0 & 0 & -s \\ 0 & p & 0 \\ s & 0 & 0 \end{pmatrix}, \tag{1.14b}$$

where $p > 1$ and $s \neq 0$.

Definition 1.1.14 (value–asymmetric templates). *The class of value–asymmetric templates is defined by template values having one of the following structural and sign patterns:*

$$A = \begin{pmatrix} 0 & r & 0 \\ 0 & p & 0 \\ 0 & -s & 0 \end{pmatrix}, \quad A = \begin{pmatrix} 0 & 0 & 0 \\ r & p & -s \\ 0 & 0 & 0 \end{pmatrix}, \quad (1.15a)$$

$$A = \begin{pmatrix} r & 0 & 0 \\ 0 & p & 0 \\ 0 & 0 & -s \end{pmatrix}, \quad A = \begin{pmatrix} 0 & 0 & -s \\ 0 & p & 0 \\ r & 0 & 0 \end{pmatrix}, \quad (1.15b)$$

where $p > 1$ and $r \cdot s > 0$.

Definition 1.1.15 (Regions α and β). *If a cell's state is such that $|x_{ij}| < 1$ then it is said to be operating in region α; otherwise it is said to be operating in region β.*

The following theorems hold for the above-defined classes of templates.

Theorem 1.1.4. *If the output nonlinear function $f(\cdot)$ is smooth and strictly monotonically increasing and the template is positive cell–linking then the CNN is completely stable except, possibly, for a set of measure zero (e.g. it is stable almost everywhere).*

Theorem 1.1.5. *A CNN with negative cell–linking template, with non-zero elements in A along a row, a column or a diagonal is completely stable.*

This result also holds in the case in which the linear term $A(i,j;k,l)y_{kl}$ is replaced by a nonlinear term $A_{i,k;l,l}(y_{kl})$ (see nonlinear templates in the following section) where $A_{i,k;l,l}(\cdot)$ is strictly monotonically increasing.

Theorem 1.1.6. *If the output nonlinear function $f(\cdot)$ is defined as in (1.1) and $\mathcal{N}$ is a CNN with opposite–sign templates then the dynamics of $\mathcal{N}$ starting from any initial state has the following properties:*

1. *if all the cells are in region α at any time $t = t_j$ then all the eigenvalues of the associated Jacobian matrix have positive real parts at $t = t_j$,*
2. *if all the cells are in region β then all the eigenvalues are -1, hence, they tend to decay to a constant if they are in the basin of attraction of an equilibrium point in this region,*
3. *if there are b variables operating in region β, while the rest are operating in region α, then the Jacobian matrix has b eigenvalues equal to -1, while all the others have positive real part.*

This last theorem does not imply that a CNN with opposite–sign template is completely stable. However, by virtue of Theorem 1.1.2 and this latter one, we can say that:

1. a CNN with opposite–sign template will never settle out of a region β,

2. any trajectory that enters or starts from region β must drift to a fixed point in region β,
3. any trajectory external to region β must tend to a region β.

Theorem 1.1.7. *A CNN with positive strictly sign symmetric template with isolated equilibria is stable almost everywhere.*

Theorem 1.1.8. *For an opposite–sign template CNN with cloning template $\left(\begin{array}{ccc} s & p & -s\end{array}\right)$, with $p > 1$, $s > 0$ and $(p-1)/2 < s < p-1$, all non-constant trajectories ϕ with $\|\phi(0)\| \geq 1$ defined on a maximal interval $[0, a)$ satisfy:*

$$\lim_{t \to a} \phi(t) = \xi, \tag{1.16}$$

where ξ is a stable fixed point.

Theorem 1.1.9. *A CNN with an acyclic digraph Γ and $A(i, j; k, l) > 1$ is completely stable.*

Theorem 1.1.10. *A template such that*

$$A(i,j;k,l) = \begin{cases} 0 & \text{if } l > j \\ 0 & \text{if } l = j \text{ and } k > i \\ \text{arbitrary} & \text{otherwise} \end{cases} \tag{1.17}$$

generates a CNN with an acyclic Γ.

The previous results can be extended to nonlinear templates[6] that are strictly increasing functions in C^1 with respect to the non local variable (i.e. with respect to y_{kl}).

Let us now suppose that the bias I and the control template B are all zero. The state equations can then be written in a compact matrix form such as:

$$\dot{\boldsymbol{x}} = \boldsymbol{x} + AF(\boldsymbol{x}). \tag{1.18}$$

Moreover we define:

Definition 1.1.16 (Diagonally dominant matrix). *The matrix $A \in \mathbb{R}^n$ is said to be diagonally dominant if*

$$|a_{ii}| \geq \sum_{j=1, j \neq i}^{n} |a_{ij}|, \quad \forall i = 1, \ldots, n. \tag{1.19}$$

Moreover, it is said to be strictly dominant if equality does not hold $\forall i = 1, \ldots, n$.

Then the following theorems hold:

---
[6] See Sect. 1.2.

Theorem 1.1.11. *There exists a fixed point x' in each saturation region iff $A - I$ is diagonally dominant.*

Theorem 1.1.12. *If there exists a row (index p) such that*

$$a_{pp} - 1 < -\sum_{j \neq p}^{n} |a_{pj}|, \quad \text{which implies } a_{pp} < 1, \tag{1.20}$$

then there does not exist a fixed point in any saturation region.

Theorem 1.1.13. *If $A - I$ is strictly dominant then a trajectory corresponding to system (1.18) converges to an equilibrium point in one of the saturation regions (i.e. the CNN is completely stable).*

1.2 Main Generalizations

As previously pointed out, several generalizations have been made to the model of Chua and Yang discussed in Sect. 1.1. The purpose of these generalizations has been to enhance the capabilities of the CNNs, broadening their field of applications or improving the efficiency of existing ones [20,21,30,36,39]. Needless to say, for every generalization there are numerous stability results.

Some of the most common generalizations are reviewed in this section. A more general and rigorous definition including most of them as particular cases will be given in the following section.

1.2.1 Nonlinear CNNs and Delay CNNs

The CNN model of Chua and Yang is indeed a nonlinear circuit because of the output function (1.1). However, some authors [21,36] refer to this model as the *linear* CNN to emphasize the linearity of the VCCSs determining the coupling, as in equations (1.3a,b).

Let us now replace the relationships (1.3a,b) by the following ones:

$$I_{xy}(i,j;k,l) = \hat{A}_{ij;kl}(v_{ykl}, v_{yij}) + A^{\tau}_{ij;kl} v_{ykl}(t-\tau), \tag{1.21a}$$

$$I_{xu}(i,j;k,l) = \hat{B}_{ij;kl}(v_{ukl}, v_{uij}) + B^{\tau}_{ij;kl} v_{ukl}(t-\tau), \tag{1.21b}$$

where $\hat{A}_{ij;kl}(\cdot,\cdot), \hat{B}_{ij;kl}(\cdot,\cdot): C(\mathbb{R}) \times C(\mathbb{R}) \to \mathbb{R}$ (i.e. they are real-valued continuous functions of, at most, two variables) while $A^{\tau}_{ij;kl}, B^{\tau}_{ij;kl} \in \mathbb{R}$, $\tau \in [0,\infty)$. Namely a *nonlinear* coupling is introduced by $\hat{A}_{ij;kl}(\cdot,\cdot)$ and $\hat{B}_{ij;kl}(\cdot,\cdot)$, while a *functional* dependence is introduced by $A^{\tau}_{ij;kl}$ and $B^{\tau}_{ij;kl}$. Equations (1.3a,b) are a particular case of (1.21a,b) and the state equation (1.8) is replaced by the following one:

$$\frac{dx_{ij}}{dt} = -x_{ij}(t) + \hat{A} * y_{ij}(t) + A^\tau * y_{ij}(t-\tau)$$
$$+ \hat{B} * u_{ij}(t) + B^\tau * u_{ij}(t-\tau) + I, \qquad (1.22)$$
$$1 \le i \le M, \ 1 \le j \le N,$$

with the consequent extension for the convolution operator. Moreover, let us observe that one of the hypotheses, namely the assumption concerning the time-invariance of the inputs, has been relaxed.

We will talk about *nonlinear CNNs* when $A^\tau_{ij;kl}, B^\tau_{ij;kl} = 0$ while we will talk about *delay-type CNNs* [40] when $\hat{A}_{ij;kl}, \hat{B}_{ij;kl} = 0$.

Moreover, other hypotheses can be relaxed. For example those of the time- or the space-invariance of the templates.

The nonlinear templates allow more sophisticated and efficient ways of processing data [20, 39]. Furthermore they permit one to mimic some biological functions such as those in the retina [41–43]. Delay-type templates find application, for instance, in detection of moving objects [44–46].

Other extensions relate to the memoryless output equation (1.1). An arbitrary *bounded* nonlinear function $f: \mathbb{R} \to \mathbb{R}$ can be used in place of (1.1). Actually, practical design considerations imply further restrictions such as continuity. Some typical output functions are shown in Fig. 1.5. One can also consider a dynamic output function:

$$\dot{v}_{yij} = -a v_{yij} + f(v_{xij}). \qquad (1.23)$$

An easy way to obtain it is to connect a capacitor in parallel with R_y in Fig. 1.1.

Another generalization involving the topology of the cell consists in substituting the circuit shown in Fig. 1.1 with a completely different one. Very

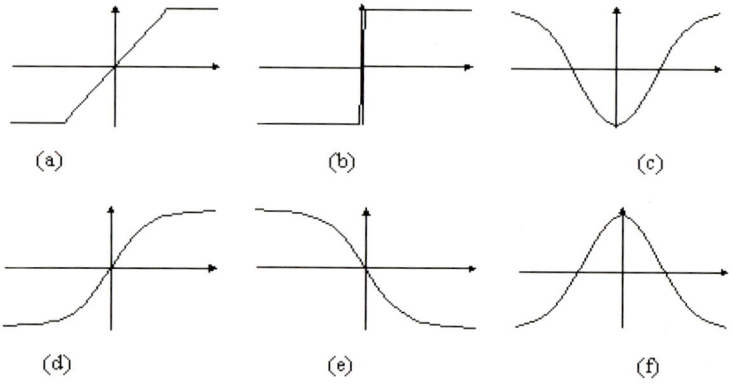

Fig. 1.5. Nonlinear output functions; (**a**) unit gain with saturation, (**b**) high gain with saturation, (**c**) inverse gaussian, (**d**) sigmoid, (**e**) inverse sigmoid, (**f**) gaussian

important is the case in which the *Chua oscillator* [47] is used as cell. This case will be considered in detail in the next chapters.

1.2.2 Nonuniform Processor CNNs and Multiple Neighborhood Size CNNs

Motivated partly by neurobiological structures, other generalizations involve *nonuniform grid* CNNs, having more than one type of cell and/or more than one size of neighborhood. These are called *nonuniform processor* CNNs (NUP-CNN) and *multiple-neighborhood-size* CNNs (MNS-CNN) respectively. Examples of grids are reported in Fig. 1.6. The adoption of a grid that is not rectangular implies a formal modification in the definition of the neighbor set and in the way in which cell's positions are specified. An example of NUP-CNN with two different kinds of cell (drawn in black and white respectively) is shown in Fig. 1.7. An example of a MNS-CNN is shown in Fig. 1.8. In this multilayer CNN there are neighbor sets of two different sizes. One layer (white cells) with a fine grid ($r = 1$) and another one (black cells) with a coarse grid ($r = 3$). This kind of architecture reflects some characteristic structures found in living visual systems [39, 42].

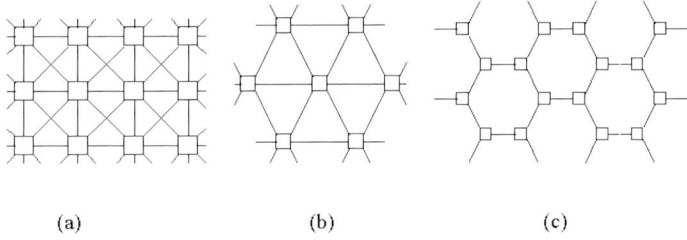

Fig. 1.6. Examples of grids; (a) rectangular, (b) triangular, (c) hexagonal

1.2.3 Discrete-Time CNNs

A very important class of CNNs is the *discrete-time CNN* (DTCNN) [48]. As the name implies, these are essentially the discrete-time version of the continuous-time models already discussed. Actually, in the original definition given in [48], some further differences (e.g. the output function is the so-called *threshold* nonlinearity and implies only *binary outputs*) exist.

A DTCNN is described by the state map

$$x_{ij}(n) = \sum_{C(k,l) \in N_r(i,j)} a(i,j;k,l) y_{kl}(n) + \sum_{C(k,l) \in N_r(i,j)} b(i,j;k,l) u_{kl} + i_{ij}$$

(1.24a)

18   1. CNN Basics

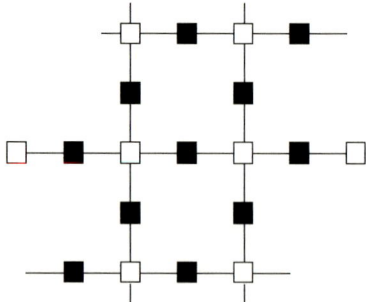

Fig. 1.7. A NUP-CNN with two kinds of cell

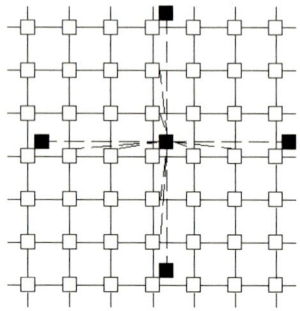

Fig. 1.8. A MNS-CNN with two kinds of neighbor sets

$$y_{ij}(n) = f(x_{ij}(n-1)) = \begin{cases} 1 & \text{if } x_{ij}(n-1) > 0 \\ -1 & \text{if } x_{ij}(n-1) < 0 \end{cases} \quad (1.24\text{b})$$

$$1 \leq i \leq M,\ 1 \leq j \leq N$$

or, in compact form, adopting the *Einstein summation*[7] convention:

$$x^c(n) = a_d^c y^d(n) + b_d^c u^d(n) + i^c, \quad (1.25\text{a})$$

$$y^c(n) = f(x^c(n-1)) = \begin{cases} 1 & \text{if } x^c(n-1) > 0 \\ -1 & \text{if } x^c(n-1) < 0, \end{cases} \quad (1.25\text{b})$$

where c is the generic cell of the DTCNN, $d \in N_r(c)$, while a_d^c, b_d^c and i^c are the cloning templates. Moreover, this latter form is also valid with non-rectangular grids. All of the above mentioned generalizations can be applied to this kind of CNN. Note that the output function is not defined at the origin but in practice there is always some noise so it does not matter.

It is immediately worth mentioning however that the DTCNN have many practical features not possessed by a conventional continuous time CNN.

---

[7] See Def. A.2.5 in Appendix A.

Among these, systematic procedures to design the templates needed for a well-defined task do exist [48–50]. A simple example explains how this is possible: just consider that, due to the output nonlinearity, the templates can be found by solving a system of inequalities obtained by imposing a desired state transition on the update equation (1.25).

Moreover DTCNNs have some appreciable robustness features. In fact, if the templates are such that:

$$\Delta = \min_{c,k} |a_d^c y^d(n) + b_d^c u^d(n) + i^c| \qquad (1.26)$$

is large enough, then the algorithm (1.25) is relatively insensitive to parameter variation in a_d^c, b_d^c or i^c smaller than Δ, because individually they cannot cause any changes in the output states [48].

Finally, the speed of the network is easily controlled by adjusting the clock frequency for the law (1.25). This is advantageous in terms of testability.

On the other hand, as is always the case when discrete-time systems and continuous-time systems are compared [51], some of the features of the continuous CNN are lost in the DTCNN. For instance, in a DTCNN, a symmetric template does not imply complete stability. Indeed, there exist examples of such systems admitting stable limit cycles [48].

1.2.4 The CNN Universal Machine

As pointed out in Sect. 1.1.5, the CNN can be used as a programmable processing device where the instructions are represented by the templates. This is the idea behind the so-called *CNN Universal Machine* (CNNUM), a further evolution of the CNN architecture, proposed by T.Roska and L.O.Chua [29, 30], in which both analog and digital circuitry coexist.

The adjective *universal* must be intended in the sense of Turing. This statement hides a rather complex definition but the essential meaning is that a Turing (universal) machine is able to realize any conceivable algorithm (recursive function). In fact, it has been proved that both the CNN [52] and the CNNUM (that we are going to briefly discuss) [53] are universal in this sense. In simple terms, it can be theoretically proved that they are able to perform any possible algorithm. This is an important result that proves the existence of an algorithm. However, it does not give an answer on *how* the CNN/CNNUM can perform a desired algorithm. This is itself a tough open problem and it is known as the *learning and design problem*[8].

On this basis, the CNNUM is a CNN in which every (analog) cell is augmented by the introduction of additional (local) analog and digital blocks. Moreover, further blocks are added to perform global tasks.

---
[8] The term *design* is used when the desired task could be translated into a set of local dynamic rules. But the term *learning* is used when the templates should be obtained so that pairs of input and output must correspond (the relationship between them may be by far too complicated for the explicit formulation of local rules) [54].

The local blocks are essentially constituted by memories storing data to be processed and sequences of template values to perform the desired processing. In addition, some local logic to govern the operation is also added. This, in analogy with cache and pipeline techniques used in many common microprocessors, allows one to improve the throughput of the network in performing template-based algorithms. Global circuitry controls the overall behavior (loading and retrieving information) of the CNN nucleus apart from the interaction of the CNNUM with the external world.

The coexistence of analog and digital sections, working together, without any data converter placed between them, suggested the introduction of the term *analogic computing* (from the contraction of "analog" and "logic") or *dual computing*. Thus the CNNUM executes *analogic* and *dual* algorithms.

Roska and Chua identified the bottleneck of the CNN as being the input/ouput with the external world. In fact, it is easily understood that while the number of cells in an integrated realization grows as N^2, the corresponding number of pins can only grow linearly. And this forces a sequential input/output from the chip.

Given that many applications of CNNs deal with the processing of sensory information, they proposed the integration of the array of sensors/antennas/coils directly on the same chip as the CNN itself [30], an idea as appealing as it is technologically challenging.

1.3 A Formal Definition

The previous sections have essentially followed the actual evolution of the CNN paradigm from its introduction to the present. It is apparent how numerous contributions and interactions between electrical engineering and other disciplines have led to different mutations and variations on the basic model of Chua and Yang.

This in turn led to successive formal definitions (first in [2], then in [20] and, more recently, in [21]) of the CNN.

In this section, the last and most general definition of cellular neural network is reported. For the definitions of *vector field*, *map*, *dynamical system* and so on, the reader should refer to the appendices.

1.3.1 The Cells and Their Coupling

Definition 1.3.1 (Cellular neural network). *A cellular neural network (CNN) is a high dimensional dynamic nonlinear circuit composed by locally coupled, spatially recurrent circuit units called cells. The resulting net may have any architecture, including rectangular, hexagonal, toroidal, spherical and so on. The CNN is defined mathematically by four specifications:*

1. Cell dynamics.
2. Synaptic law.
3. Boundary conditions.
4. Initial conditions.

Definition 1.3.2 (Cell dynamics). *The internal circuit core of the cell can be any dynamical system. The cell dynamics is defined by an evolution equation. In the case of continuous-time lumped circuits, the dynamics is defined by the state equation:*

$$\dot{\boldsymbol{x}}_\alpha = -\boldsymbol{g}(\boldsymbol{x}_\alpha, \boldsymbol{z}_\alpha, \boldsymbol{u}_\alpha(t), \boldsymbol{I}^s_\alpha), \qquad (1.27)$$

where $\boldsymbol{x}_\alpha, \boldsymbol{z}_\alpha, \boldsymbol{u}_\alpha \in \mathbb{R}^m$ *are the state vector, threshold (DC bias) and input vector of the cell* n_α *at position* α *respectively.* $\boldsymbol{I}^s_\alpha$ *is a synaptic law and* $\boldsymbol{g}\colon \mathbb{R}^m \times \mathbb{R}^m \times \mathbb{R}^m \times \mathbb{R}^m \to \mathbb{R}^m$ *is a vector field.*
For a discrete-time circuit, the dynamics is defined by the state update law:

$$\dot{\boldsymbol{x}}_\alpha = -\boldsymbol{G}(\boldsymbol{x}_\alpha, \boldsymbol{z}_\alpha, \boldsymbol{u}_\alpha(t), \boldsymbol{I}^s_\alpha), \qquad (1.28)$$

where $\boldsymbol{G}\colon \mathbb{R}^m \times \mathbb{R}^m \times \mathbb{R}^m \times \mathbb{R}^m \to \mathbb{R}^m$ *is a map.*

Note the explicit dependence of $\boldsymbol{x}_\alpha$ on $\boldsymbol{x}_\alpha$ itself (self-feedback), on the threshold $\boldsymbol{z}_\alpha$ and its inputs $\boldsymbol{u}_\alpha(t)$, and also on the synaptic law defined below.

Definition 1.3.3 (Sphere of influence). *In this section, the sphere of influence* S_α *of the cell* n_α *coincides with the previously defined neighbor set* $N_r(n_\alpha)$ *without* n_α *itself:*

$$S_\alpha \doteq N_r(n_\alpha) - \{n_\alpha\}. \qquad (1.29)$$

Definition 1.3.4 (Synaptic law). *The synaptic law defines the coupling between the considered cell* n_α *and all the cells* $n_{\alpha+\beta}$ *within a prescribed sphere of influence* S_α *of* n_α *itself:*

$$\boldsymbol{I}^s_\alpha = \hat{A}^\beta_\alpha \boldsymbol{x}_{\alpha+\beta} + A^\beta_\alpha * \boldsymbol{f}_\beta(\boldsymbol{x}_\alpha, \boldsymbol{x}_{\alpha+\beta}) + B^\beta_\alpha * \boldsymbol{u}_{\alpha+\beta}(t), \qquad (1.30)$$

where we use the Einstein summation rule.
The first term $\hat{A}^\beta_\alpha \boldsymbol{x}_{\alpha+\beta}$ *is the linear feedback of the states of the neighboring cells* $n_{\alpha+\beta}$. $\hat{A}^\beta_\alpha$ *is the state template*[9].
The second term $A^\beta_\alpha * \boldsymbol{f}_\beta(\boldsymbol{x}_\alpha, \boldsymbol{x}_{\alpha+\beta})$ *defines the arbitrary nonlinear coupling.* A^β_α *is the nonlinear feedback template.*
The last term $B^\beta_\alpha * \boldsymbol{u}_{\alpha+\beta}(t)$ *accounts for the contribution of external inputs.* B^β_α *is the feedforward or control template.*

---

[9] The introduction of the direct contribution from the state itself was first introduced in [55].

Before proceeding any further, it is worth making a few remarks about the common case in which a two-dimensional rectangular-grid CNN is considered. In this circumstance the contributions given in equations (1.27, 1.28, 1.30) can be written in explicit form. First of all, the notation regarding the sphere of influence has to be understood as follows:

$$kl \in S_\alpha \iff k, l \in \{-r, \ldots, 0, \ldots, r\}, (k, l) \neq (0, 0). \tag{1.31}$$

The simplest example for (1.27) is the first-order cell

$$\dot{x}_{ij} = -g(x_{ij}) + I_{ij}^s, \tag{1.32}$$

with $g \colon \mathbb{R} \to \mathbb{R}$.

The state contribution is written as

$$\hat{A}_\alpha^\beta \boldsymbol{x}_{\alpha+\beta} = \sum_{kl \in S_\alpha} \hat{a}_{ij;kl} \boldsymbol{x}_{i+k, j+l}. \tag{1.33}$$

The nonlinear (output) coupling expresses an arbitrary nonlinear interaction law between the cells in a neighborhood. The same is valid for the feedforward template. Therefore both the nature of $\boldsymbol{f}_\beta$ and that of the nonlinear feedback A_α^β and feedforward B_α^β templates can be chosen freely. Some illustrative examples for the nonlinear feedback follow:

$$A_\alpha^\beta * \boldsymbol{f}_\beta(\boldsymbol{x}_\alpha, \boldsymbol{x}_{\alpha+\beta}) = \sum_{kl \in S_\alpha} a_{ij;kl} \boldsymbol{f}(\boldsymbol{x}_{i+k, j+l}), \tag{1.34a}$$

$$A_\alpha^\beta * \boldsymbol{f}_\beta(\boldsymbol{x}_\alpha, \boldsymbol{x}_{\alpha+\beta}) = \sum_{kl \in S_\alpha} a_{ij;kl} \boldsymbol{f}(\boldsymbol{x}_{i+k, j+l}(t - T) - \boldsymbol{x}_{i,j}(t - T)), \tag{1.34b}$$

$$A_\alpha^\beta * \boldsymbol{f}_\beta(\boldsymbol{x}_\alpha, \boldsymbol{x}_{\alpha+\beta})$$
$$= \sum_{kl \in S_\alpha} a_{ij;kl} \int_0^t \boldsymbol{h}(t - \tau) \cdot (\boldsymbol{x}_{i+k, j+l}(\tau) - \boldsymbol{x}_{i,j}(\tau)) \mathrm{d}\tau, \tag{1.34c}$$

$$A_\alpha^\beta * \boldsymbol{f}_\beta(\boldsymbol{x}_\alpha, \boldsymbol{x}_{\alpha+\beta})$$
$$= \sum_{kl \in S_\alpha} a_{ij;kl} \sum_{n=1}^M \int_0^t \cdots \int_0^t \int_0^t \boldsymbol{h}_n(t - \tau_1, t - \tau_2, \ldots, t - \tau_n)$$
$$\cdot \boldsymbol{x}_{i+k, j+l}(\tau_1) \cdots \boldsymbol{x}_{i+k, j+l}(\tau_n) \mathrm{d}\tau_1 \mathrm{d}\tau_2 \cdots \mathrm{d}\tau_n. \tag{1.34d}$$

The reader will recognize these as the memoryless output coupling, the nonlinear delay, the convolution and the Volterra series operator, respectively.

1.3.2 Boundary Conditions

In the previous sections the importance of the boundary conditions has been pointed out. In fact, these define the way the boundary cells work and so, by indirect propagation, may affect the whole network behavior. It is worth noticing that in some cases (especially in some theoretical derivations, because of the formal simplification that this hypothesis sometimes implies) it is assumed the CNN is extended to infinity.

Some of the most common boundary conditions are now defined. In order to give mathematical definitions a formal artifice is introduced.

Definition 1.3.5 (Missing cell). *Let $\mathcal{C}$ be the set of cells of a finite CNN and $\mathcal{C}^\infty$ its extension to infinity ($\mathcal{C} \subset \mathcal{C}^\infty$). Let $n_\alpha \in \mathcal{C}$ be a boundary cell and S_α its sphere of influence. A missing cell (of n_α) $n_{\alpha+\beta}$ is defined as:*

$$n_{\alpha+\beta} \in \mathcal{C}^\infty \; : \; n_{\alpha+\beta} \in S_\alpha \; \text{and} \; n_{\alpha+\beta} \notin \mathcal{C}. \tag{1.35}$$

Definition 1.3.6 (Fixed (Dirichlet) boundary condition). *Let $n_{\alpha+\beta}$ be a missing cell and $\boldsymbol{E}^x_{\alpha+\beta}, \boldsymbol{E}^u_{\alpha+\beta} \in \mathbb{R}^m$ two constant vectors. Then the Dirichlet condition is defined by*

$$\boldsymbol{x}_{\alpha+\beta} = \boldsymbol{E}^x_{\alpha+\beta}, \; \boldsymbol{u}_{\alpha+\beta} = \boldsymbol{E}^u_{\alpha+\beta}. \tag{1.36}$$

In other words, the CNN is clamped at its ends to some fixed (possibly space-invariant) potential. For instance, this is the case of the CNN model of Chua and Yang, where the potential is uniformly at ground.

Definition 1.3.7 (Zero–flux (Neuman) boundary condition). *Let $n_{\alpha+\beta}$ be a missing cell of n_α. The zero–flux condition is defined by*

$$\boldsymbol{x}_{\alpha+\beta} = \boldsymbol{x}_\alpha, \; \boldsymbol{u}_{\alpha+\beta} = \boldsymbol{u}_\alpha. \tag{1.37}$$

In this case the state and input of the missing cell match those of the boundary cell.

Definition 1.3.8 (Periodic (toroidal) boundary condition). *Let n_α and n_γ be two boundary cells placed at the corresponding opposite ends of a CNN. Let $n_{\alpha+\beta}$ be the missing cell of n_α lying on the symmetry axis uniquely defined by n_α and n_γ. The toroidal condition is defined by*

$$\boldsymbol{x}_{\alpha+\beta} = \boldsymbol{x}_\gamma, \; \boldsymbol{u}_{\alpha+\beta} = \boldsymbol{u}_\gamma. \tag{1.38}$$

In this case the network behaves as if it is joined onto itself forming a torus.

1.4 Summary

In this chapter the basic concepts and definitions of cellular neural networks have been reviewed.

The original Chua and Yang model and its most important generalizations have been summarized, introducing the reader to this continuously evolving field.

Finally a broad and formal definition of CNN has been given.

2. Some Applications of CNNs

In this chapter some original applications of cellular neural networks to real-world problems are discussed. A library of templates and dual algorithms is included for reference in Appendix B.

In Sect. 2.1 a new image processing technique based on CNNs for improving the automatic classification of fruits (in particular, oranges) is introduced. It allows the digitized orange images to be processed in order to highlight some peculiarities of the fruits. In this way the subsequent classification step is greatly simplified and improved. Moreover, the real-time processing characteristic of CNNs is a great advantage over the traditional computing resources commonly used in this kind of processing. The proposed task is accomplished by the choice of suitable templates in a simple Chua and Yang CNN model. These templates are described and some examples are reported.

In Sect. 2.2 a new method to filter 2D NMR spectra against uncertainties arising from experiments and data acquisition machinery is proposed. The method introduced is explained and is applied to the filtering of a real 2D NMR spectrum of a protein.

In Sect. 2.3 the simulation of some environmental models for air quality is accomplished by suitable CNNs. These have been derived from the partial differential equations describing the concentrations of pollutants under wind action. In particular three different cases have been considered: the one-dimensional advection, the one-dimensional diffusion, and the two-dimensional advection.

For each one, the proper CNN structure is derived. In addition, some examples are presented.

2.1 CNN-Based Image Pre-processing for the Automatic Classification of Fruits

One of the main problems in the automation of modern farming production concerns the selection of good fruits from the crop as a whole. This is necessary for commercial reasons and because a bad (rotten) fruit placed among good ones can cause the whole lot to deteriorate. In particular we are interested in the selection of oranges. Currently this task is mainly done manually.

Some attempts at automatic classification using traditional computing resources and algorithms have been made. Unfortunately the average classification time for each orange using this equipment is too long for an efficient real-time application. It is clear that the classification of fruits based on their digitized images can be improved and simplified if redundant details are removed from these images. This can be accomplished by preliminary image filtering. Again, this has been attempted by digital computers but has proved too time-consuming to be actually applied.

In [56] an alternative solution that uses CNNs for a fast processing of orange images is proposed. The choice of suitable templates allows the desired pre-filtering to be achieved even by a simple Chua and Yang CNN model.

Some words must be said about automatic orange image classification. For each orange six different views are available (top, bottom, etc.). The basis of the classification is the recognition of possible spots on the surface of the orange: If the fruit presents a spot it must be rejected. For a correct operation it is important to distinguish correctly between spots and other parts of the fruit such as the stalk.

If, in fact, the latter is recognized as a spot, every fruit will be rejected. However, stalks and spots present some peculiarities that can be exploited in order to distinguish between them. More specifically, stalks present a star-like shape while spots present an almost elliptical shape (Fig. 2.1). Some stalks, nevertheless, have a rounded shape so they can be confused with spots. In this case the irregular surface of stalks can be used to distinguish them from the smooth surface of spots.

2.1.1 The Pre-filtering

It is now clear that, roughly speaking, the classification is actually based on the presence/absence of spots in the orange being examined and that the main requirement is the ability to avoid confusion between spots and stalks. Therefore the required filtering strategy must be able to clean the images of useless particulars such as shadows, small irregularities on the surface of the fruit, noise, and so on, leaving just the interesting elements, i.e. spots and stalks. Moreover, the classification will be enhanced if the different peculiarities of these two elements are further stressed.

2.1 CNN-Based Image Pre-processing for the Automatic Classification of Fruits

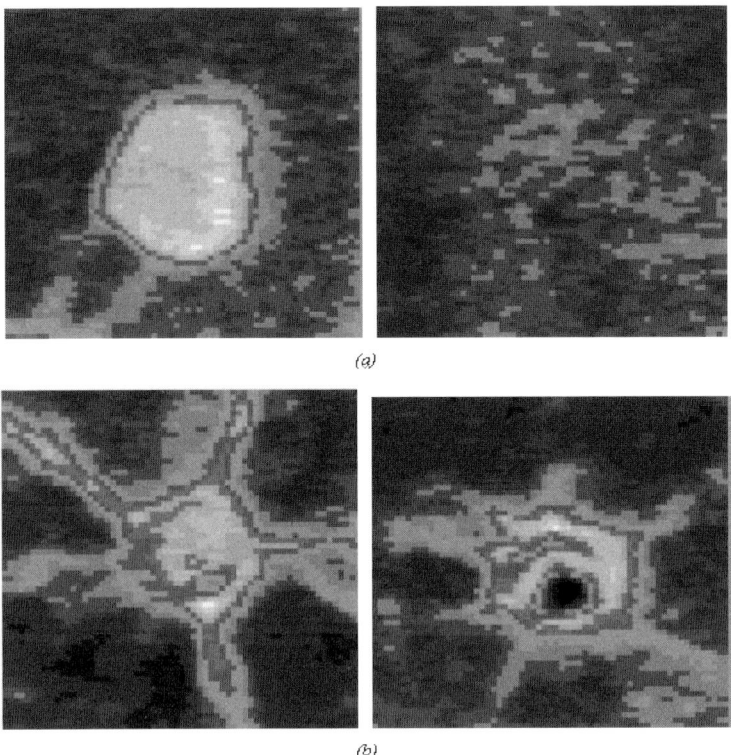

Fig. 2.1. (a) Two spots in the orange's surface. (b) Two stalks

Let us consider the following template:

$$A = \begin{pmatrix} 0 & 0 & 0 & 0 & 0 & 0 & 0 \\ 0 & 0 & 0 & 0 & 0 & 0 & 0 \\ 0 & 0 & 0 & 0 & 0 & 0 & 0 \\ 0 & 0 & 0 & 2 & 0 & 0 & 0 \\ 0 & 0 & 0 & 0 & 0 & 0 & 0 \\ 0 & 0 & 0 & 0 & 0 & 0 & 0 \\ 0 & 0 & 0 & 0 & 0 & 0 & 0 \end{pmatrix}, \qquad (2.1a)$$

$$B = \begin{pmatrix} +1 & -1 & -1 & +1 & -1 & -1 & +1 \\ -1 & +1 & -1 & +1 & -1 & +1 & -1 \\ -1 & -1 & +1 & +1 & +1 & -1 & -1 \\ +1 & +1 & +1 & +2 & +1 & +1 & +1 \\ -1 & -1 & +1 & +1 & +1 & -1 & -1 \\ -1 & +1 & -1 & +1 & -1 & +1 & -1 \\ +1 & -1 & -1 & +1 & -1 & -1 & +1 \end{pmatrix}, \quad I = 0.2. \qquad (2.1b)$$

In (2.1a,b) we can immediately note some peculiarities:

1. $A(i,j;i,j) = 2 > 1$ so steady-state output saturation is guaranteed, according to Theorem 1.1.1.
2. The feedback template is reciprocal so the convergence (complete stability) is guaranteed as well, in agreement with what was mentioned in Sect. 1.1.3.

Let us note its resemblance to some well-known templates used to extract particular connected configurations in images [15]. On the other hand, one of the design guidelines was the need to extract the above-mentioned configurations of spots and stalks. In particular let us consider the control template B. It contains the value 1.0 along the eight principal directions (north, south, etc.) and the value -1.0 in the remaining positions in order to reveal the possible connected components along these directions.

The bias current acts as a threshold while the feedback template drives the system to the steady state.

It is worth noting that a small neighbor radius template implies poor efficiency in the connected line extraction task because it reveals thin lines but encounters some problems with wide and/or broken lines. A big radius template implies a complex practical realization and it is not consistent with the local interactions of CNNs.

A trade-off between complexity and efficiency was found with a radius $r = 3$ in the proposed templates.

In Fig. 2.1 some images of spots and stalks are shown, while Fig. 2.2 gives the results of the proposed processing.

From these figures it can be noted that, as anticipated, just the required elements have been preserved. Moreover, the inner irregularities of the spots have been removed while the circular irregularities inside the central part of the stalks and their star-like structure have been further highlighted.

By varying the bias value I and the central control coefficient $B(i,j;i,j)$, some peculiarities of the resulting image processing (e.g. noise selectivity, resolution etc.) can be enhanced according to particular requirements. This leads to a family of templates.

2.2 Processing of NMR Spectra

The problem of noise removal from corrupted images is a topic of great interest in a variety of research areas, ranging from robotics and automation to medicine and rehabilitation. In particular, as regards the biological and medical field, experiments using nuclear magnetic resonance (NMR) spectra are very helpful because of the meaningful information that derives from this kind of experiment performed on living tissues. Suitable filtering of these spectra is required but quite difficult to obtain, owing to disturbances arising from the fact that the sample is analyzed while moving in solution. Moreover, additional noise sources deriving from data acquisition equipment and instrument resolution make the filtering process even harder.

2.2 Processing of NMR Spectra

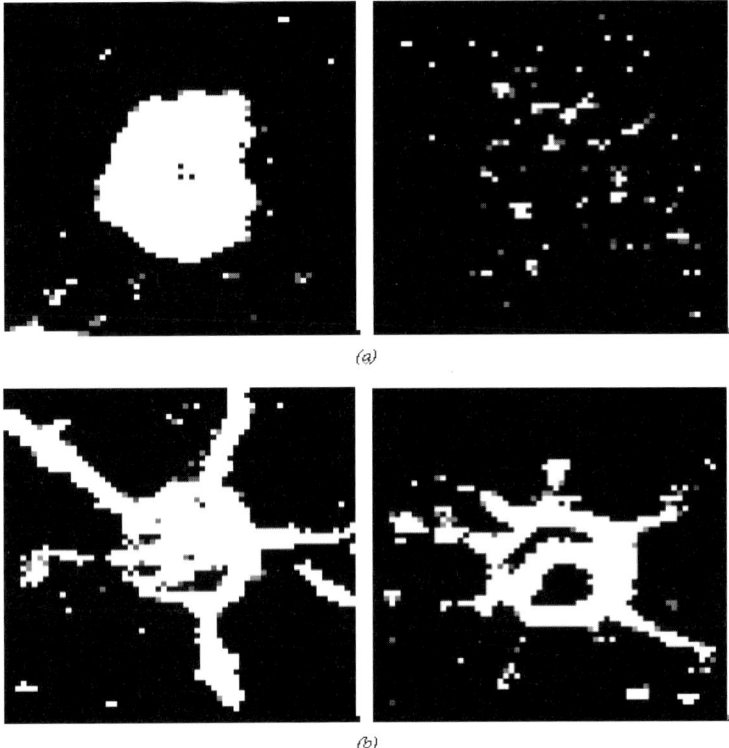

Fig. 2.2a,b. The results of the proposed processing on the images shown in Fig. 2.1

In this section the problem of noise removal in 2D NMR spectra is efficiently solved by using a CNN approach. This is achieved in four steps, where the CNN is used in each phase like an analog processor whose operation is fixed by a proper choice of its templates.

2.2.1 Two-Dimensional NMR Spectra

Among the wide variety of applications of 2D NMR spectra, they are used in the biochemical field to study and reconstruct protein structure, i.e. it is possible to identify which aminoacids make up a protein, to establish their sequence in the protein, and how they are located in three-dimensional space [57]. The power of NMR processing lies in its capability to analyze biopolymers in solution, i.e. in a conformational state very close to their living conditions. Actually, owing to NMR's feature of analyzing proteins in solution, NMR data referring to the same molecule may vary a lot, due to the vibrations and conformational motions of the molecule which is free to move around an equilibrium position. Such a situation causes some problems when performing spectral analysis. The presence of noise makes the interpretation of spectra more difficult.

The proposed procedure was applied to filter a particular 2D NMR spectrum, the TOCSY (TOtal Correlation SpectroscopY) spectrum [58], although it is quite general and can be applied to any kind of 2D NMR spectrum.

The TOCSY spectrum allows us to identify the correlations among the spins constituting the protein being examined by means of the presence/absence of resonance peaks in some particular positions on a two-dimensional plane. For our aims, the presence of a peak is more important than its height so the three-dimensional spectral surface can be reduced to a two-dimensional array called a map. The peaks are ordered in sets called lists. Every list is a row or a column of peaks that permits a particular residue in the protein to be characterized.

The processing of these maps (in order to identify the protein) is a time-consuming task, especially if it is done in a traditional way, i.e. by human experts. Some artificial intelligence tools have been developed in order to improve and to automate this job [59]. Due to the presence of noise, the peaks belonging to a list are not exactly aligned in a straight line as they should be in the ideal case. This is nothing special for a human expert because he/she expects to find the peaks around particular regions, but it can obviously be an important problem for an automatic recognition method. It is clear that if the map is appropriately pre-filtered (rebuilt), before the classification process, the simplicity and the reliability of the identification is greatly improved.

2.2.2 Processing of NMR Spectra with CNNs

Let us show how a simple reconstruction of the noisy lists can be achieved by a dual CNN algorithm. This algorithm is composed of a sequence of four cloning templates for the model of Chua and Yang. It is assumed that the execution of every step of the algorithm is completed when the network reaches the steady-state which always consists of settling at an equilibrium point.

In Fig. 2.3 an example of a TOCSY bidimensional spectrum is shown. This is a very wide array (typically 1024×1024 or 800×800) so for a practical application the proposed processing must be repeated *window by window*, i.e. the CNN is used to process a small rectangular region (a window) of the map; subsequently the window is moved to another position and the processing is repeated and so on.

The processed windows should not overlap and it is possible for them to be processed concurrently. In this work 61×61 windows were considered.

As can be seen in Fig. 2.3, the spectrum is symmetric with respect to the diagonal of the hydrogen resonance peaks. Moreover, the lists are horizontal in the region to the right of the diagonal, while they are vertical in the region above the diagonal. We concentrated on the former region. In agreement with the above definitions, the map is actually an array of pixels. A black pixel represents the presence of a peak while a white pixel represents its absence.

2.2 Processing of NMR Spectra

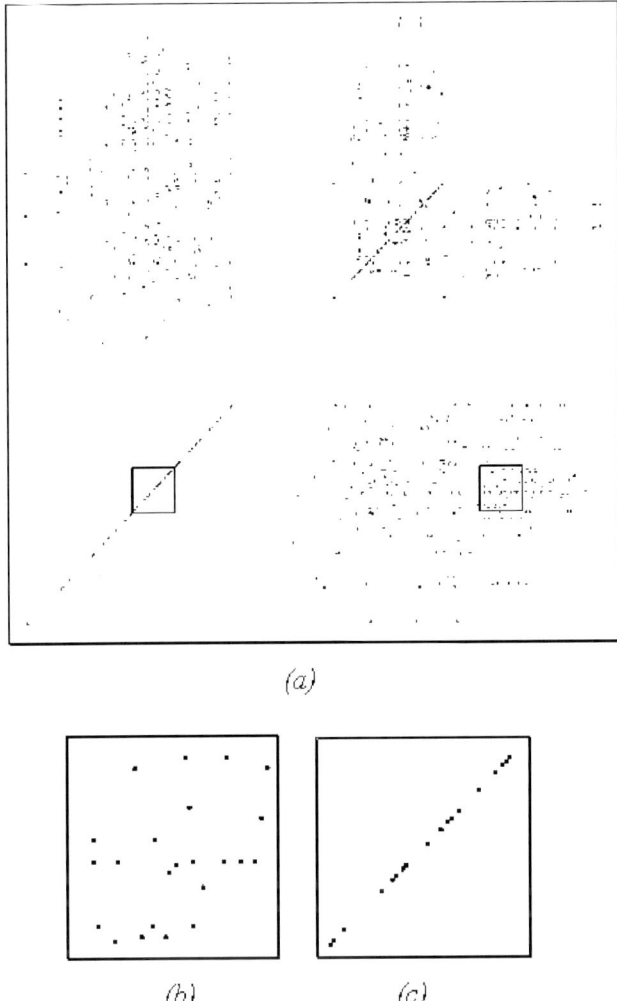

Fig. 2.3. (a) The TOCSY spectrum, (b) the selected window, and (c) its corresponding hydrogen peaks

An important and useful point is that the peaks on the diagonal determine the straight horizontal line identifying the lists. More precisely, every list has hydrogen resonance peak as an element in the diagonal [57]. Therefore a reliable way to reconstruct the lists is to exploit the virtual lines determined by the peaks in the diagonal; so for any processed window it is necessary to consider a new window containing the corresponding hydrogen peaks (as is shown in Fig. 2.3).

In order to describe the proposed algorithm it is necessary to describe the four instructions/templates used. A name is assigned to each template, like computer instructions. The first instruction is called STRETCH. Its effect is to add one new black pixel below any black one formerly present and the corresponding template is:

$$A = \begin{pmatrix} 0 & 0 & 0 \\ 0 & 2 & 0 \\ 0 & 0 & 0 \end{pmatrix}, \quad B = \begin{pmatrix} 0 & 1 & 0 \\ 0 & 0 & 0 \\ 0 & 0 & 0 \end{pmatrix}, \quad I = 1. \tag{2.2}$$

STRETCH requires that data are fed into the inputs and processed as the initial state of the CNN.

The second one is called LINE and it generates a horizontal line of black pixels in any row containing a black pixel. The chosen template is:

$$A = \begin{pmatrix} 0 & 0 & 0 \\ 1 & 2 & 1 \\ 0 & 0 & 0 \end{pmatrix}, \quad B = 0, \quad I = 1.9. \tag{2.3}$$

LINE requires its own data as the initial state (inputs are insignificant). The third one is the well-known template AND. This is the only already-known template and its effect is the logical AND between the inputs and the initial state.

$$A = \begin{pmatrix} 0 & 0 & 0 \\ 0 & 1.5 & 0 \\ 0 & 0 & 0 \end{pmatrix}, \quad B = \begin{pmatrix} 0 & 0 & 0 \\ 0 & 1.5 & 0 \\ 0 & 0 & 0 \end{pmatrix}, \quad I = -1. \tag{2.4}$$

The last instruction is SHRINK. Its effect is to delete any black pixel that has another black pixel above it. The corresponding template is:

$$A = \begin{pmatrix} 0 & 0 & 0 \\ 0 & 1.5 & 0 \\ 0 & 0 & 0 \end{pmatrix}, \quad B = \begin{pmatrix} 0 & -1 & 0 \\ 0 & 1 & 0 \\ 0 & 0 & 0 \end{pmatrix}, \quad I = -1. \tag{2.5}$$

SHRINK requires its own data in the inputs and initial state.

2.2.3 Description of the Dual Algorithm

In this section the algorithm for filtering 2D NMR spectra is described and applied to an experiment performed on the BPTI protein. It consists of the following steps:

1. Choose the window; assign the value 1.0 to the black pixels and -1.0 to the white pixels; set these values in the corresponding inputs and initial state of the CNN; execute STRETCH (Fig. 2.4).
2. Analogously, set the corresponding "hydrogen window" as the initial state of the CNN and execute LINE (Fig. 2.5).

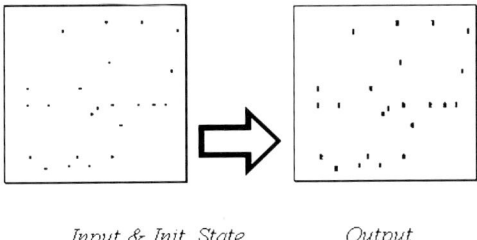

Input & Init. State   *Output*

Fig. 2.4. The execution of STRETCH

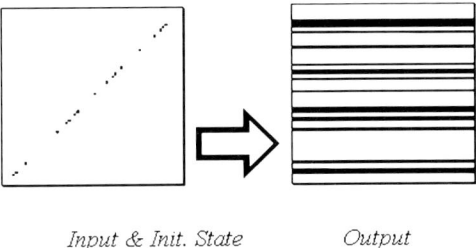

Input & Init. State   *Output*

Fig. 2.5. The execution of LINE

3. Set the result of step 1 (STRETCH) as the inputs and set the result of step 2 (LINE) as the initial state; execute AND (Fig. 2.6).
4. Set the result of step 3 (AND) as the inputs and initial state and execute SHRINK (Fig. 2.7).

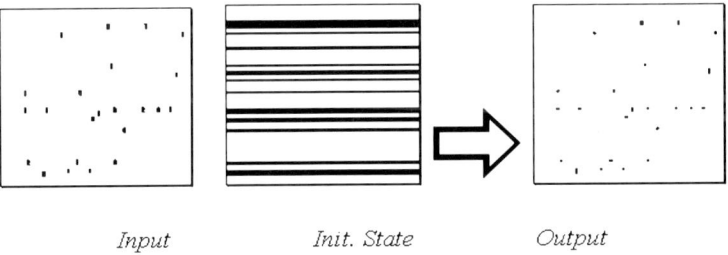

Input   *Init. State*   *Output*

Fig. 2.6. The execution of AND

It is worth noting that steps 1 and 2 are independent so they can be executed concurrently. Now let us compare the ideal window (i.e. how the peak should be in the ideal case when there is no noise) shown in Fig. 2.8a, the noisy window (i.e. the window just processed) shown in Fig. 2.8b, and the result of the CNN processing (rebuilt window) shown in Fig. 2.8c. Fig. 2.9 makes this comparison simpler.

34   2. Some Applications of CNNs

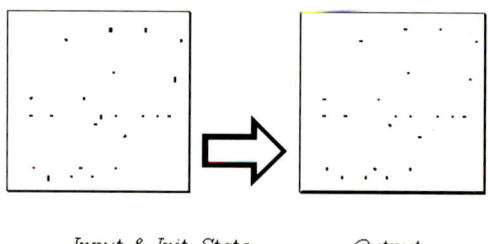

Input & Init. State   Output

Fig. 2.7. The execution of SHRINK

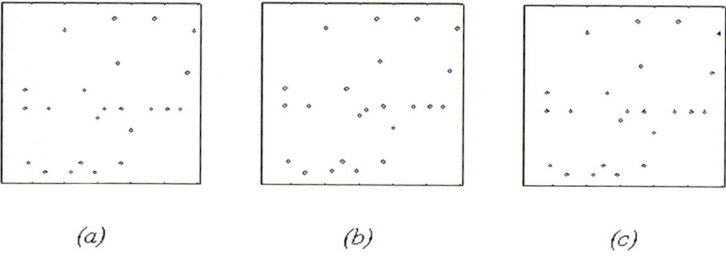

(a)   (b)   (c)

Fig. 2.8. (a) The ideal window, (b) the noisy window, (c) the rebuilt window

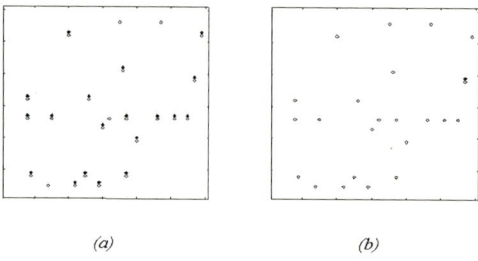

(a)   (b)

Fig. 2.9. (a) Overlapping of the ideal and noisy windows. (b) Overlapping of the ideal and rebuilt windows. The white peaks belong to the ideal window while the black peaks belong to the other overlapped windows. Therefore, in (a) 19 peaks are in the wrong position while in (b) just one peak is in the wrong position

In Fig. 2.9a the ideal and noisy window have been overlapped while in Fig. 2.9b the ideal and rebuilt window have been overlapped. The ideal window contains 23 peaks. The noisy window differs from the ideal one by 19 peaks (i.e. the error is $19/23 \approx 83\%$) while the rebuilt window differs from the ideal one by only one peak (i.e. the error is $1/23 \approx 4\%$).

The error has been significantly reduced. However, the efficiency of this filtering varies. A number of different cases were examined under application of the proposed algorithm and, in the worst case, a maximum final error of around 15% was obtained, even with very noisy windows (more than 80%

initial noise). Nevertheless the reconstruction is often perfect (0% error). This happens when the lists are less crowded than the presented example (in particular it happens if the result of the LINE is a set of divided thin stripes).

It is worth considering that if the lists are too close, even a human expert will find it difficult to decide whether a peak is not in its exact position and, if this is the case, to decide the right list to assign it to.

2.3 Air Quality Modeling

Much attention is currently devoted to environmental problems; among these air quality modeling is of particular interest. Ozone air quality modeling, for example, is of great concern, especially in the United States [60]. Air quality models are mathematical descriptions of atmospheric transport, diffusion and chemical reaction of pollutants [60]. Usually, the concentrations of chemical species in the air are the unknown variables. These models are used to predict how peak concentrations will change in response to prescribed changes in meteorology and in the source of pollution.

In this section the various models are described in terms of partial differential equations (PDEs) and suitable CNNs are introduced in order to solve these PDEs[1] [61,62]. In this way simulations and predictions of the advection and diffusion of chemical pollutants in the air can be efficiently accomplished. Three cases have been considered: the advection and the diffusion in a single dimension and the advection in a two-dimensional plane.

2.3.1 Models

Air quality models can be divided in two classes: *diagnostic* models and *prognostic* models. The former are statistical descriptions of the observed data, while the latter are based on physicochemical principles governing air pollution [60]. In this section, only prognostic models are taken into account.

The general model describing pollution in a 3-D time-dependent domain Ω_t is a coupled system of nonlinear parabolic partial differential equations [60]:

$$\frac{\partial c_i}{\partial t} + \nabla \cdot (\boldsymbol{u} c_i) = \nabla \cdot (K \nabla c_i) + f_i(c_1, \ldots, c_p) \tag{2.6}$$

$$x \in \Omega_t, \quad t > 0, \quad 1 \leq i \leq p.$$

---

[1] In this section, some of the symbols (e.g. x, y, i, y and so on) are replicated with different meaning. This is due to the fact that these are well-established notations in the specialized literature of the different disciplines involved. Changing one or the other one would create confusion to some readers with no real improvement in the accessibility of the document. Therefore we prefer to leave the original nomenclature while the meaning is fully clear from the context.

where $\boldsymbol{u}$ is the air velocity field (wind field), K is a diffusion matrix, f_i is the chemical formation (or depletion) rate of species i, c_i are the concentrations of chemical species and the model (2.6) is derived as the conservation of mass equations for these species in Ω_t.

Here the concentrations of pollutants are unknown while the wind field and diffusion matrix are known. Of course, the equation (2.6) must be complemented by the initial conditions

$$c_i(\boldsymbol{x},0) = c_{i0}(\boldsymbol{x}), \quad \text{for} \quad \boldsymbol{x} \in \Omega_0, \quad 1 \le i \le p, \tag{2.7}$$

and the boundary conditions

$$ac_i + b(\nabla \cdot c_i) \cdot \boldsymbol{\nu} = g_i, \quad \boldsymbol{x} \in \partial\Omega_t, \quad t > 0, \quad 1 \le i \le p, \tag{2.8}$$

where $a, b, g \colon (\boldsymbol{x},t) \to \mathbb{R}$ and $a \ge 0$, $b \ge 0$, $a^2 + b^2 > 0, \forall (\boldsymbol{x},t) \in \Omega_t \times \mathbb{R}^+$; while $\boldsymbol{\nu}$ is the outward unit normal to $\partial\Omega_t$.

Important particular cases derived from this general model are the *advection* and *diffusion*. In the former case the diffusion matrix K vanishes (or its contribution is negligible in comparison to wind effects), that is, the pollutants are transported by the wind without diffusion.

In the diffusion case of course K cannot be neglected.

2.3.2 CNNs for Air Quality Modeling

These PDEs are traditionally solved by numerical finite difference methods [60]. Of course it is often time-consuming because a huge amount of elementary computations are needed. However, CNNs have often been used in the solution of many PDEs [21, 63–65]. In this section a new approach for air quality prediction is presented [61, 62]. Attention is focused on the diffusion and transport of a single pollutant (e.g. the ozone) ignoring the various underlying processes.

In particular the following cases have been considered:

1. 1D advection,
2. 1D diffusion,
3. 2D advection.

Let us begin with the one-dimensional advection. Let us suppose that the wind field is uniform and that diffusion can be neglected ($k_{ij} = 0$). For the sake of simplicity the reference system can be chosen so that $\boldsymbol{u} = (U,0,0)$ with $U \in \mathbb{R}$. With these hypotheses the model (2.6) simplifies to the one-dimensional model:

$$\frac{\partial c}{\partial t} + U \frac{\partial c}{\partial x} = 0. \tag{2.9}$$

It can be proved that with this model and with the above hypothesis about the wind field, the peak of an initial distribution of the pollutant cannot grow as the system evolves [60]. The shape of the initial distribution will be simply transported by the wind field.

In order to obtain a CNN model, let us introduce a uniform discretization grid for the spatial variable x. Moreover the spatial derivative can be approximated by the simple formula:

$$\left.\frac{\partial c}{\partial x}\right|_{x=j\Delta x} \cong \frac{c_j - c_{j-1}}{\Delta x}, \quad \text{where} \quad c_j \equiv c(j\Delta x, t). \tag{2.10}$$

Substituting (2.9) into (2.10) the following model is obtained:

$$\frac{\partial c_j}{\partial t} = -\frac{U}{\Delta x}[c_j - c_{j-1}]. \tag{2.11}$$

As Δx tends to zero this model approaches the model (2.9). Therefore it can be easily proved that if the discretization length Δx is sufficiently small then, again, the initial peak value of c cannot grow but it will be just translated.

This consideration allows one to map this equation into the following linear CNN model:

$$\frac{dx_j}{dt} = -x_j + A * y_j \tag{2.12a}$$

with

$$A = \left(\frac{U}{\Delta x}, \; 1 - \frac{U}{\Delta x}, \; 0\right), \tag{2.12b}$$

where x_j corresponds to c_j and differs just by a scale factor that must be chosen such that the initial concentration peak corresponds to 1 (or less). In this way y_j will always correspond to x_j because saturation will never occur.

Let us now consider the one-dimensional diffusion. In this case, the hypothesis on the wind field is the same as in the previous case but now diffusion in the x direction is present. More precisely:

$$\boldsymbol{u} = (U, 0, 0), \quad U \in \mathbb{R}, \quad K = \begin{pmatrix} k & 0 & 0 \\ 0 & 0 & 0 \\ 0 & 0 & 0 \end{pmatrix}. \tag{2.13}$$

So the general model (2.6) is reduced to the one-dimensional model:

$$\frac{\partial c}{\partial t} + U \frac{\partial c}{\partial x} = k \frac{\partial^2 c}{\partial x^2}. \tag{2.14}$$

In this case the initial distribution of the pollutant will be transported by the wind field and will diffuse at the same time. For these reasons it is evident that the concentration peak cannot grow but, possibly, it will decrease.

Analogously to the previous case let us introduce a uniform discretization grid into the one-dimensional spatial variable and let us approximate the spatial derivatives with the following formulas:

$$\left.\frac{\partial c}{\partial x}\right|_{x=j\Delta x} \simeq \frac{c_j - c_{j-1}}{\Delta x}, \qquad \left.\frac{\partial^2 c}{\partial x^2}\right|_{x=j\Delta x} \simeq \frac{c_{j+1} - 2c_j + c_{j-1}}{\Delta x^2}. \qquad (2.15)$$

Therefore the model (2.14) is approximated by the following spatio-discrete model:

$$\frac{\partial c_j}{\partial t} = -\left(\frac{U}{\Delta x} + \frac{2k}{\Delta x^2}\right)c_j + \left(\frac{k}{\Delta x^2} + \frac{U}{\Delta x}\right)c_{j-1} + \frac{k}{\Delta x^2}c_{j+1}. \qquad (2.16)$$

This, again, can be mapped on a simple linear CNN model with:

$$\frac{dx_j}{dt} = -x_j + A * y_j \qquad (2.17a)$$

where

$$A = \left(\frac{U}{\Delta x} + \frac{k}{\Delta x^2}, \; 1 - \frac{U}{\Delta x} - \frac{2k}{\Delta x^2}, \; \frac{k}{\Delta x^2}\right), \qquad (2.17b)$$

which differs from (2.12b) just for the template. It is evident that, in this case too, y_j cannot saturate.

The third case, the two-dimensional advection, is surely the most interesting. Here, the wind field is no longer uniform but depends on the spatial position. However, as in the first case considered, the diffusion can be disregarded. So the general model (2.6) reduces to the following one:

$$\frac{\partial c}{\partial t} + \frac{\partial(Uc)}{\partial x} + \frac{\partial(Vc)}{\partial y} = 0, \qquad (2.18)$$

where the wind field is stationary and has been described by $\mathbf{u} = (U(x,y), V(x,y))$, x and y being the spatial coordinates. This is a nonlinear PDE and can be rewritten as

$$\frac{\partial c}{\partial t} + U\frac{\partial c}{\partial x} + V\frac{\partial c}{\partial y} + c\frac{\partial U}{\partial x} + c\frac{\partial V}{\partial y} = 0. \qquad (2.19)$$

Similarly to previous cases, a 2D spatial grid $(i\Delta x, j\Delta y)$ is introduced and spatial derivatives are approximated by finite difference terms:

$$\begin{aligned}\left.\frac{\partial c}{\partial x}\right|_{(i,j)} &\simeq \frac{c_{ij} - c_{(i-1)j}}{\Delta x}, & \left.\frac{\partial U}{\partial x}\right|_{(i,j)} &\simeq \frac{U_{ij} - U_{(i-1)j}}{\Delta x}, \\ \left.\frac{\partial c}{\partial y}\right|_{(i,j)} &\simeq \frac{c_{ij} - c_{i(j-1)}}{\Delta y}, & \left.\frac{\partial V}{\partial y}\right|_{(i,j)} &\simeq \frac{V_{ij} - V_{i(j-1)}}{\Delta y}\end{aligned} \qquad (2.20)$$

leading to a time-continuous, spatio-discrete model. For the sake of simplicity it can be supposed that $\Delta x = \Delta y \doteq h$ and with trivial algebra, analogously the the two previous cases, the model can be represented by the following nonlinear three-layer CNN:

$$\frac{d\boldsymbol{x}_{ij}}{dt} = -\boldsymbol{x}_{ij} + A * \mathbf{y}_{ij}, \qquad \boldsymbol{x}_{ij} = \begin{pmatrix} V_{ij} \\ U_{ij} \\ c_{ij} \end{pmatrix},$$

$$A = \begin{pmatrix} A_{11} & 0 & 0 \\ 0 & A_{22} & 0 \\ 0 & 0 & A_{33} \end{pmatrix}, \quad A_{11} = A_{22} = \begin{pmatrix} 0 & 0 & 0 \\ 0 & 1 & 0 \\ 0 & 0 & 0 \end{pmatrix}, \tag{2.21}$$

$$A_{33} = \frac{1}{h} \begin{pmatrix} 0 & U_{ij} & 0 \\ V_{ij} & (h - 2U_{ij} - 2V_{ij} + U_{(i-1)j} + V_{i(j-1)}) & 0 \\ 0 & 0 & 0 \end{pmatrix}.$$

It can be seen from these that A_{11} and A_{22} ensure the time-invariance of the wind field and A_{33} is not a constant coefficients matrix but its elements are functions of the state variables of the layers one and two, so the CNN of equation (2.21) is a nonlinear CNN.

2.3.3 Examples

Here we describe two examples of prediction using the CNN models introduced above.

In the first example the case of $1D$ advection is examined. A one-dimensional CNN array of 50 cells is used. For simplicity we choose $U/\Delta x = 1$. So the feedback template is simply reduced to $A = [1\ 0\ 0]$.

In Fig. 2.10 the initial distribution of the pollutant (represented by means of the state variables) is shown. A classic Gaussian shape has been chosen. In Fig. 2.11 the distribution is reported at different time instants; it is clear how the shape is simply translated in the wind direction.

However the most interesting case is the 2D advection that is the topic of the second example where a 25×25 CNN is considered. It has been supposed that:

1. the initial distribution of the pollutant is Gaussian and centered at $(12, 12)$; it is shown in Fig. 2.12;
2. the stationary wind field has unit magnitude but is not uniformly distributed; it is shown in Fig. 2.13.

Under these conditions the pollutant is transported by the wind. The initial distribution of the concentration c is put out of shape by the different directions of the wind in the various part of its spatial distribution.

Figure 2.14 shows the results of the simulation of the CNN in (2.21) at different time instants. It shows how a constant level section of the pollutant distribution is transported and modified by the wind field.

40   2. Some Applications of CNNs

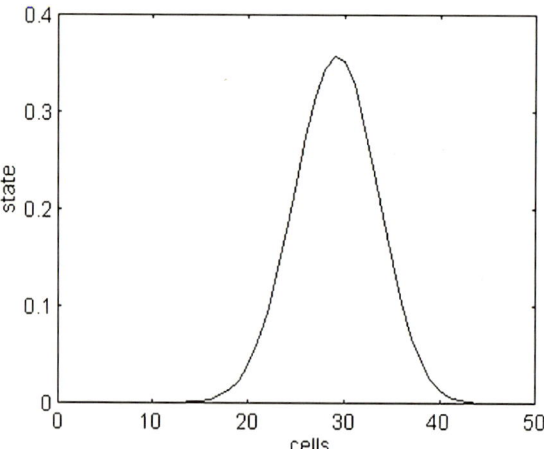

Fig. 2.10. Initial pollutant concentration distribution

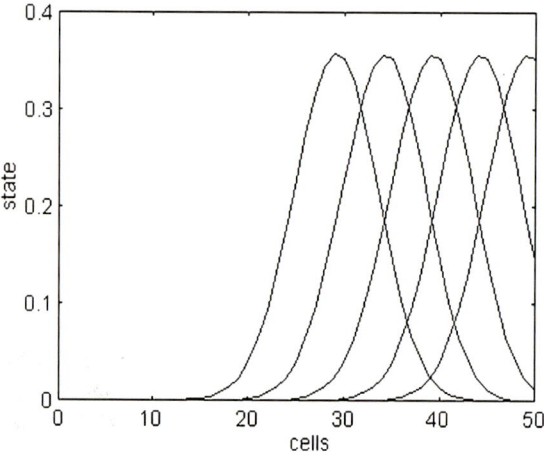

Fig. 2.11. The distribution at subsequent time instants

2.4 Conclusions

In this chapter some new applications of CNNs to real-world problems have been presented.

In Sect. 2.1 a new image processing technique for filtering orange images has been presented. It has been developed in order to enhance the automatic classification of these fruits. In fact, from the above considerations it follows that recognition and classification is greatly simplified, freeing the artificial classifier from redundant information.

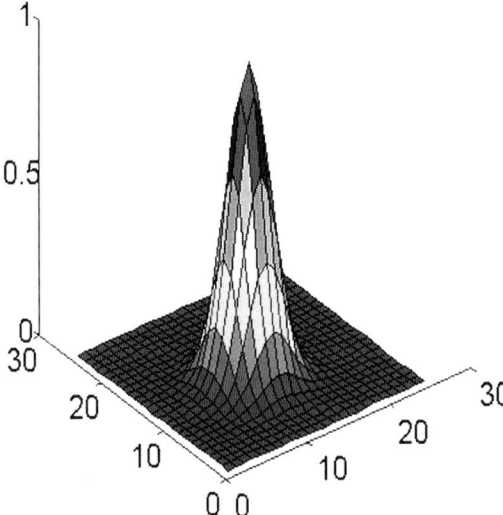

Fig. 2.12. Initial pollutant concentration distribution

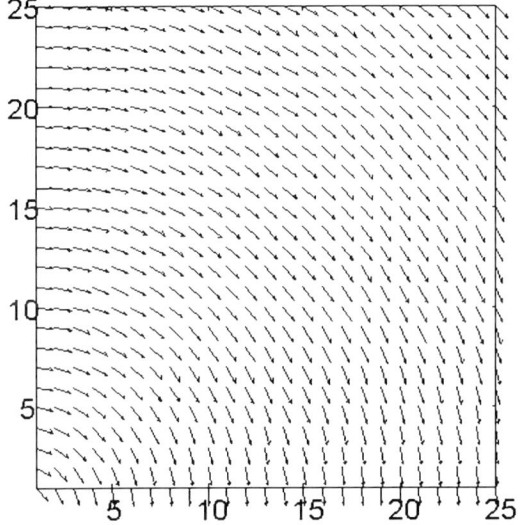

Fig. 2.13. The wind field

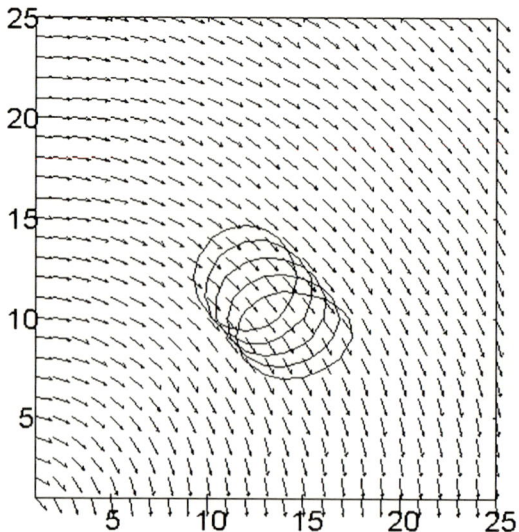

Fig. 2.14. Contour of the pollutant concentration surface transported and modified by the wind at five different time instants, obtained by the simulation of the multi-layer CNN of equations (2.21)

Moreover, the approach introduced only requires the simplest CNN model. This is also important in view of possible future actual hardware implementations.

In Sect. 2.2 a novel strategy to filter 2D NMR spectra efficiently has been introduced The example reported shows the efficiency of the noise removal. The particular spectrum taken into consideration belongs to the class of the so-called TOCSY spectra and represents resonance peaks corresponding to BPTI protein residues. However, this strategy can be directly applied to other 2D NMR spectra. This method is applied in a phase prior to the true analysis of the spectrum. Clearly, if the filtering phase is accurate then the following recognition step is made easier and more reliable. Residue recognition has already been performed employing artificial neural networks [66].

In Sect. 2.3 some CNN models for the prediction of air quality are proposed. These are derived analytically from traditional nonlinear PDE models. Simulation results of some cases are also reported. The CNN approach has many advantages over traditional finite-elements methods; among these the time-continuity of the CNN models and the natural computing parallelism of this analogic architecture.

3. The CNN as a Generator of Nonlinear Dynamics

Many researchers would claim that the main advances that have revolutionized science in this century are Einstein's theory of relativity, quantum mechanics, and the chaos theory. In fact, the chaotic behavior of nonlinear dynamical systems, although not explicitly mentioned, was already known in its essence by Henri Poincaré and other mathematicians at the end of the last century [67, 68].

Nevertheless a rapidly growing interest in chaotic and complex systems has involved several scientific disciplines in the last two or three decades; these, of course, include electrical and electronic engineering [69].

Indeed, as engineers, having acknowledged that chaotic oscillations show up in strongly nonlinear systems such as electronic circuits [69–71], we are interested in their study in view of applications.

To cite a few examples, chaos has been applied to broaden the capture range of phase-locked loops, in pseudo-random number generators, in secure communication systems, to suppress the injection of harmonics back into the power distribution network due to switched-mode power supplies and so on [69, 71]. Moreover, in some cases, it is of interest to control a chaotic system to follow a desired (e.g. periodic) trajectory [72].

Cellular neural networks are nonlinear dynamic circuits and so it is natural to expect that chaotic motions can show up. In fact, F.Zou and J.A.Nossek reported [73] and studied [74, 75] the first chaotic attractor in a CNN composed of three cells[1].

On these basis, however, given that a CNN is a programmable circuit[2] and so extremely flexible, it is natural to ask "what kind of nonlinear dynamics can be obtained from a CNN?". This question is harder to answer than might be expected at first sight.

The reasons for these difficulties are numerous. But just two of them are enough to give the reader a glimpse of the challenge that this question implies:

1. It well-known that *general* methods for the study of nonlinear systems do not exist and the analytic study of a third-order system[3] can be often extremely hard if not an intractable problem [67, 68, 76, 77].
2. Every cell in a CNN increases the system order by one unit.

---

[1] Specifically, a Chua and Yang model.
[2] By changing its templates.
[3] The minimum order required to have chaos in a continuous-time system.

44   3. The CNN as a Generator of Nonlinear Dynamics

As previously stated however, several electronic circuits have been found to behave in a chaotic way under certain choices of their parameters. Moreover, analytical studies concerning the nature of their dynamics and strange attractors have been carried out for some of them.

A more accessible question could then be: "is it possible to obtain the dynamics belonging to many different circuits and systems on a cellular neural network?".

What would be the consequence of a positive answer to this question? First of all a theoretical consequence: the CNN would represent a general model for the class of circuits for which it reproduces the dynamics. In other words, the cell would be the *primitive* of a wide class of dynamic circuits.

Secondly, but no less important, a practical consequence: given that chaos, nowadays, is actively used in engineering applications, being able to obtain several different dynamics from the same unique circuit would boost the capabilities of existing applications and open the way to new and powerful ones [78].

In this chapter it is proved that a CNN is able to exactly reproduce the dynamics of several well-known nonlinear oscillators. The main emphasis is on chaotic and hyperchaotic dynamics but more exotic behaviors, such as the so-called *canards*, are also considered. In particular, eight different cases are considered, including both theoretical and experimental results. Thereafter a general analysis of the conditions under which the dynamics of a circuit can be reproduced by such a CNN is presented.

A wider discussion about the consequences of this result will be given as conclusion of this chapter, while Chap. 4 will discuss some of the applications.

3.1 The State Controlled CNN Model

In Chua and Yang's original definition of CNN discussed in Sect. 1.1 the coupling of the cells is obtained through the inputs and nonlinear outputs only.

In [55] however this restriction has been removed by the introduction of a direct dependence on the neighbors' state.

In Sect. 1.3.1, Def. 1.3.4, it was emphasized that a direct dependence of the cell dynamics on the state vector of its neighbors is introduced by the *linear feedback template* (also called *state template*) $\hat{A}$.

Strictly speaking it can be easily proved that this dependence is implicitly obtained even without the need of the state template if the output nonlinearity $y = f(x)$ is such that $\partial f / \partial x \neq 0$ almost everywhere[4].

Regardless of this it is useful to introduce a particular sub-class of cellular neural networks.

---

[4] This is not true in the Chua and Yang model.

3.1 The State Controlled CNN Model

Definition 3.1.1 (SC-CNN [55, 79]). *A state-controlled cellular neural network (SC-CNN[5]) is a CNN with non-zero linear feedback template $\hat{A}$.*

In this chapter we will consider SC-CNNs with a small number of cells. Therefore, we will often refer to the simple one-dimensional linear SC-CNN with the following state equation:

$$\dot{x}_i = -x_i + \sum_{C(j) \in N_r(i)} \left[\hat{A}_{i;j} x_j + A_{i;j} y_j + B_{i;j} u_j \right] + I \qquad (3.1)$$
$$1 \leq i, j \leq N$$

or to its multi-layer version.

The theoretical propositions of this chapter will be complemented by experimental results. For this purpose, given that the order of the circuits considered is relatively low, simple implementations using off-the-shelf discrete components have been assembled.

3.1.1 Discrete Components Realization of SC-CNN Cells

If a circuit prototype for a SC-CNN with a limited number of cells is needed, an op-amp-based circuit is often a good choice on account of its simplicity and reliability. Here two possible circuit realizations for a SC-CNN cell are given. The first one is shown in Fig. 3.1; it is essentially constituted by three blocks.

The block $B1$ implements the output nonlinear function by exploiting the natural output saturation of amplifiers. It basically consists of an inverting

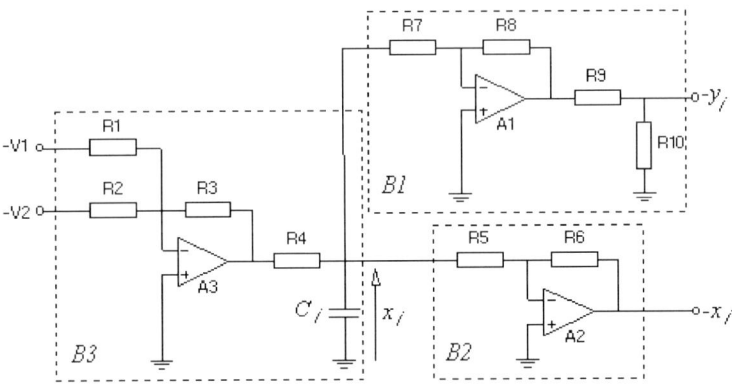

Fig. 3.1. First circuit realization for a SC-CNN cell

---

[5] The acronym SC should not be confused with the abbreviation used for Switched Capacitor circuits.

amplifier stage in which the gain has to be chosen so that the output saturates when the input voltage reaches the breakpoints (i.e. when $|x_{ij}| \geq 1$). This amplifier is followed by a voltage divider that is used to scale the output voltage in the $[-1,1]$ range. From these considerations the following design equations hold:

$$R_8/R_7 = V_{\text{sat}A}/V_{\text{sat}x} \tag{3.2a}$$
$$R_7/R_8 = R_{10}/(R_9 + R_{10}), \tag{3.2b}$$

where $V_{\text{sat}A}$ is the output saturation voltage of $A1$, while $V_{\text{sat}x}$ is its corresponding input voltage (i.e., in our case $V_{\text{sat}x} = 1$). The input and output impedances of $B1$ are R_7 and the parallel of R_9 and R_{10}, respectively.

The second block, $B2$, is a simple unit gain inverting amplifier (so $R_5 = R_6$) with an input impedance equal to R_5.

The block $B3$ is the fundamental core of the cell and is constituted by an inverting summing amplifier followed by a simple RC network. The impedances seen from the two inputs $-V_1$ and $-V_2$ are R_1 and R_2. If the parallel of the input impedances of blocks $B2$ and $B1$ is very high compared with the output impedance of block $B3$ (that is, $R_4/(1 + j\omega R_4 C_j)$), then blocks $B2$ and $B1$ do not strongly influence the capacitor voltage. If this hypothesis is satisfied the following state equation holds:

$$C_j \dot{x}_j = -\frac{x_j}{R_4} + \frac{R_3}{R_1 R_4} V_1 + \frac{R_3}{R_2 R_4} V_2. \tag{3.3}$$

This equation is formally equivalent (apart from a constant multiplying coefficient) to the SC-CNN model (3.1); in fact the inputs can be fed by the signals corresponding to contributions from the other cells.

However, this circuit can be simplified, reducing the number of op-amps needed, if an algebraic summing amplifier is used instead of the summing inverting amplifier of block $B3$ [80]. In this way the inverting amplifier of block $B2$ can be avoided because each of the possible signs for the desired

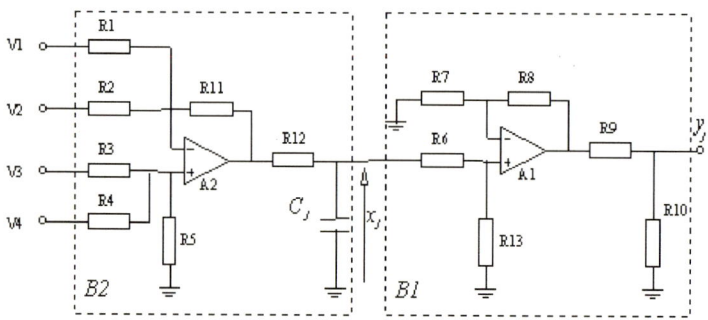

Fig. 3.2. Second circuit realization for a SC-CNN cell

gains can be obtained easily. This circuit is shown in Fig. 3.2 and is made up of just two blocks: the algebraic summing amplifier of block $B2$ and the non-inverting amplifier block $B1$ used to implement the nonlinear output function (so, again, this is designed to have a gain corresponding to the $V_{\text{sat}A}/V_{\text{sat}x}$ ratio). Here again, the input impedance of $B1$ is chosen so that it is very high with respect to the output impedance of $B2$. Both blocks have been completed with offset compensation resistors (R_5 and R_{13}) that must be chosen in accordance with the following relations:

$$\frac{1}{R_5} = \frac{1}{R_{11}} + \frac{1}{R_1} + \frac{1}{R_2} - \frac{1}{R_3} - \frac{1}{R_4}, \qquad (3.4a)$$

$$\frac{1}{R_{13}} = \frac{1}{R_8} + \frac{1}{R_7} - \frac{1}{R_6}. \qquad (3.4b)$$

The corresponding state equation for the single cell is therefore:

$$C_j \dot{x}_j = -\frac{x_j}{R_{12}} + \frac{R_{11}}{R_1 R_{12}} V_1 + \frac{R_{11}}{R_2 R_{12}} V_2 - \frac{R_{11}}{R_3 R_{12}} V_3 - \frac{R_{11}}{R_4 R_{12}} V_4. \qquad (3.5)$$

Both the schemes presented can easily be extended to an arbitrary number of inputs.

3.2 Chua Oscillator Dynamics Generated by the SC-CNN

3.2.1 Main Result

The Chua oscillator[6], shown in Fig. 3.3, is the simplest autonomous third-order nonlinear electronic circuit with a rich variety of dynamical behaviors including chaos, stochastic resonance, 1/f noise spectrum, and chaos–chaos intermittency [47, 81]. Its state equation is

$$\frac{dv_1}{d\tau} = \frac{1}{C_1}\left[G(v_2 - v_1) - g(v_1)\right],$$

$$\frac{dv_2}{d\tau} = \frac{1}{C_2}\left[G(v_1 - v_2) + i_3\right], \qquad (3.6a)$$

$$\frac{di_3}{d\tau} = -\frac{1}{L}\left[v_2 + R_0 i_3\right],$$

where

$$g(v_1) = G_b v_1 + 0.5 \cdot (G_a - G_b) \cdot [|v_1 + E| - |v_1 - E|] . \qquad (3.6b)$$

The piecewise-linear resistor N_r of $i - v$-characteristic $g(v)$ is known as the *Chua's diode*.

---
[6] Also known as the *unfolded Chua circuit*.

48   3. The CNN as a Generator of Nonlinear Dynamics

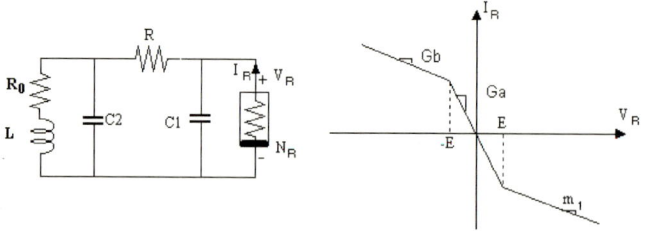

Fig. 3.3. Chua's oscillator

Taking

$$\begin{aligned}
x &= v_1/E; & y &= v_2/E; & z &= i_3/(EG); \\
t &= (\tau G)/C_2; & m_0 &= (G_a/G)+1; & m_1 &= (G_b/G)+1; \\
\alpha &= C_2/C_1; & \beta &= C_2/(LG^2); & \gamma &= (C_2 R_0)/(GL);
\end{aligned} \quad (3.7)$$

equation (3.6) can be rewritten in the more convenient dimensionless form

$$\begin{aligned}
\dot{x} &= \alpha\left[y - h(x)\right], \\
\dot{y} &= x - y + z, \\
\dot{z} &= -\beta y - \gamma z,
\end{aligned} \quad (3.8a)$$

where

$$h(x) = m_1 x + 0.5 \cdot (m_0 - m_1) \cdot [|x+1| - |x-1|], \quad (3.8b)$$

x, y, and z being the state variables and α, β, γ, m_0, m_1 the system parameters.

Chua proved that the state equation (3.8), which is uniquely determined by 5 parameters, is topologically conjugate[7] to a 21-parameter family of continuous odd-symmetric piecewise-linear equations in $\mathbb{R}^3$ [81].

On account of this result Chua's oscillator is considered the canonical circuit for studying chaos.

A zoo of more than 30 strange attractors generated by these equations can be found in [81]. It is nevertheless worth noting that the only strange attractor admissible on Chua's oscillator for positive circuit elements is that known as the *double scroll*. All of the other attractors have been obtained with negative capacitors/inductors.

Let us now consider the following result [55].

Proposition 3.2.1 (Chua's oscillator dynamics in a SC-CNN). *The dynamics of Chua's Oscillator with state equation (3.8) can be obtained in a SC-CNN with state equation (3.1) $\forall \alpha, \beta, \gamma, m_0, m_1 \in \mathbb{R}$.*

---
[7] Refer to Def. A.8.2 in Appendix A.

3.2 Chua Oscillator Dynamics Generated by the SC-CNN

Proof. Chua's oscillator is a third-order nonlinear dynamic system, and therefore if an equivalent dynamic system is to be obtained by using SC-CNN cells, three cells are needed: one for each state variable of the given circuit.

Let us assume the correspondence $x_1 = x$, $x_2 = y$, $x_3 = x$.

Observe (3.8) and notice that the first state equation (associated with x) does not contain any contribution from z; the second one (associated with y) contains contributions from x (the previous one) and z (the following one); finally, the third equation (associated with z) does not contain any contribution from x. This implies that a unit neighborhood radius r can be chosen. Moreover, Chua's oscillator is an autonomous circuit so the control templates in (3.1) are equal to zero.

So, finally, by direct comparison of (3.8) and (3.1), the following identities provide the templates for the SC-CNN (3.1) and prove the equivalence of the two systems:

$$\begin{aligned}
&A_{1;2} = A_{1;3} = A_{2;2} = A_{2;3} = A_{3;2} = A_{3;3} = 0; \\
&A_{2;1} = A_{3;1} = 0; \quad \hat{A}_{1;3} = \hat{A}_{3;1} = \hat{A}_{2;2} = 0; \\
&I_1 = I_2 = I_3 = 0; \quad A_{1;1} = \alpha \cdot (m_1 - m_0); \\
&\hat{A}_{3;3} = 1 - \gamma; \quad \hat{A}_{2;1} = \hat{A}_{2;3} = 1; \\
&\hat{A}_{1;1} = 1 - \alpha \cdot m_1; \quad \hat{A}_{1;2} = \alpha; \quad \hat{A}_{3;2} = -\beta.
\end{aligned} \qquad (3.9)$$

□

It can be seen, however, that the SC-CNN with templates (3.2.1) is not space-invariant because each of the three cells has its own template set. This is fully compatible with the definition of CNNs and it does not create technical problems because of the rather limited size of the network. However, although it does not need analytic proof, it is worth noting that a three-layer linear SC-CNN is again able to reproduce the above dynamics.

Hence a simple three-layer linear SC-CNN is able to fully reproduce the dynamics of a CNN having the Chua oscillator as cell, a case quite commonly found in the literature [14, 20, 82, 83].

3.2.2 Experimental Results

The SC-CNN provided by Proposition 3.2.1 can be immediately implemented by using for instance one of the approaches discussed in Sect. 3.1.1.

Let us consider a few examples.

A *double scroll attractor* is observed in Chua's circuit dynamics if $\alpha = 9$, $\beta = 14.286$, $\gamma = 0$, $m_0 = -1/7$ and $m_1 = 2/7$. The simulated phase portrait in the $x - y$ plane for the Chua's oscillator equation (3.8) is shown in Fig. 3.4. Inserting these parameter values into (3.2.1) the corresponding SC-CNN is obtained. A circuit implementing it is shown in Fig. 3.5 [80]. The corresponding experimental phase portrait in the $x_1 - x_2$ plane is shown in

50   3. The CNN as a Generator of Nonlinear Dynamics

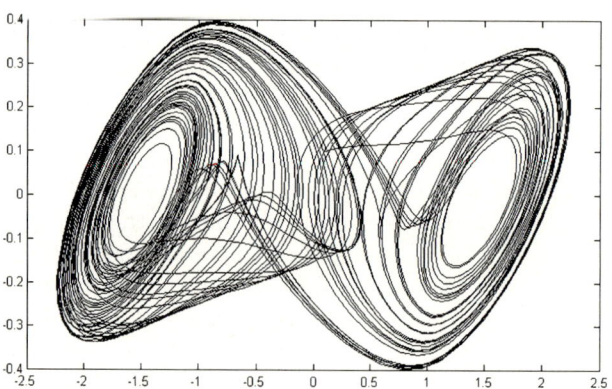

Fig. 3.4. The double scroll attractor in the $x - y$ plane

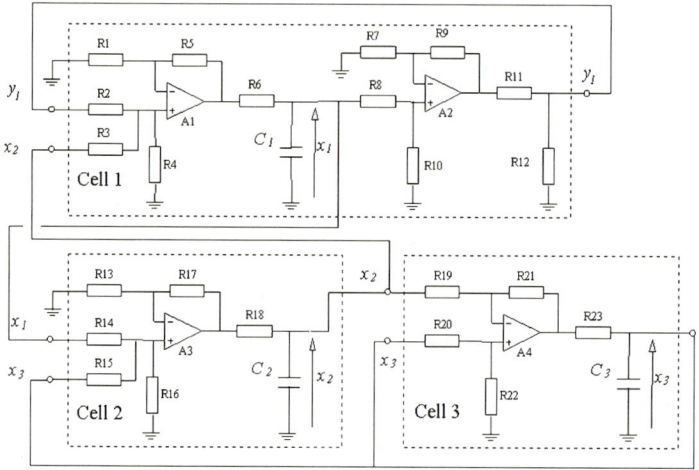

Fig. 3.5. A SC-CNN implementation for the double scroll dynamics

Fig. 3.6, while the part list is reported in Table 3.1. It is interesting to note that with this approach, given that templates have no restriction concerning signs, the SC-CNN can implement in a straightforward way any of the strange attractors that in Chua's oscillators would have required negative elements[8].

With regard to this let us consider two examples. In the first one the following set of parameters is considered: $\alpha = -4.08685$, $\beta = -2$, $\gamma = 0$, $m_0 = -1/7$, $m_1 = 2/7$. The simulated attractor obtained by system (3.8) is shown in Fig. 3.7. The associated experimental phase portrait obtained with the SC-CNN is shown in Fig. 3.8. In the second example the following

---

[8] Implementations of negative resistances, capacitances, and inductances do exist but they involve complex and often band-limited additional circuitry.

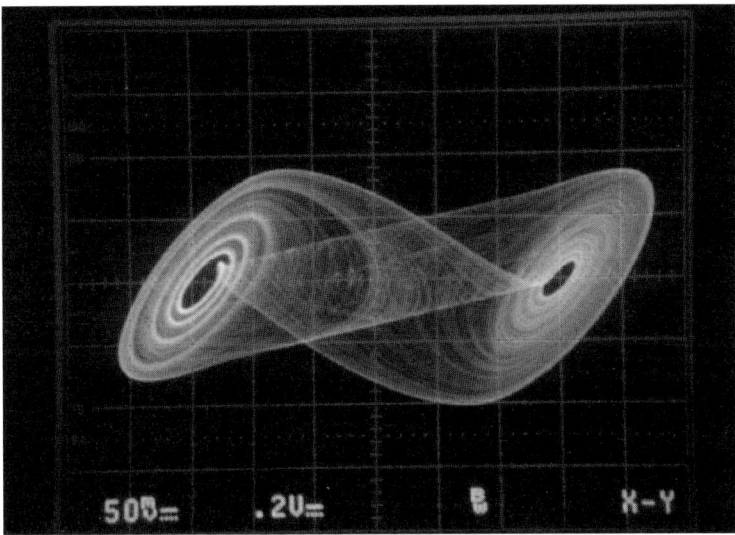

Fig. 3.6. The experimental phase portrait in $x_1 - x_2$ obtained by the SC-CNN

Table 3.1. Part list for the circuit shown in Fig. 3.5

Cell 1				
R1=4K;	R2=13.2K;	R3=5.7K;	R4=20K;	R5=20K;
R6=1K;	R7=75K;	R8=75K;	R9=1M;	R10=1M;
R11=12.1K;	R12=1K;	C1=100n;		
Cell 2				
R13=51.1K;	R14=100K;	R15=100K;	R16=100K;	R17=100K;
R18=1K;	C2=100n;			
Cell 3				
R19=8.2K;	R20=100K;	R21=100K;	R22=7.8K;	R23=1K;
C3=100n;				

set is considered: $\alpha = -6.69191$, $\beta = -1.52061$, $\gamma = 0$, $m_0 = -1/7$, $m_1 = 2/7$. Figure 3.9 shows the corresponding simulation obtained by the Chua oscillator, while Fig. 3.10 shows the experimental phase portrait.

3.3 Chaotic Dynamics of a Colpitts Oscillator

It has been recently shown that a Colpitts oscillator, for a particular choice of its parameters, shows a strange attractor in its dynamics [84]. In addition, it has been proved that this attractor is actually topologically conjugated to a member of the Chua circuit family [85, 86].

52   3. The CNN as a Generator of Nonlinear Dynamics

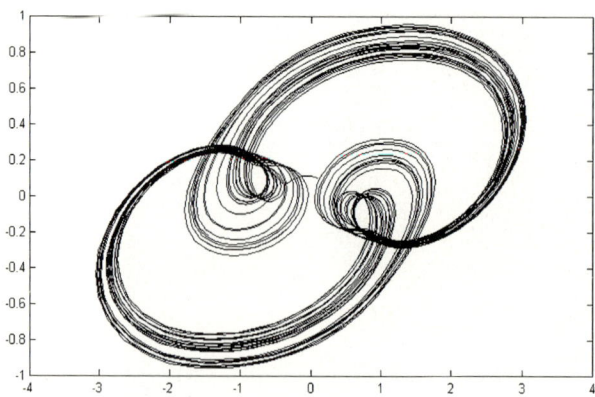

Fig. 3.7. Strange attractor for the first set of parameters in the $x - y$ plane

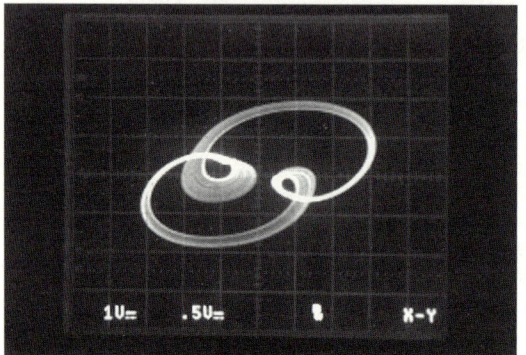

Fig. 3.8. The experimental phase portrait in $x_1 - x_2$ obtained by the SC-CNN for the first set of parameters

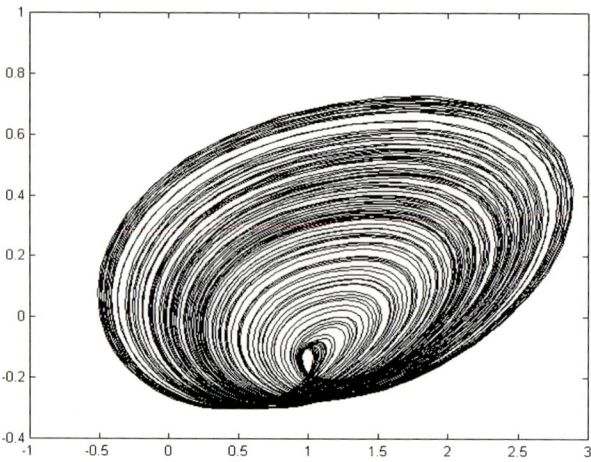

Fig. 3.9. Strange attractor for the second set of parameters in the $x - y$ plane

3.3 Chaotic Dynamics of a Colpitts Oscillator

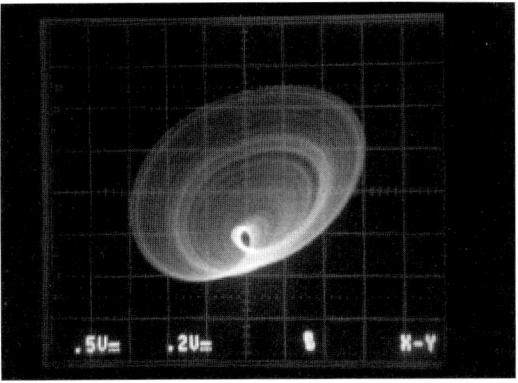

Fig. 3.10. The experimental phase portrait in $x_1 - x_2$ obtained by the SC-CNN for the second set of parameters

A piecewise linear (PWL)[9] model of the Colpitts oscillator is [84]

$$C\frac{dV_{CE}}{d\tau} = I_L - I_C,$$
$$C\frac{dV_{BE}}{d\tau} = -\frac{V_{EE} + V_{BE}}{R_{EE}} - I_L - I_B, \quad (3.10a)$$
$$L\frac{dI_L}{d\tau} = V_{CC} - V_{CE} + V_{BE} - I_L R_L,$$

with

$$I_B = \begin{cases} 0 & \text{if } V_{BE} \le V_{TH} \\ \frac{V_{BE} - V_{TH}}{R_{ON}} & \text{if } V_{BE} > V_{TH}, \end{cases} \quad (3.10b)$$

$$I_C = \beta_F I_B, \quad (3.10c)$$

where V_{CE}, V_{BE} and I_L are the state variables. The strange behavior is observed in the dynamics of the circuit if the following values are assumed for the model parameters:

$$V_{TH} = 0.75\,\text{V}; \quad R_{ON} = 200\,\Omega; \quad R_L = 35\,\Omega; \quad L = 98.5\,\mu\text{H};$$
$$C = 54\,\text{nF}; \quad R_{EE} = 400\,\Omega; \quad \beta_F = 256; \quad V_{EE} = -5\,\text{V}; \quad V_{CC} = 5\,\text{V}. \quad (3.11)$$

The phase portrait obtained by PSPICE simulation of the Colpitts oscillator is depicted in Fig. 3.11. This strange attractor can be reproduced in a SC-CNN composed of three cells [87, 88].

Proposition 3.3.1. *The dynamics of the Colpitts oscillator with state equation (3.10) can be obtained in a SC-CNN with state equation (3.1).*

---
[9] See Def. A.2.4 of Appendix A.

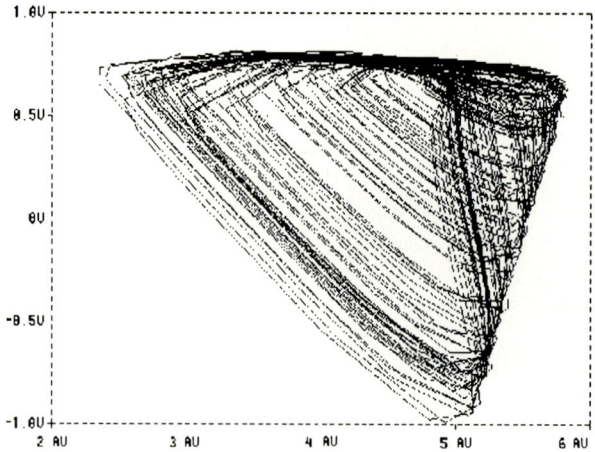

Fig. 3.11. PSPICE simulation of the Colpitts attractor in the $V_{ce} - V_{be}$ plane

Proof. The following normalized variables are assumed to represent the system given in (3.10):

$$\begin{aligned}
x_1 &= \frac{V_{CE}}{V_{CC}}; \\
x_2 &= \frac{aV_{BE} + b}{k}; \\
x_3 &= \frac{R_L I_L}{V_{CC}}; \\
t &= \frac{\tau}{R_L C};
\end{aligned} \tag{3.12a}$$

with

$$k = aV_{TH} + b, \tag{3.12b}$$

a and b being suitable real constants used to obtain the required nonlinearity (a possible choice is $a = b = 4$). Moreover, the nonlinear function (3.10b) can be substituted by the following function without changes in the system behaviour

$$I_B = \frac{k}{aR_{ON}} [x_2 - 0.5(|x_2 + 1| - |x_2 - 1|)] \tag{3.13}$$

because the independent variable x_2 never assumes values less than -1, so, in the range of values actually assumed by x_2 and with the linear transformation described above, the two functions are equivalent.

Therefore the following dimensionless state representation is obtained:

$$\dot{x}_1 = -x_1 + \left[x_1 - \frac{\beta_F k R_L}{a R_{ON} V_{CC}} x_2 + x_3\right] + \frac{\beta_F k R_L}{a R_{ON} V_{CC}} y_2,$$

$$\dot{x}_2 = -x_2 + \left[(1 - \frac{R_L}{R_{EE}} - \frac{R_L}{R_{ON}})x_2 - \frac{aV_{CC}}{k}x_3\right]$$
$$+ \frac{R_L}{R_{ON}} y_2 + \frac{R_L(b - aV_{EE})}{kR_{EE}}, \quad (3.14a)$$

$$\dot{x}_3 = -x_3 + \left[-\frac{R_L^2 C}{L} x_1 + \frac{kR_L^2 C}{aLV_{CC}} x_2 + (1 - \frac{R_L^2 C}{L})x_3\right]$$
$$+ \frac{R_L^2 C}{L} - \frac{bR_L^2 C}{aLV_{CC}},$$

where

$$y_2 = 0.5 \cdot (|x_2 + 1| - |x_2 - 1|). \quad (3.14b)$$

But this is the equation of a SC-CNN with 3 cells and the following template coefficients:

$$\hat{A}_{1;1} = 1; \quad \hat{A}_{1;2} = -\frac{\beta_F k R_L}{a R_{ON} V_{CC}}; \quad \hat{A}_{1;3} = 1; \quad A_{1;2} = \frac{\beta_F k R_L}{a R_{ON} V_{CC}};$$

$$\hat{A}_{2;2} = 1 - \frac{R_L}{R_{EE}} - \frac{R_L}{R_{ON}}; \quad \hat{A}_{2;3} = -\frac{aV_{CC}}{k} \quad A_{2;2} = \frac{R_L}{R_{ON}}; \quad (3.15)$$

$$\hat{A}_{3;1} = -\frac{R_L^2 C}{L}; \quad \hat{A}_{3;2} = \frac{kR_L^2 C}{aLV_{CC}}; \quad \hat{A}_{3;3} = 1 - \frac{R_L^2 C}{L};$$

$$I_2 = \frac{R_L(b - aV_{EE})}{kR_{EE}}; \quad I_3 = \frac{R_L^2 C}{L} - \frac{bR_L^2 C}{aLV_{CC}}.$$

This proves the proposition. □

The experimental phase portrait obtained by the SC-CNN is shown in Fig. 3.12.

3.4 Hysteresis Hyperchaotic Oscillator

In continuous-time systems, when the system order is higher than 3, it is possible to have more than one positive Lyapunov exponent[10]. If this is the case, the corresponding behavior is known as *hyperchaos*.

T. Saito recently introduced a four-dimensional nonlinear autonomous circuit that shows hyperchaos [89]. It is called the *Saito hysteresis chaos generator* (for simplicity we will refer to it as the SHCG).

---

[10] See Sect. A.7.3 in Appendix A.

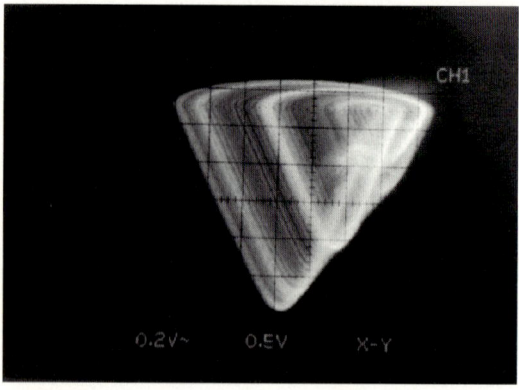

Fig. 3.12. The experimental phase portrait in $x_1 - x_2$ obtained by the SC-CNN for the Colpitts dynamics

The dimensionless state equation of the SHCG is

$$\begin{aligned} \dot{x} &= -z - w, \\ \dot{y} &= \gamma(2\delta y + z), \\ \dot{z} &= \rho(x - y), \\ \epsilon \dot{w} &= x - h(w), \end{aligned} \qquad (3.16a)$$

where

$$h(w) = w - (|w + 1| - |w - 1|), \qquad (3.16b)$$

x, y, z and w being the state variables and γ, δ, ρ and ϵ the system parameters. The SHCG is shown in Fig. 3.13. N_r is a nonlinear resistor and is responsible for $h(x)$. Letting $\epsilon \to 0$, the nonlinear resistor N_r and the inductor L_0 lead to the *jump phenomenon* and *hysteresis* [90].

Moreover, by varying the parameters of the SHCG it is possible to generate a wide variety of dynamic behaviors (periodic solutions, quasi-periodic solutions, chaos, hyperchaos).

This fourth-order oscillator is, again, a member of the SC-CNN family. In fact, the following proposition holds [91]:

Proposition 3.4.1. *The dynamics of the SHCG circuit with state equation (3.16) can be obtained in a SC-CNN with state equation (3.1), $\forall \gamma, \delta, \rho, \epsilon \in \mathbb{R}$.*

Proof. The proof is immediately achieved if in equation (3.1), for a SC-CNN composed of 4 cells, the following template coefficients are taken

3.4 Hysteresis Hyperchaotic Oscillator

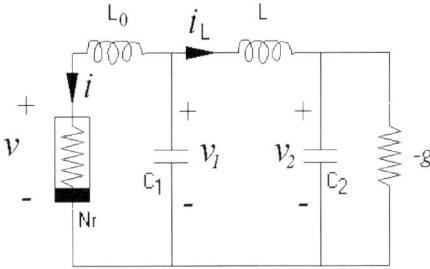

Fig. 3.13. Saito hysteresis chaos generator

$$A_{k;k} = 0; \text{ for } k = 1, 2, 3 \quad A_{j;k} = 0; \text{ for } j, k = 1, \ldots, 4, (j \neq k)$$
$$\hat{A}_{1;2} = \hat{A}_{2;1} = \hat{A}_{2;4} = \hat{A}_{3;4} = \hat{A}_{4;2} = \hat{A}_{4;3} = 0; \quad i_j = 0; \text{ for } j = 1, \ldots, 4$$
$$\hat{A}_{1;1} = 1; \quad \hat{A}_{1;3} = \hat{A}_{1;4} = -1; \quad \hat{A}_{2;2} = 1 + 2\gamma\delta; \quad \hat{A}_{2;3} = \gamma;$$
$$\hat{A}_{3;1} = \rho; \quad \hat{A}_{3;2} = -\rho; \quad \hat{A}_{3;3} = 1; \quad A_{4,4} = 2/\epsilon;$$
$$\hat{A}_{4;1} = 1/\epsilon; \quad \hat{A}_{4;4} = 1 - 1/\epsilon;$$
(3.17)

and assuming the correspondence $x_1 = x$, $x_2 = y$, $x_3 = z$ and $x_4 = w$. □

The experimentally observed hyperchaotic attractor, corresponding to the parameter set $\gamma = 1$, $\rho = 14$, $\delta = 1$ and $\epsilon \to 0$ (in practice $\epsilon = 10^{-2}$ is sufficient) obtained in the four-cells SC-CNN is reported in Fig. 3.14.

Incidentally, this was the very first example of true hyperchaotic attractor ever observed in a CNN [91].

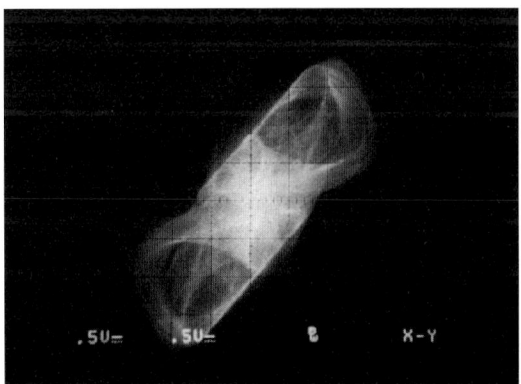

Fig. 3.14. The experimental phase portrait in $x_1 - x_2$ obtained by the SC-CNN for the SHCG hyperchaotic attractor

3.5 n-Double Scroll Attractors

In Sect. 3.2 we already encountered the well-known *double-scroll attractor*[11] of the Chua oscillator. Indeed, this attractor and its geometric structure are probably among the most extensively studied [47, 92, 93] in electrical engineering.

In [94] J.Λ.K. Suykens and J. Vandewalle introduced a new family of circuits derived from the traditional Chua's circuit by modifying the characteristic of the Chua diode. The corresponding attractors of this generalized Chua circuit have been called *n-double scroll attractors*. One of these looks like more double scroll attractors, of different sizes, one contained inside another like a Russian matrioska. In this framework, the classic double scroll corresponds to the *1-double scroll*.

The generalized Chua diode of Suykens and Vandewalle, however, had a quite involved $i-v$ characteristic presenting intervals with polynomial nonlinearity and singular points. Indeed, a circuit implementation of that specific family has never been presented.

In this section we show that n-double scroll attractors[12] can also be obtained if a PWL continuous $i-v$ characteristic is chosen for the Chua diode.

Moreover, the n-double scroll family obtained will be realized by a SC-CNN with appropriate output nonlinearity.

3.5.1 A New Realization of the n-Double Scroll Family

A general expression for a $(n+1)$-segments scalar PWL function $f \colon \mathbb{R} \to \mathbb{R}$ is given by

$$f(x) = a_0 + a_1 x + \sum_{j=1}^{n} b_j |x - E_j|, \qquad (3.18)$$

where $E_1 < E_2 < \cdots < E_n$ are the n break-points and $a_0, a_1, b_1, b_2, ..., b_n \in \mathbb{R}$ are related to the segment slopes and to $f(0)$ by the following expressions:

$$a_1 = \frac{1}{2}(m_0 + m_n), \quad b_j = \frac{1}{2}(m_j - m_{j-1}), \quad a_0 = f(0) - \sum_{j=1}^{n} b_j |E_j| \qquad (3.19)$$

in which m_0 is the first linear segment slope and m_j is the $(j+1)$-th linear segment slope (Fig. 3.15).

Let us then introduce the following generalized Chua oscillator:

---

[11] That has been proved to be a true chaotic attractor in the sense of Šilnikov [92]. See also Sect. A.9 in Appendix A.
[12] That, most likely, are topologically conjugated with the n-double scrolls in [94].

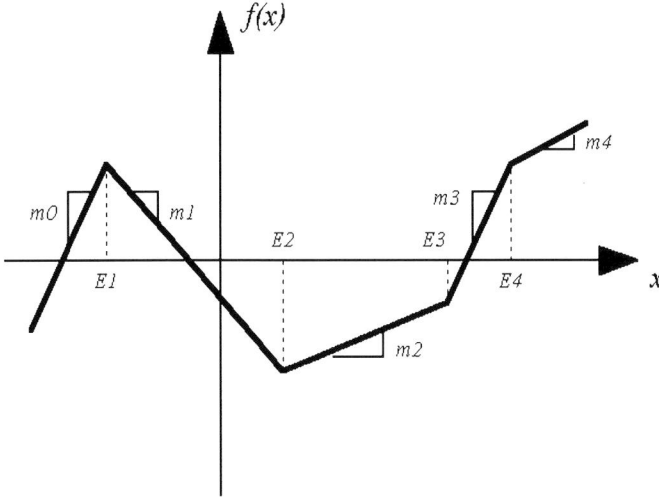

Fig. 3.15. A generic PWL continuous function

Definition 3.5.1 (PWL n-double scroll family [95]). *The state equations*

$$\dot{x} = \alpha \left[y - h(x) \right]$$
$$\dot{y} = x - y + z \quad (3.20\mathrm{a})$$
$$\dot{z} = -\beta y - \gamma z,$$

where

$$h(x) = m_{2n-1} x + \frac{1}{2} \sum_{k=1}^{2n-1} (m_{k-1} - m_k) \left[|x + b_k| - |x - b_k| \right] ; \quad (3.20\mathrm{b})$$

b_k *are the $2n-1$ break points and m_k are the $2n$ corresponding slopes, define a family of circuits, systems or vector fields that will be referred to as the PWL n-double scroll family.*

Observe that $b_1 = 1$, $h(0) = 0$, $h(x) = -h(-x)$ and the restriction of (3.20) to $-b_2 \le x \le b_2$ coincide with the classic Chua circuit but the nonlinear resistor has been modified maintaining its PWL nature.

Moreover, (3.20) is invariant under the transformation $s \colon \mathbb{R}^3 \to \mathbb{R}^3$:

$$s(x, y, z) \doteq \begin{pmatrix} -x \\ -y \\ -z \end{pmatrix} \quad (3.21)$$

and so the geometry of the state space of (3.20) has a symmetry about the origin.

Now, to understand what follows it is important to recall a few facts regarding the 1-double scroll.

1. $\gamma = 0$.
2. The state space $\mathbf{X}$ of the Chua oscillator can be partitioned into three subspaces:
$$D_0 = \{[x\ y\ z]' \in \mathbf{X} : |x| \leq 1\}, \tag{3.22a}$$
$$D_1 = \{[x\ y\ z]' \in \mathbf{X} : x > 1\}, \tag{3.22b}$$
$$D_{-1} = \{[x\ y\ z]' \in \mathbf{X} : x < -1\}. \tag{3.22c}$$

3. There exist three fixed points (saddle foci):
$$\boldsymbol{x}_1 = \begin{pmatrix} -\kappa \\ 0 \\ \kappa \end{pmatrix} \in D_1, \quad \boldsymbol{x}_0 = \begin{pmatrix} 0 \\ 0 \\ 0 \end{pmatrix} \in D_0, \quad \boldsymbol{x}_{-1} = \begin{pmatrix} \kappa \\ 0 \\ -\kappa \end{pmatrix} \in D_{-1}, \tag{3.23a}$$

where
$$k \doteq \frac{m_1 - m_0}{m_1}. \tag{3.23b}$$

4. The eigenvalues of the linearized systems corresponding to these fixed points are
$$\begin{array}{c} \lambda(\boldsymbol{x}_0) = \{\gamma_0, \sigma_0 \pm j\omega_0\}; \ \lambda(\boldsymbol{x}_1) = \lambda(\boldsymbol{x}_{-1}) = \{\gamma_1, \sigma_1 \pm j\omega_1\}; \\ \gamma_0 > 0; \ \sigma_0 < 0; \ \omega_0 > 0; \ \gamma_1 < 0; \ \sigma_1 > 0; \ \omega_1 > 0 \end{array} \tag{3.24}$$

and the dimensions of the corresponding stable and unstable eigenspaces are
$$\begin{array}{l} \dim E^s(\boldsymbol{x}_{\pm 1}) = \dim E^u(\boldsymbol{x}_0) = 1, \\ \dim E^u(\boldsymbol{x}_{\pm 1}) = \dim E^s(\boldsymbol{x}_0) = 2. \end{array} \tag{3.25}$$

5. Roughly speaking, the shape of the double scroll is constituted by two outer discs (due to $\boldsymbol{x}_{\pm 1}$), linked by inner (almost) straight trajectories (due to $\boldsymbol{x}_0$).

The proof in [92] of the true chaotic nature of the double scroll is based on the existence of an odd-symmetrically related pair of homoclinic trajectories $\mathcal{H}^\mp$ based at the origin $\boldsymbol{x}_0$ and consequent application of the Šilnikov theorem.

Let us now discuss an algorithm to generate n-double scroll from (3.20). From a geometrical point of view, once the *double scroll order* n has been fixed, the goal is to find the proper set of slopes for (3.20b) such that

1. the origin x_0 keeps its saddle focus nature;
2. a new pair of homoclinic trajectories $\mathcal{H}_n^{\mp}$ (still based at x_0) are formed.

Fortunately, a sufficient condition to fulfill the first requirement is to keep m_0 of (3.20b) at the value known for the 1-double scroll (i.e. $m_0 = -1/7$). Regarding the second condition, it can be noticed that the horseshoe chaos, according to the Šilnikov theorem, is structurally stable.

Exploiting the above cited symmetry, it is necessary to consider only the sub-space for $x > 0$.

Let us first obtain the 2-double scroll, starting from the 1-double scroll. For this purpose, it is necessary to add two new discs to the sides of the classical double scroll.

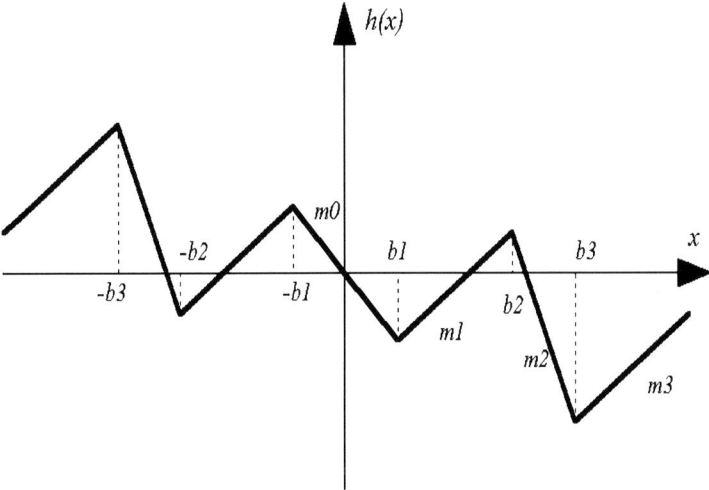

Fig. 3.16. The function $h(x)$ for the 2-double scroll

So let us add two new break points (b_2 and b_3 respectively) to the right hand side of b_1 in order to add a new segment with negative slope and another one with positive slope. The same modification will be made on the negative side in order to maintain the odd-symmetry of $h(x)$ (see Fig. 3.16). The new segment's slopes are m_2 and m_3.

As in the case of Chua's circuit, it is useful to partition the state space into different regions determined by the breakpoints of the PWL nonlinearity:

$$R_1 = \{[x\,y\,z]' \in \mathbf{X} \,:\, x \leq -b_3\},$$
$$R_2 = \{[x\,y\,z]' \in \mathbf{X} \,:\, -b_3 < x \leq -b_2\},$$
$$R_3 = \{[x\,y\,z]' \in \mathbf{X} \,:\, -b_2 < x \leq -b_1\},$$
$$R_4 = \{[x\,y\,z]' \in \mathbf{X} \,:\, -b_1 < x < b_1\}, \qquad (3.26)$$
$$R_5 = \{[x\,y\,z]' \in \mathbf{X} \,:\, b_1 \leq x < b_2\},$$
$$R_6 = \{[x\,y\,z]' \in \mathbf{X} \,:\, b_2 \leq x < b_3\},$$
$$R_7 = \{[x\,y\,z]' \in \mathbf{X} \,:\, x > b_3\}.$$

What we want to do is to introduce the new break points b_2 and b_3 close enough to the focus $\boldsymbol{x}_1$ so that some of the external trajectories in R_5, following a outward spiral path, can encounter the boundary imposed by b_2. When this happens, these trajectories are "snatched" because, due to m_2, the vector field changes direction in R_6. Hence, in R_6, the vector field accelerates the trajectories towards the new boundary b_3. At this point, the state enters R_7, where a new unstable focus (the stability of which depends on m_3), of the same nature of $\boldsymbol{x}_1$, forces an outward spiral motion. This would create the desired outer disc. These growing spiral trajectories will eventually again hit the boundary fixed by b_3. Depending on the region of this intersection, these ones are redirected again inside the outer region or accelerated towards the other inner regions (and hence towards the inner double scroll again).

This simple and intuitive strategy suggests some guidelines:

1. choose the break point b_2 near the right edge of the 1-scroll disc, i.e. around $x = 2$;
2. choose the slope of the outer segment (m_3 in this case) corresponding to the new disc, to equal the slope associated to the inner disc, i.e. $m_3 = m_1$;
3. if the slope m_2 is chosen equal to m_0 then too few trajectories will be snatched from the inner disc, therefore a higher slope is necessary;
4. choose the remaining parameters equal to the Chua's double scroll parameters, i.e. $\alpha = 9$, $\beta = 14.286$, $\gamma = 0$, $m_0 = -1/7$ and $m_1 = 2/7$.

With the above criteria in mind, a feasible choice is rapidly obtained by simulation. For instance:

$$b_2 = 2.15; \qquad b_3 = 3.6; \qquad m_2 = -4/7; \qquad m_3 = 2/7. \qquad (3.27)$$

With these parameters the attractor of Fig. 3.17 is obtained. The same attractor together with the 7 regions of (3.26) is shown in Fig. 3.18.

If a 3-double scroll is desired the above outlined strategy can be repeated. In fact, starting from the 2-double scroll, two other break points b_4 and b_5 must be added to $h(x)$ with corresponding negative and positive slopes.

In Fig. 3.19 the 3-double scroll is shown. It has been obtained by choosing $b_4 = 8.2$, $b_5 = 13$, $m_4 = m_2 = -4/7$ and $m_5 = m_3 = 2/7$.

It is now clear that this procedure obtains the n-double scroll from the $(n-1)$-double scroll. In order to further simplify the tuning in this procedure

3.5 n-Double Scroll Attractors 63

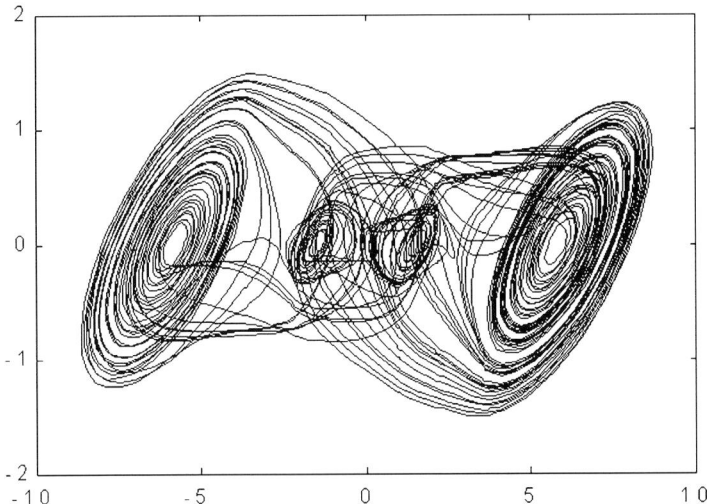

Fig. 3.17. The 2-double scroll attractor

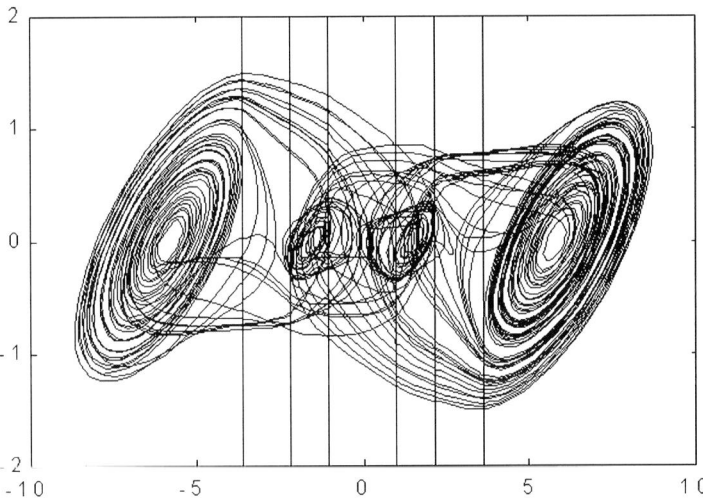

Fig. 3.18. The 2-double scroll attractor and the seven regions in which the state space can be partitioned

64   3. The CNN as a Generator of Nonlinear Dynamics

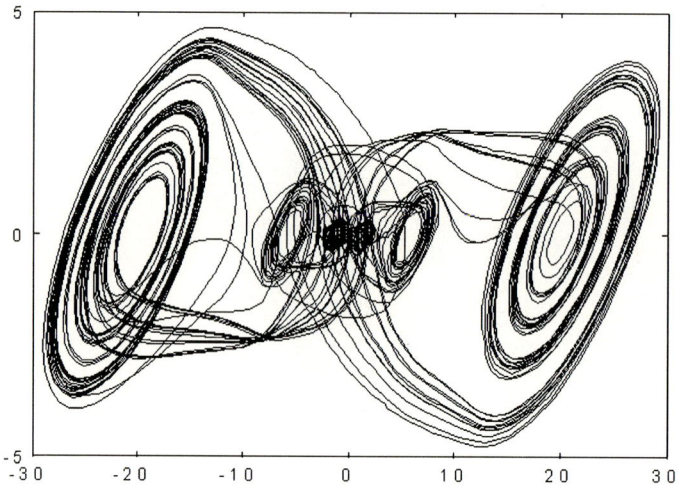

Fig. 3.19. The 2-double scroll attractor

we suggest taking the negative slope and the positive slope, respectively, equal to $-4/7$ and $2/7$ as in the case of the 3-double scroll; the break-point values can then be found using the above algorithm.

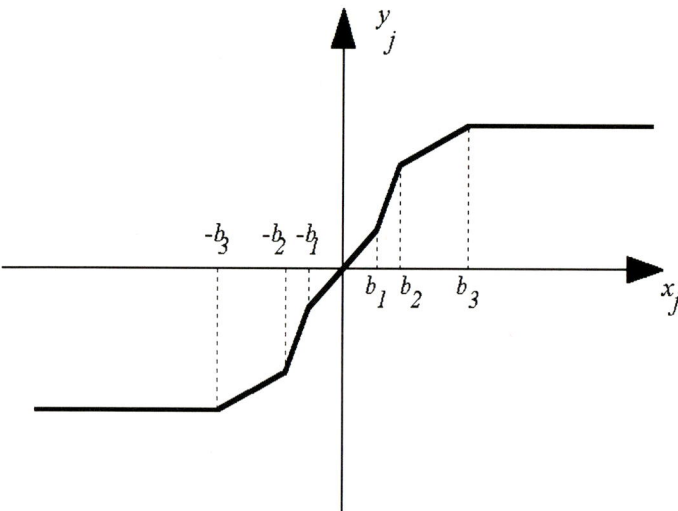

Fig. 3.20. The output nonlinearity of the SC-CNN admitting a 2-double scroll attractor

3.5.2 n-Double Scrolls in SC-CNNs

Once n-double scrolls have been obtained by means of the family of PWL systems (3.20), the circuit implementation using SC-CNN is straightforward.

A modification of the output function $y = f(x)$ is required however. The classic saturation function (1.1) is replaced by its PWL generalization

$$y_j = \frac{1}{2} \sum_{k=1}^{2n-1} n_k(|x + b_k| - |x - b_k|), \qquad (3.28)$$

where b_k are the above fixed break-points and n_k are related to the slopes of the segments composing (3.28). This function is shown in Fig. 3.20 for the 2-double scroll case. The template set for this three-cell SC-CNN will be the same as the one seen for the case of Chua's Oscillator (refer to equation (3.2.1) in Proposition 3.2.1) with the only exception of $\hat{A}_{1;1}$ now equal to $1 - \alpha m_{2n-1}$. The coefficients n_k in (3.28) are given by:

$$n_k = \alpha(m_k - m_{k-1}), \qquad k = 1, \ldots, 2n-1. \qquad (3.29)$$

In relation to the circuit implementation, again, the only difference from the other cases concerns the circuit for the output function (3.28). The circuit is shown in Fig. 3.21. This equation is obtained by a weighted sum of three nonlinear terms. These terms can be separately realized by inverting/non-inverting amplifiers whose outputs are saturated when their inputs reach the desired break-point (B1, B2, and B3). Afterwards, these three outputs are

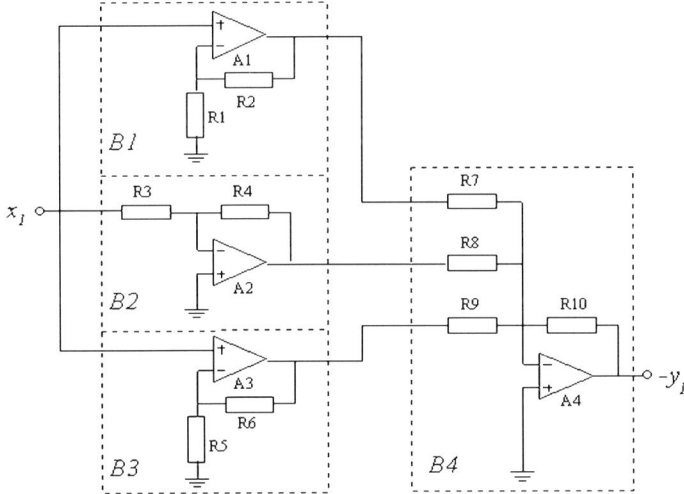

Fig. 3.21. A circuit implementation for the output output nonlinearity of the SC-CNN admitting a 2-double scroll attractor

added with appropriate weights by a summing amplifier (B4). Consider the blocks B1, B2, B3. If b_1, b_2 and b_3 are the corresponding desired break-points then the following equations hold:

$$1 + \frac{R_2}{R_1} = \frac{E_{sat}}{b_1}, \quad \frac{R_4}{R_3} = \frac{E_{sat}}{b_2}, \quad 1 + \frac{R_6}{R_5} = \frac{E_{sat}}{b_3} \qquad (3.30)$$

where E_{sat} is the saturation voltage of the differential amplifiers. The following summing stage is easily designed for the proper weights, taking into account that the various outputs of B1, B2, B3 are saturated when x_j reaches the corresponding break-points. Further details about the actual implementation and the values of the components can be found in [95]. The experimental phase portraits of the 2-double scroll obtained with this SC-CNN are shown in Fig. 3.22.

3.6 Nonlinear Dynamics Potpourri

From what we have seen shown so far, it may appear that only autonomous circuits with nonlinear resistors, whose $i - v$ characteristic is a function of a single state variable, belong to the SC-CNN family.

In reality there is no such limitation. In fact, in this section, we see how a further group of nonlinear circuits with particular features are realized by SC-CNNs. In particular, two non-autonomous circuits, one of which includes a nonlinear inductor, a circuit admitting a *Canard* and a circuit with a non-linearity function of the sum of two state variables are considered.

3.6.1 A Non-autonomous Second Order Chaotic Circuit

We now consider a second order non-autonomous (driven) circuit as described in [96]. The state equations of the circuit are

$$\begin{aligned} \epsilon \dot{x} &= y - f(x) \\ \dot{y} &= -x + u + b \end{aligned} \qquad (3.31a)$$

with $\epsilon > 0$; $k > 0$; $b > 0$; and

$$f(x) = x + 0.5(1 + k)(|x - 1| - |x + 1|), \qquad (3.31b)$$

where x and y are the state variables and u is an external input. These two state equations can be realized by a two-cell SC-CNN as stated by the following proposition [79]:

Proposition 3.6.1. *The forced oscillator (3.31) can be realized by a two-cell non-autonomous linear SC-CNN with the following templates:*

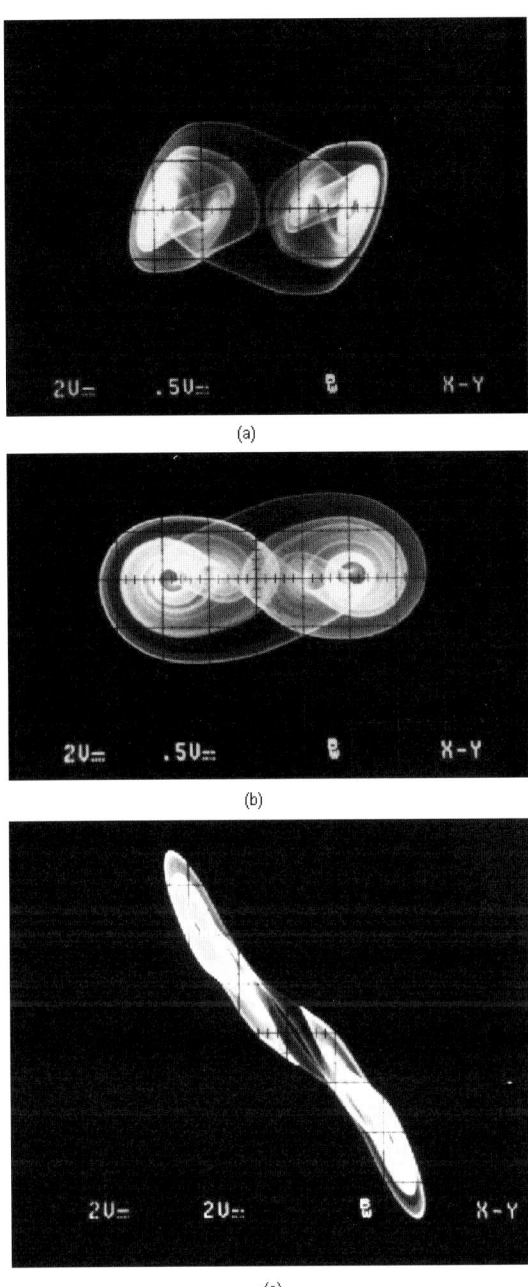

Fig. 3.22. Observed phase portraits of the 2-double scroll obtained in the SC-CNN.
(**a**) $x_1 - x_2$, (**b**) $x_2 - x_3$, (**c**) $x_1 - x_3$

$$A_{1;1} = \frac{1+k}{\epsilon}, \qquad \hat{A}_{1;1} = 1 - \frac{1}{\epsilon}, \qquad \hat{A}_{2;1} = -1,$$
$$\hat{A}_{1;2} = \frac{1}{\epsilon}, \qquad B_{2;2} = 1, \qquad \hat{A}_{2;2} = 1, \qquad I_2 = b, \qquad (3.32)$$

the remaining coefficients being zero.

The behavior of the system (3.31) is very sensitive to parameter variations. In particular, for the case of a sinusoidal input signal $u = a\cos\omega t$ and the following choice for the parameters: $\epsilon = 0.2$, $a = 0.2$, $\omega = 2.0\pi$, $k = 0.885$, the parameter b was varied within the narrow interval $[1, 1.0154]$ [96]. One of the attractors obtained with the SC-CNN based circuit is shown in Fig. 3.23.

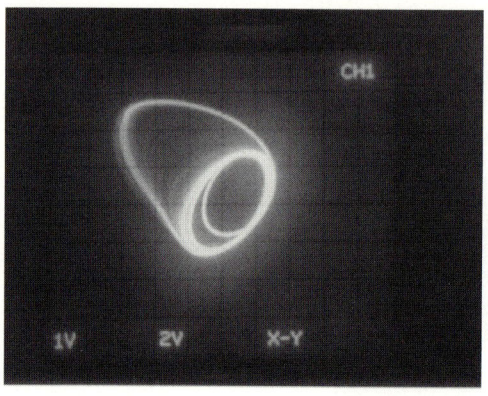

Fig. 3.23. Observed phase portrait of the non-autonomous SC-CNN in $x_1 - x_2$

3.6.2 A Circuit with a Nonlinear Reactive Element

Let us now consider the SC-CNN dynamic realization of a non-autonomous circuit containing a nonlinear inductor [97]. The electrical scheme is reported in Fig. 3.24. In [97] the following state representation was given:

$$\dot{V}_1 = V_2;$$
$$C\left(\frac{d\phi(i)}{di}\right)\dot{V}_2 = -CRV_2 - V_1 + E\cos\omega t; \qquad (3.33a)$$

with

$$\phi(i) = \begin{cases} L_0 i & \text{if } i \le i_0 \\ L_1 i + (\phi_m - L_1 i_0) & \text{if } i > i_0, \end{cases} \qquad (3.33b)$$

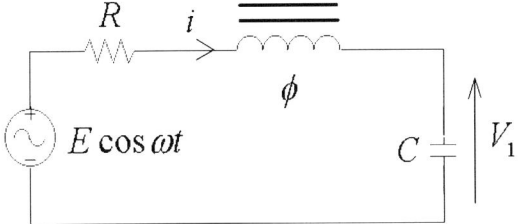

Fig. 3.24. Circuit with nonlinear inductor

where V_2 is defined as $V_2 \equiv i/C$, i and C being the inductor current and the capacitor value respectively. A strange behavior appears if the circuit parameters are assumed [97] as $k = 0.032$, $\nu = 0.7$, $B = 0.725$, $\alpha \to 0$, where $\alpha = \frac{L_1}{L_0}$, $\nu = \sqrt{L_0 C}\omega$, $k = R\sqrt{C/L_0}$, $B = E\sqrt{L_0 C}/\phi_m$. From these one obtains

$$L_0 = 1\,\text{H}; \quad L_1 = 0.5 \times 10^{-2}\,\text{H}; \quad \omega = 1\,\text{rad/s};$$
$$E = 1\,\text{V}; \quad C = 0.49\,\text{F}; \quad \phi_m = 0.9655\,\text{Wb}; \quad R = 4.571 \times 10^{-2}\,\Omega. \tag{3.34}$$

However, using the model (3.33) the SC-CNN realization is not so straightforward. Therefore, an alternative equivalent mathematical description is required.

Proposition 3.6.2. *The circuit shown in Fig. 3.24 can be described by the following state representation:*

$$\dot{\phi} = e_s - Ri(\phi) - V_1,$$
$$\dot{V_1} = \frac{1}{C}i(\phi), \tag{3.35a}$$

with

$$i(\phi) = \begin{cases} L_0^{-1}\phi & \text{if } \phi \leq \phi_m \\ L_1^{-1}(\phi - \phi_m) + i_0 & \text{if } \phi > \phi_m \end{cases} \tag{3.35b}$$

and this system can be realized by a single-layer two-cell non-autonomous SC-CNN with the following templates: $\hat{A}_{1;1} = 1 - RL_1^{-1}$, $\hat{A}_{1;2} = -1/\phi_m$, $B_{1;1} = 1/\phi_m$, $A_{1;1} = R(L_1^{-1} - L_0^{-1})$, $\hat{A}_{2;2} = 1$, $\hat{A}_{2;1} = \phi_m/CL_1$, $A_{2;1} = \phi_m(L_0^{-1} - L_1^{-1})/C$, *the remaining ones being zero.*

Proof. The alternative state representation (3.35) is obtained by choosing the capacitor voltage V_1 and the magnetic induction flux ϕ as state variables. With this choice, the circuit can be described by the following equations:

$$e_s = Ri + \dot{\phi} + V_1$$
$$C\dot{V_1} = i \tag{3.36}$$

and $\phi_m = L_0 i_0$, $e_s = E\cos\omega t$. From these relationships (3.35) is obtained. However, taking into account that real inductor nonlinearities are of odd-symmetry, this nonlinear characteristic can be replaced by the following one:

$$i(\phi) = L_1^{-1}\phi + \frac{(L_0^{-1} - L_1^{-1})}{2}(|\phi + \phi_m| - |\phi - \phi_m|). \tag{3.37}$$

This substitution does not affect the dynamics of the system because the values actually assumed by ϕ make them equivalent. It is now clear that the model (3.35) with (3.37) can be suitably realized by a two-cell linear non-autonomous SC-CNN. In fact taking: $x_1 = \phi/\phi_m$, $x_2 = V_1$, $u_1 = e_s$, as cell state variables and input, and with the following template coefficients:

$$\hat{A}_{1;1} = 1 - RL_1^{-1}, \ \hat{A}_{1;2} = -\frac{1}{\phi_m}, \ B_{1;1} = \frac{1}{\phi_m}, \ A_{1;1} = R(L_1^{-1} - L_0^{-1}),$$
$$\hat{A}_{2;2} = 1, \ \hat{A}_{2;1} = \frac{\phi_m}{CL_1}, \ A_{2;1} = \frac{\phi_m(L_0^{-1} - L_1^{-1})}{C}, \tag{3.38}$$

the desired dynamic can be reproduced as in the other propositions reported. □

The particular case in which the parameter set (3.34) has been adopted can obviously be reproduced with the SC-CNN realization and the experimental phase portrait observed in the x_1-x_2 plane is shown in Fig. 3.25.

Let us note that (3.35) is an alternative model of (3.33), both of them describe the same physical system. There thus exists a bijective correspondence between the orbits of the two representations. Moreover, the nonlinear functions involved are all invertible and differentiable almost everywhere. Finally, the same time variable is chosen. Therefore, a straightforward consequence of Proposition 3.6.2 is that:

Corollary 3.6.1. *The strange attractors of system (3.35) and (3.33) are topologically conjugated.*

3.6.3 Canards and Chaos

Canards are nonlinear phenomena that can be observed in the so-called slow–fast systems [98,99]. They are certain singular solutions of these nonlinear systems which, at first, move with a slow dynamic; then, the state goes through the trajectory faster and, subsequently, it slows down again and so on until it vanishes while a irregular oscillation or a cycle appears. The main feature of Canards is that they are extremely sensitive to parameter variations; in fact they are structurally unstable. Our attention has been devoted to the Bonhoeffer–Van der Pol (BVP) system [98] mathematically described by the following state space equations:

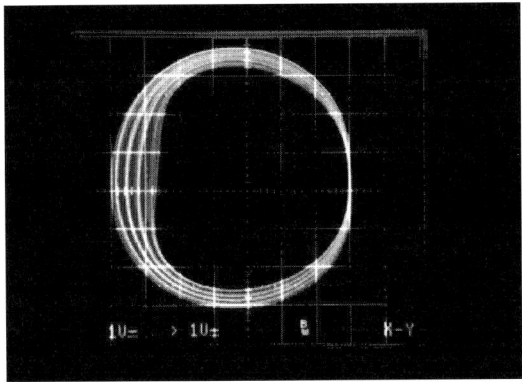

Fig. 3.25. The Experimental phase portrait in $x_1 - -x_2$ obtained by the SC-CNN

$$\dot{x} = -\frac{y + f(x)}{\epsilon}, \quad (3.39\text{a})$$
$$\dot{y} = x - \alpha y - \beta,$$

with $\alpha > 0$, $\epsilon > 0$, $\epsilon \sim 0$, $\beta = 2.8$ and

$$f(x) = \frac{x^3 - 27x}{18}. \quad (3.39\text{b})$$

Furthermore, the previous circuit has been generalized leading to a third-order autonomous PWL system with the following state representation [99]:

$$\begin{aligned}\dot{x} &= z - y, \\ \dot{y} &= \alpha(x + y), \\ \dot{z} &= -\frac{x + n(z)}{\epsilon},\end{aligned} \quad (3.40\text{a})$$

with

$$n(z) = \beta z + 0.5(\gamma - \beta)(|z + 1| - |z - 1|), \quad (3.40\text{b})$$

and $\alpha > 0$, $\beta > 0$, $\gamma < 0$. It is worth noting that in the BVP system (3.39) the nonlinear function $f(x)$ in (3.39b) is a cubic continuous nonlinearity while in the third-order generalized BVP (3.40) the nonlinearity $n(z)$ in (3.40b) is a PWL one. It has been proved that the BVP system has the Canard trajectories while the third-order BVP cannot possess them because of the nature of its nonlinearity [98, 99].

Both of the previously assumed systems can be realized with SC-CNN as already shown. While in the three-cell SC-CNN realizing the third-order BVP system a linear SC-CNN can be used, in contrast, in the two cell realization of the BVP system, a linear SC-CNN with a smooth output function (instead of the stiff PWL saturation function) must be used. This smooth output nonlinearity was obtained by using a circuit with an appropriate diode network. The diode nonlinearity permits us to obtain the required smoother shape.

Proposition 3.6.3. *The state equations of the BVP system (3.39) can be realized by a two-cell SC-CNN with the defined templates: $\hat{A}_{1;1} = 1$, $\hat{A}_{1;2} = -1/\epsilon$, $A_{1;1} = -1/\epsilon$, $\hat{A}_{2;1} = 1$, $\hat{A}_{2;2} = 1 - \alpha$, $I_2 = -\beta$, the other template coefficients being zero and taking the nonlinear function $f(x) = (x^3 - 27x)/18$ as the output nonlinearity.*

Proposition 3.6.4. *The state equations of the third-order generalized BVP system (3.40) can be realized by a three-cell linear SC-CNN with the following templates: $\hat{A}_{1;1} = 1$, $\hat{A}_{1;2} = -1$, $\hat{A}_{1;3} = 1$, $\hat{A}_{2;1} = \alpha$, $\hat{A}_{2;2} = 1 + \alpha$, $\hat{A}_{3;1} = -1/\epsilon$, $\hat{A}_{3;3} = 1 - \beta/\epsilon$, $A_{3;3} = (\beta - \gamma)/\epsilon$, the other template coefficients being zero.*

The experimentally observed phase portraits, referring to the SC-CNN realizations, are shown in Fig. 3.26a,b.

3.6.4 Multimode Chaos in Coupled Oscillators

In [100], Nishio and Ushida describe some interesting nonlinear phenomena that were observed in coupled oscillators. The state equations of a single oscillator are

$$L_1 \frac{di_1}{d\tau} = -v - Ri_1 - v_d(i_1 + i_2),$$
$$L_2 \frac{di_2}{d\tau} = -v_r(i_2) - v_d(i_1 + i_2), \quad (3.41a)$$
$$C \frac{dv}{d\tau} = i_1,$$

where i_1, i_2 and v are the state variables and v_d and v_r are nonlinear functions defined as follows:

$$v_r(i_2) = \begin{cases} V - ri_2 & \text{if } i_2 > J_A, \\ (\frac{R_d r_d}{2R_d + r_d} - r)i_2 & \text{if } |i_2| \leq J_A, \\ -V - ri_2 & \text{if } i_2 < -J_A, \end{cases} \quad (3.41b)$$

where $J_A = (2R_d + r_d)/R_d r_d \cdot V$,

$$v_d(i_1 + i_2) = \begin{cases} 3V & \text{if } i_1 + i_2 > J_B, \\ \frac{3}{2} r_d(i_1 + i_2) & \text{if } |i_1 + i_2| \leq J_B, \\ -3V & \text{if } i_1 + i_2 < -J_B, \end{cases} \quad (3.41c)$$

where $J_B = (2 \cdot V)/r_d$.

3.6 Nonlinear Dynamics Potpourri 73

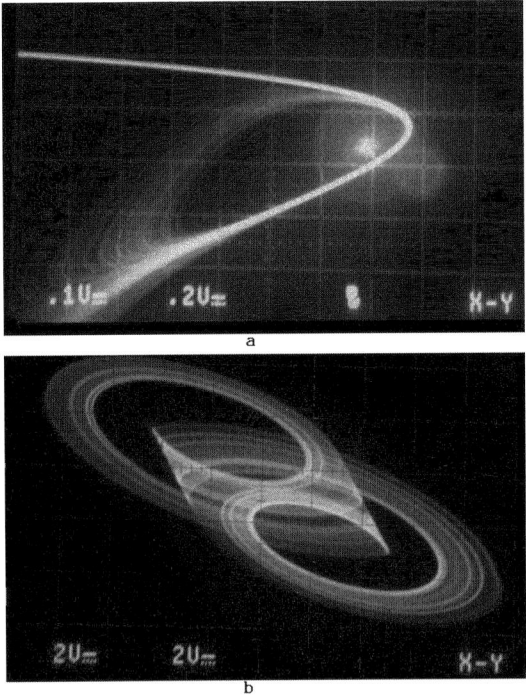

Fig. 3.26. (a) The canard observed in the SC-CNN realization of the BVP system. (b) The chaotic attractor observed in the SC-CNN realization of the generalized BVP system

This circuit becomes chaotic if the following choice is made for its parameters: $L_1 = 100\,\text{mH}$; $L_2 = 200\,\text{mH}$; $C = 0.068\,\mu\text{F}$; $R = 10\,\Omega$; $R_d = 560\,\Omega$; $r = 1\,\text{k}\Omega$; $r_d = 15\,\text{M}\Omega$; $V = 0.7\,\text{V}$.

It can be noted that, in this case (see (3.41)), a nonlinear function of two state variables is present.

Proposition 3.6.5. *The nonlinear dynamic system (3.41) is topologically conjugated to a three-cell linear SC-CNN with the following templates:*

$$\hat{A}_{1;1} = 1 - \frac{R}{L_1}, \quad \hat{A}_{1;2} = -\left(\frac{R_1}{L_1} + \frac{r}{L_2}\right)\frac{(2R_d + r_d)}{2VR_d},$$

$$\hat{A}_{1;3} = -\frac{r_d}{2VL_1}, \quad A_{1;1} = -\left(\frac{1}{L_1} + \frac{1}{L_2}\right)\frac{3r_d}{2},$$

$$\hat{A}_{2;2} = 1 - \frac{r}{L_2}, \quad A_{2;1} = -\frac{3VR_dr_d}{L_2(2R_d + r_d)}, \quad A_{2;2} = -\frac{VR_dr_d}{L_2(2R_d + r_d)},$$

$$\hat{A}_{3;3} = 1, \quad \hat{A}_{3;1} = \frac{2V}{Cr_d}, \quad \hat{A}_{3;2} = -\frac{2V}{Cr_d},$$

(3.42)

all the remaining template coefficients being zero.

Proof. Let us consider the following linear state transformation for system (3.41):

$$x_1 = r_d(i_1 + i_2)/(2V),$$
$$x_2 = R_d r_d i_2 / (2R_d + r_d), \qquad (3.43)$$
$$x_3 = v.$$

Therefore, (3.41) can be rewritten as follows:

$$\begin{aligned}
\dot{x}_1 &= -\frac{R}{L_1} x_1 - \left(\frac{R_1}{L_1} + \frac{r}{L_2}\right) \frac{(2R_d + r_d)}{2VR_d} x_2 - \frac{r_d}{2VL_1} x_3 \\
&\quad - \left(\frac{1}{L_1} + \frac{1}{L_2}\right) \frac{3r_d}{2} y_1, \\
\dot{x}_2 &= -\frac{r}{L_2} x_2 - \frac{VR_d r_d}{L_2(2R_d + r_d)} y_2 - \frac{3VR_d r_d}{L_2(2R_d + r_d)} y_1, \\
\dot{x}_3 &= -x_3 + \frac{2V}{Cr_d}(x_1 - x_2) + x_3.
\end{aligned} \qquad (3.44)$$

These state equations represent the model of a three-cell linear SC-CNN if the template coefficients (3.42) are assumed. This proves the proposition. □

Once the SC-CNN realization is accomplished, the original variables of the classical realization can be obtained by inverting the transformation[13] (3.43). In this case the phase portrait observed, referring to the variables i_1 and i_2 is shown in Fig. 3.27.

3.6.5 Coupled Circuits

Two of the previously considered oscillators can be coupled by using a capacitor in series with the existing ones [100]. From a mathematical point of view this implies that the first equation in system (3.41) is modified as follows:

$$L_1 \frac{di_1}{d\tau} = -v_1 - Ri_1 - v_d(i_1 + i_2) - v_0, \qquad (3.45a)$$

with

$$C_0 v_0 = C v_1 + C v_2, \qquad (3.45b)$$

where v_0 is the branch voltage of the coupling capacitor C_0, and v_1 and v_2 are the branch voltages of the two capacitors in the two Nishio–Ushida chaotic circuits.

---

[13] Any linear transformation and/or its inverse can be obtained in straightforward way by means of op-amp-based arrays of algebraic adders.

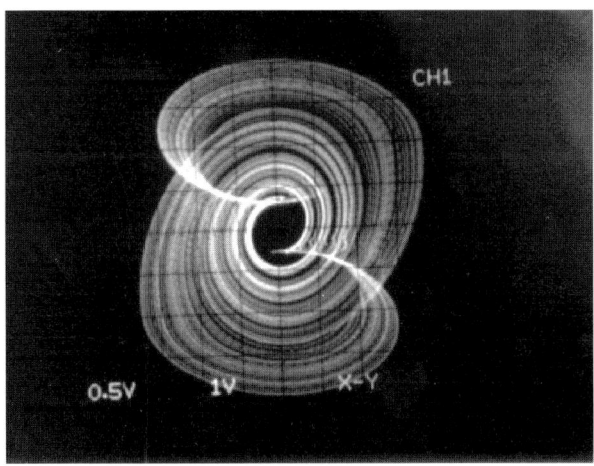

Fig. 3.27. The attractor of the single oscillator (3.41) obtained by a SC-CNN and then mapped into $i_1 - i_2$

A six state equation system for the coupled oscillators is derived after some algebra from the expressions (3.45b) and (3.45a). This system can be realized by using six SC-CNN cells in the same way as proved above for the autonomous circuit.

The attractors observed, corresponding to a variety of parameter sets (mainly the capacitors of the six cells), are shown in Fig. 3.28.

3.7 General Case and Conclusions

As seen from previous sections a wide variety of well-known nonlinear circuits belong to the family of SC-CNN. The list of examples could continue further but it is preferable to seek a general formulation.

To this end the following sufficient condition is presented.

Proposition 3.7.1. *Let us consider an n-th order circuit, system or vector field with the state representation*

$$\dot{\boldsymbol{x}} = H\boldsymbol{x} + G\boldsymbol{f}(L\boldsymbol{x}) , \quad (3.46a)$$

where $\boldsymbol{x} \in \mathbb{R}^n$, $\boldsymbol{f} \colon \mathbb{R}^n \to \mathbb{R}^n$, $H, G, L \in \mathbb{R}^{n,n}$ *and* $\mathbf{L}$ *is nonsingular (i.e.* $\exists L^{-1}$*). Moreover:*

$$\left| \frac{\partial f_i}{\partial x_j} \right| = \begin{cases} 0 & \text{if } |i - j| > r \\ \geq 0 & \text{otherwise} , \end{cases} \quad (3.46b)$$

where r is the neighbor radius. Then (3.46a) is topologically conjugated to the SC-CNN family (3.1).

76   3. The CNN as a Generator of Nonlinear Dynamics

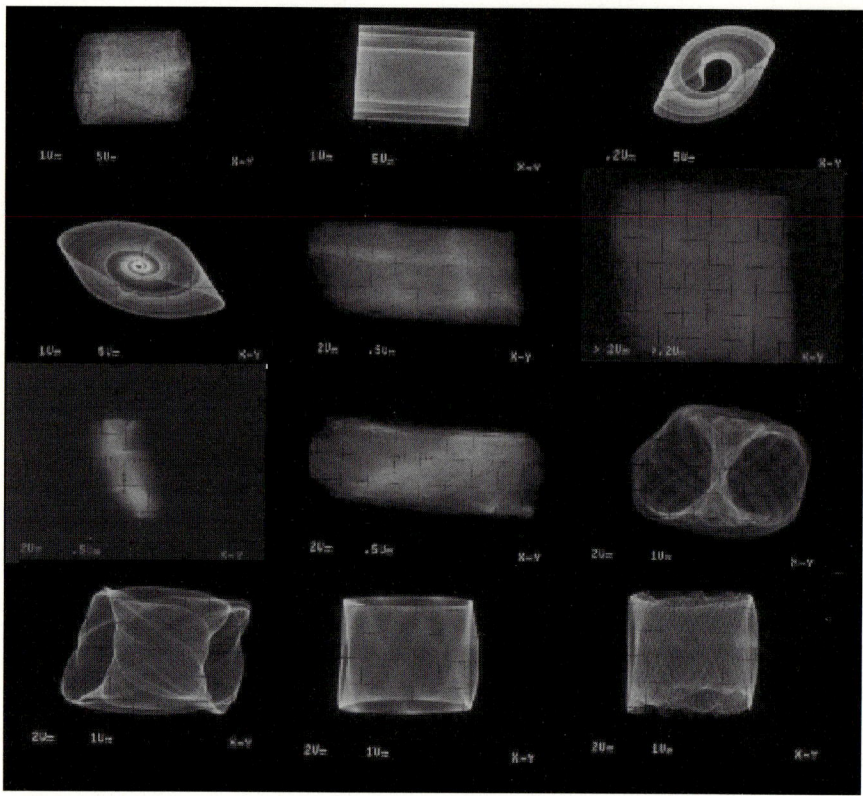

Fig. 3.28. A gallery of attractors from the six cell SC-CNN

Proof. Let us consider the nonsingular state variable linear transformation $z = Lx$. Hence $L^{-1}z = x$ and $L^{-1}\dot{z} = \dot{x}$. By substitution in (3.46a) the following system is obtained:

$$L^{-1}\dot{z} = HL^{-1}z + Gf(z) \tag{3.47}$$

namely:

$$\dot{z} = LHL^{-1}z + LGf(z). \tag{3.48}$$

This last equation is nothing other than the matrix form of (3.1) where LHL^{-1} defines the state templates and LG defines the feedback templates. Condition (3.46b) is necessary to satisfy the definition of local interaction imposed by the synaptic law (1.30) in Def. 1.3.4. In this regard it is important to observe that re-ordering the scalar equations composing system (3.46a) (and so re-numbering the components of x or z) can satisfy (3.46b). □

Condition (3.46b) must not be confused with the condition for *cooperative systems*[14] and it is actually less restrictive.

Corollary 3.7.1. *If all the hypotheses of Prop. 3.7.1 are satisfied and, moreover, $\boldsymbol{f}(\boldsymbol{x})$ is*

1. *Piecewise linear and also such that*
2. $\forall i, j = 1, \ldots, n$ *and* $\forall \tilde{\boldsymbol{x}} \in \mathbb{R}^n$ *constant vector:*

$$f_i(\boldsymbol{x})\big|_{\boldsymbol{x}=[\tilde{x}_1,\ldots,x_j,\ldots,\tilde{x}_n]'} = a_0 + a_1 x_j + \frac{1}{2} a_2 (|x_j + b_k| - |x_j - b_k|) \tag{3.49}$$

then system (3.46a) is realized by a SC-CNN with saturation output functions.

The extension of these results to non-autonomous circuits and systems is trivial.

Let us now discuss some of the many implications of the various results in this chapter.

3.7.1 Theoretical Implications

First of all, if it is true that Chua's circuit is the simplest electronic circuit able to show a wide variety of nonlinear dynamic behaviors, then from what we saw in Sect. 3.2 it can be stated that the primitive circuit for obtaining such behaviors is the SC-CNN cell.

Moreover, CNN arrays based on the Chua oscillator as basic cell can actually be substituted by multilayer (3 layer) SC-CNNs with simpler and more homogeneous circuit topology.

The results shown provide a common electrical model for the study of a wide class of nonlinear circuits. The approach introduced unites the fields of "conventional" CNNs and of nonlinear oscillators.

The SC-CNN becomes a general programmable generator of nonlinear dynamics.

3.7.2 Practical Implications

It is worth noting that in a SC-CNN:

1. there are no inductors,
2. all the capacitors are identical,
3. the actual absolute value of the capacitors does not really matter[15] but the ratio of the capacitors is important,
4. saturation-type nonlinearity is a natural characteristic of any amplifier.

---

[14] See Def. A.7.13 in Appendix A.
[15] It affects the whole time-scale of the dynamics but not its shape or nature.

These features are of paramount importance for VLSI implementation. In fact [101–103]:

1. inductors are very difficult to realize in IC technology,
2. the tolerance on the absolute value of integrated capacitors is very high (on the order of ±20% or more),
3. but, on the other hand, the ratios between capacitors on the same chip can be very accurate (±1% or even ±0.1%) if supported by careful layout work,
4. amplifiers with saturation-type nonlinearity are ubiquitous,
5. feedback amplifiers allow very accurate gains to be realized.

At this point it is important to stress that although chaos and other nonlinear behaviors are relatively robust with respect to parameter variations, the need for accuracy cannot be underestimated.

This means, for example, that a direct[16] IC implementation of a traditional Chua oscillator, a SHCG or one of the other above reported circuits[17], is programmed to be a failure.

A way to keep control of (or to recover from) parameter variations is absolutely indispensable[18].

However the SC-CNN approach offers a natural solution to all of these problems. Moreover it provides all the different dynamics in one circuit completely on demand of the user.

Some good reasons for providing a *programmable chaos generator* (*PCG*) can be found in applications as pseudo-random sequences/numbers generators [69] and for secure communication/chaos-based cryptography [104–107]. In all of these applications any chaotic attractor corresponds, roughly speaking, to a "seed", a code or a "key". A PCG offers a fast re-programmable multi-key system [78].

More discussion of this subject can be found in Chap. 4.

As a final remark we emphasize that, is some situations, spatio-invariant SC-CNNs can be a valuable alternative to the Chua and Yang CNN. Some further theoretical results on the stability of SC-CNNs can be found in [108].

---

[16] Component by component without any device able to control the value of the circuit elements.
[17] In fact, realized with discrete components.
[18] Namely, compulsory introduction of feedback circuitry is required.

4. Synchronization

One recent and interesting research topic in the circuit area is the synchronization of nonlinear circuits (SNC) [104–107,109,110]. Some techniques have been developed to force two (or more) identical nonlinear dynamic circuits, starting from different initial conditions, to synchronize, namely to follow identical trajectories (asymptotically, at least). This is particularly interesting if the circuits behave in a chaotic way because of their sensitive dependence on initial conditions. However, if certain conditions are satisfied [104, 105], then these circuits can be successfully synchronized.

Synchronization principles have been applied to realize analog masking systems for secure communication [105, 107].

In this chapter we describe an experimental study of a chaotic transceiver using SC-CNNs and non-ideal communication channels [111, 112]. Moreover a new method to identify the parameters of a chaotic circuit using synchronization and genetic algorithms is discussed [113–115].

4.1 Background

In this Section we will briefly recall some of the concepts of synchronization that are strictly necessary for what follows. A good introduction to this subject can be found in [105].

The synchronization of nonlinear systems is defined as follows [105].

Definition 4.1.1 (Synchronization). *Let us consider two (or more) nonlinear systems ($N \geq 2$):*

$$\dot{\boldsymbol{x}}_i = \boldsymbol{f}_i(\boldsymbol{x}_i), \tag{4.1a}$$

where $\boldsymbol{x}_i \in \mathbb{R}^n$, $\boldsymbol{f}_i \colon \mathbb{R}^n \to \mathbb{R}^n$ and $1 \leq i \leq N$; if:

$$\lim_{t \to \infty} |\boldsymbol{x}_i(t) - \boldsymbol{x}_j(t)| = 0, \tag{4.1b}$$

with $i \neq j$ then the N systems are synchronized.

In this chapter just two systems will always be considered ($N = 2$). In order to obtain synchronization, many different approaches are possible. When the two systems are coupled such that the first one is independent from the other one, while the dynamics of the second one is influenced by the first one (one way coupling) then we have a *master-slave* configuration[1].

4.1.1 Pecora–Carroll Approach

Let us consider a dynamic system [104]:

$$\dot{u} = f(u), \qquad (4.2)$$

and partition it into two subsystems, $u = (v, w)$:

$$\begin{aligned} \dot{v} &= g(v, w), \\ \dot{w} &= h(v, w), \end{aligned} \qquad (4.3)$$

where $v = [u_1, ..., u_m]'$, $g = [f_1(u), ..., f_m(u)]'$, $w = [u_{m+1}, ..., u_n]'$, $h = [f_{m+1}(u), ..., f_n(u)]'$.

The partition is arbitrary since the equations can be previously reordered. A new system w' can now be considered; it is created by duplicating the w system; moreover the set of variables $v'$are replaced by their corresponding v:

$$\dot{w}' = h(v, w'). \qquad (4.4)$$

In this way the w' system is forced by the u system by means of the v variables; w' is called the *response* system. If, as time elapses, $w'(t) \to w(t)$ then the master and slave synchronize.

For this purpose one can consider the system obtained as the difference between these two. In other words, coherent with Def. 4.1.1 we want $\Delta w(t) \doteq w'(t) - w(t)$ to converge to zero as $t \to \infty$. This leads to the variational equation[2]:

$$\frac{d\Delta w}{dt} = D_w h(v, w) \Delta w + O\left((\Delta w)^2\right), \qquad (4.5)$$

where $D_w h$ is the Jacobian of the w subsystem. In the limit of small Δw, higher order terms can be neglected and the variational equation for the w subsystem remains.

The Lyapunov exponents resulting from this variational equation (i.e. the Lyapunov exponents of the difference between w and w') are the *conditional Lyapunov exponents*. If all the conditional Lyapunov exponents are less than zero, then the response systems will synchronize with the master.

This synchronization scheme can be further completed as follows. The v subsystem can also be reproduced to create the v'' subsystem which is driven with the w' variables. The complete slave system is therefore:

---

[1] Where, obviously, the master is the independent system.
[2] See Sect. A.7.2 in Appendix A.

$$\dot{v}'' = g(v'', w'),$$
$$\dot{w}' = h(v, w'').$$
(4.6)

Again, if all the corresponding conditional Lyapunov exponents are negative, then the master (4.3) and the slave (4.6) will synchronize and, in particular, $v'' \longrightarrow v$ as $t \longrightarrow \infty$. This last set-up is known as the *cascaded synchronization* scheme [106].

An interesting result concerning synchronization is the fact that the two systems may be synchronized even in the presence of noise or if the driving signal has been altered by a filter [106].

4.1.2 Inverse System Approach

Consider another master–slave set-up. Let the master be forced by an external signal $s(t)$ and let $y(t)$ be the corresponding response. $s(t)$ can have a regular waveform while $y(t)$, assuming that the master is a chaotic circuit, is taken to be chaotic[3].

The equations of the master and its output can be written as:
$$\dot{x} = f(x, s),$$
$$y = g(x, s),$$
(4.7)

where $x \in \mathbb{R}^n$ is the state, $f: \mathbb{R}^{n+1} \to \mathbb{R}^n$ the vector field, $s \in \mathbb{R}$ the forcing signal, and $g: \mathbb{R}^{n+1} \to \mathbb{R}$ the output function.

The output signal $y(t)$ is used to drive the slave system:
$$\dot{\hat{x}} = \hat{f}(\hat{x}, y),$$
$$\hat{s} = h(\hat{x}, y),$$
(4.8)

where $\hat{x} \in \mathbb{R}^n$ is the state, $\hat{f}: \mathbb{R}^{n+1} \to \mathbb{R}^n$ the vector field, $\hat{s} \in \mathbb{R}$ the response to y and $h: \mathbb{R}^{n+1} \to \mathbb{R}$ the output function.

If, for equal initial conditions $x(0) = \hat{x}(0)$, it happens that $\hat{s}(t) = s(t)$, $\forall t \geq 0$ then the slave is said to work as *inverse system*.

However, given that there is no real control over initial conditions[4], a real inverse system set-up is assumed to be able to synchronize (at least asymptotically) regardless of the initial conditions.

4.2 Experimental Signal Transmission Using Synchronized SC-CNN

In this section, a simple architecture based on two state controlled cellular neural networks, is proposed for transmitting signals by applying the chaos

---
[3] Although an external forcing signal can, in some circumstances, cause a bifurcation and so the system can lose the chaotic behavior.
[4] And this is particularly critical in the case of chaotic circuits because of their sensitive dependence on initial conditions.

82   4. Synchronization

synchronization principles. In particular, the inverse system approach is considered.

Experimental and simulation results will be reported. Moreover, an experimental study of the consequences of the introduction of a non-ideal transmission channel between transmitter and receiver is included.

4.2.1 Circuit Description

The circuit considered is composed by the four blocks shown in Fig. 4.1.

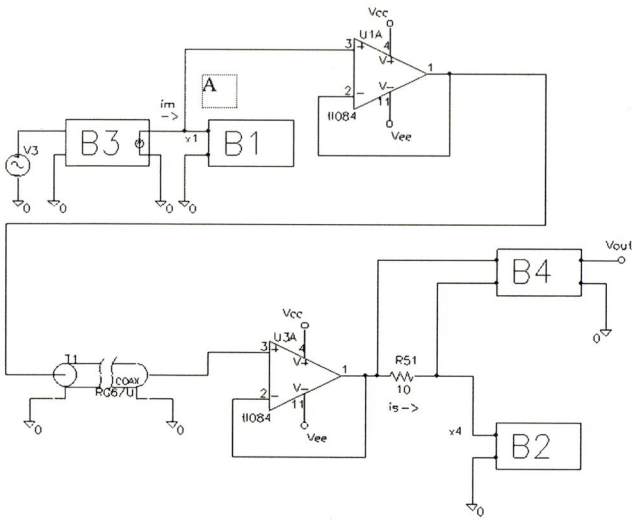

Fig. 4.1. Block diagram of the master–slave set-up

Blocks B1 and B2 are two identical three-cell SC-CNNs realizing the dynamics of the double-scroll as shown in Chap. 3. The corresponding circuit schematic, already seen in Fig. 3.5, is shown again in Fig. 4.2 for convenience.

Block B3 (see Fig. 4.3) is used to convert the voltage signal $V3$ representing the message to be encrypted in the current i_m. This current is added to the node A (see Fig. 4.1) and B1 responds with a chaotic modulated signal x_1 (the transmitted signal). Two buffers have been put between the channel input and the chaotic modulator B1 and between the channel output and the de-modulator B2 in order to avoid possible loads to the SC-CNN circuits.

The synchronization signal is imposed at the node corresponding to x_4 (the homolog of x_1 for the slave system B2) and B2 will respond drawing the current i_s. It has been recalled in Sect. 4.1.2 that iff B1 and B2 are synchronized then $i_s = i_m$. The current i_s will be transformed into the voltage V_{out} by the block B4 shown in Fig. 4.4.

4.2 Experimental Signal Transmission Using Synchronized SC-CNN

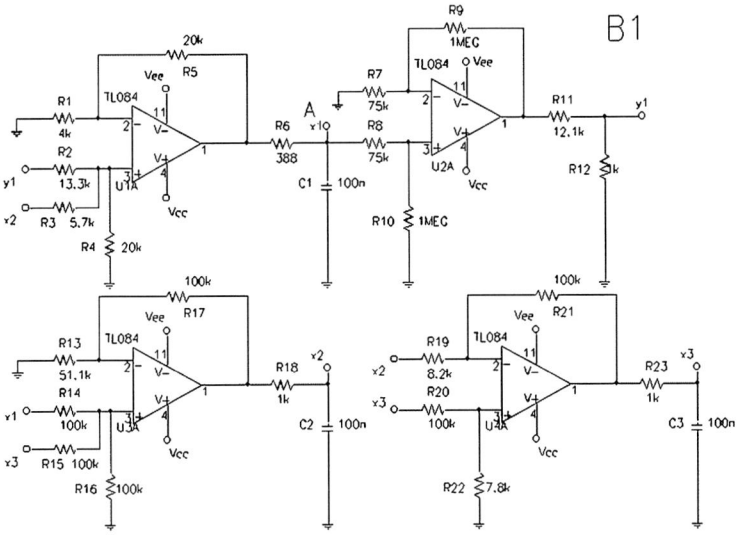

Fig. 4.2. SC-CNN realization of the Chua oscillator

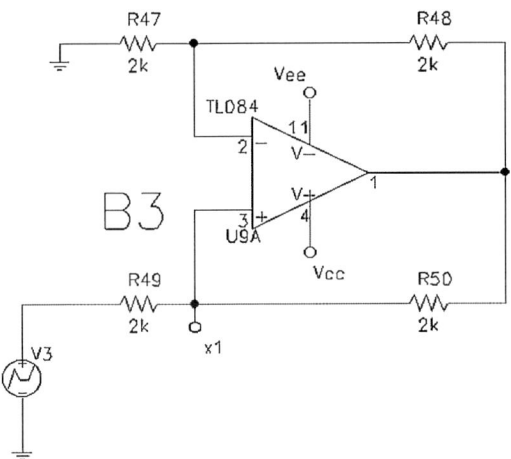

Fig. 4.3. Voltage-to-current converter realizing block B3

4.2.2 Synchronization: Results of Experiment and Simulation

Let us consider the ideal case in which the master and slave are directly coupled. Figure 4.5 shows the not-coded[5] and decoded messages superposed.

It is seen that apart from a brief transient, due to different initial conditions, the two waveforms are in a good agreement. This case will be consid-

---

[5] The original message before the chaotic cryptography.

84   4. Synchronization

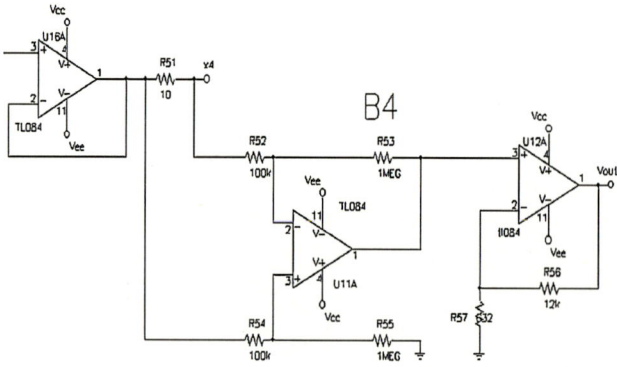

Fig. 4.4. Current sensing stage realizing block B4

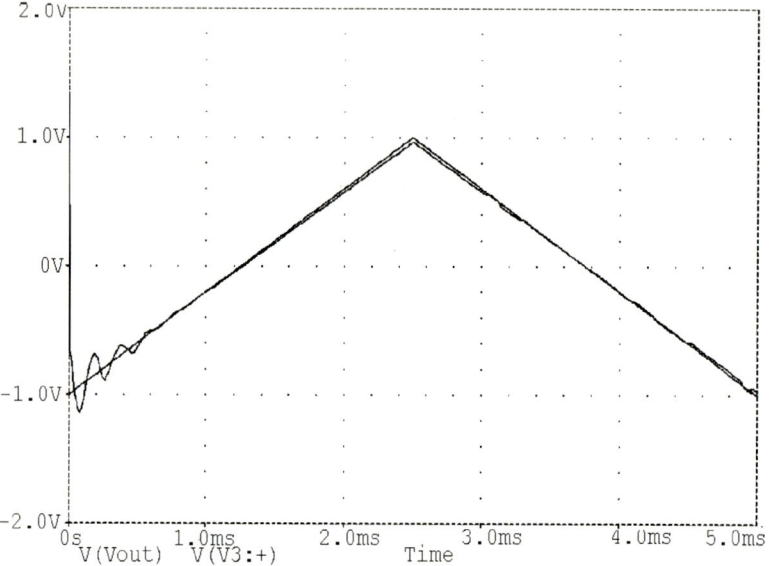

Fig. 4.5. Ideal case: Transmitted (V_3) and decoded (V_{out}) signal with direct coupled circuits

ered as a reference for the next discussion on the introduction of a non-ideal channel.

A triangular waveform is considered (as message) instead of a sinusoidal one because it is well known [105] that the behavior of this system is dependent on the frequency and on the amplitude of the tone.

The corresponding experimentally observed waveforms are shown in Fig. 4.6b.

4.2 Experimental Signal Transmission Using Synchronized SC-CNN

The quality of the experimental synchronization can be appreciated by looking at Fig. 4.6a which shows the variable x_2 versus the variable x_5, corresponding to one another, in the master and slave systems. Finally, the observed phase portrait in the $x_1 - x_2$ plane when the forcing signal is applied has been reported in Fig. 4.6c. In Fig. 4.6b, the presence of a small ripple super-imposed onto the decoded signal can be observed.

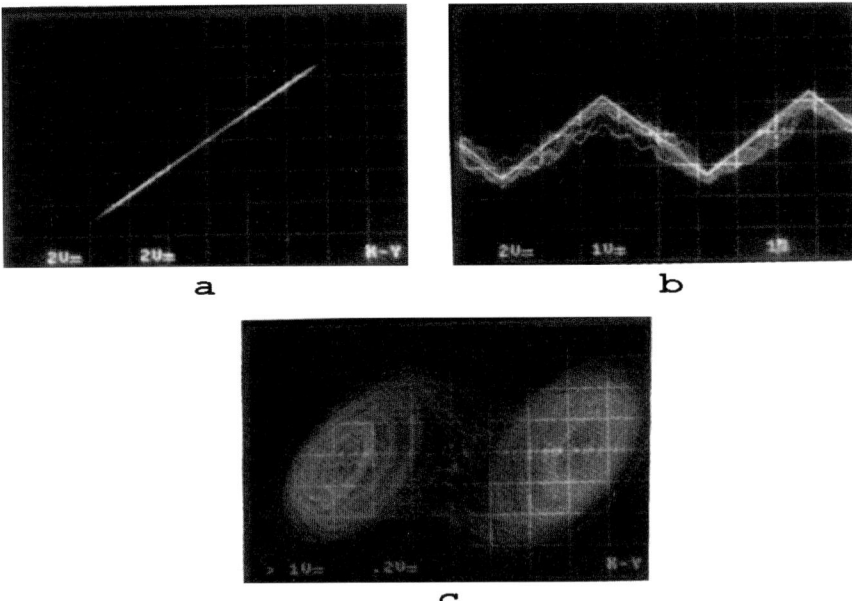

Fig. 4.6. Experimentally observed waveforms. (**a**) x_2 (*second variable of the master*) vs x_5 (*second variable of the slave*); (**b**) V_{out} and V_3 superposed; (**c**) Observed phase portrait in the $x_1 - x_2$ plane when the forcing signal is applied

This could lead to the erroneous conclusion that the two circuits are not exactly synchronized; however, Fig. 4.6a excludes this case. The correct explanation of this fact is immediately obtained by measuring the actual current fed by the block B3. In fact, this current is slightly corrupted by the master itself due to the non-ideal features of the actual current source generator.

Apart from the triangular wave, other different waveforms have also been employed (e.g. sine waves, square waves, speech and musical signals and so on) with successful results.

Finally, we note that the above-mentioned ripple is always present independently of the nature of the transmitted signals considered. Therefore it could be significantly reduced by using a better current source or by filtering the decoded signal by means of a low-pass filter.

4.2.3 Non-ideal Channel Effects

We now discuss the effects of coupling the transmitter and the receiver with a commercial coaxial cable. In particular the model RG6/U with characteristic impedance $Z_0 = 75\,\Omega$ is considered. Two different cases have been investigated. In the first case the transmission channel is adapted with its proper characteristic impedance, while in the second, the line is not adapted.

The circuit corresponding to the latter case is the one shown in Figs. 4.1–4.4. Otherwise if the channel has to be adapted then a $75\,\Omega$ resistor must be inserted between the output of the buffer U1A and the input of the line T1 in Fig. 4.1; furthermore, another $75\,\Omega$ resistance must be inserted between the positive input of buffer U3A and the ground in the same figure.

Moreover, due to the voltage divider effect at the input of the line T1, in this case, the buffer U3A must be replaced by a non-inverting stage with voltage gain $G = 2$.

Let us refer to the ideal case of directly coupled circuits considered in the previous section. The spectrum of the synchronization signal x_1, when the master is not forced by any message, is displayed in Fig. 4.7.

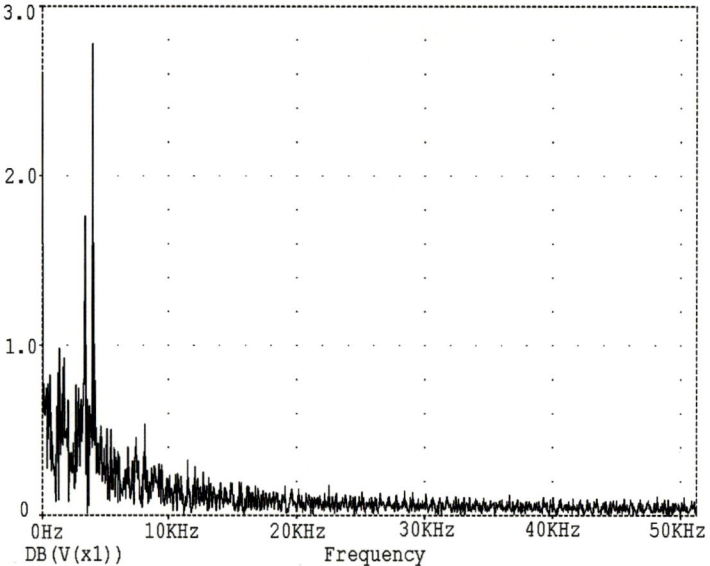

Fig. 4.7. Spectrum of the coupling signal x_1

It can be seen that the majority of the signal power is located in a band below 50 kHz, therefore in the following it will be assumed that the signal is band-limited to this range.

4.2 Experimental Signal Transmission Using Synchronized SC-CNN

The Bode plots of the transfer function of the adapted line for a 1 km cable and 10 km cable are reported in Figs. 4.8 and 4.9 respectively.

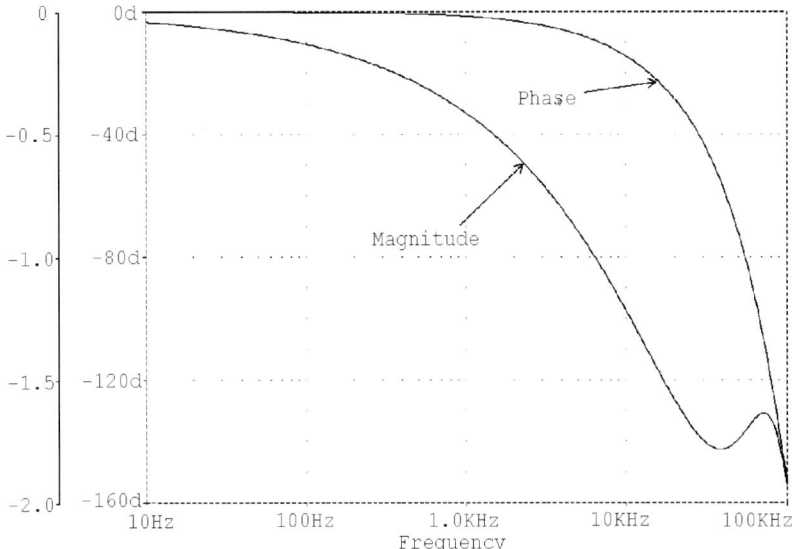

Fig. 4.8. Bode plot of a 1 km transmission line

Let us consider the case of the adapted channel. In a first instance a 1 km long line is considered. The corresponding transmitted and decoded waveforms are reported in Fig. 4.10, while in Fig. 4.11 the case of a 10 km line is depicted.

Of course, while in the first case there is still a good agreement between the two signals, in the case of the 10 km cable the degradation due to the line is excessive.

If the line is not adapted then the corresponding waveforms are those reported in Figs. 4.12 and 4.13.

From these it follows that while for the 1 km cable the degradation is still acceptable (and comparable to that of the adapted line), in the 10 km case the distortion of the decoded message is greater than that observed for the adapted line.

4.2.4 Effects of Additive Noise and Disturbances on the Channel

Here, the effects of additive noise and disturbances in the signal x_1 on the "quality" of the synchronization are considered. An ideal coupling is assumed.

The block diagram shown in Fig. 4.14 represents the experimental set-up.

In particular the block B5 represents the noise generator embedded into the spectrum analyzer used for the measurements (HP 35665A).

88   4. Synchronization

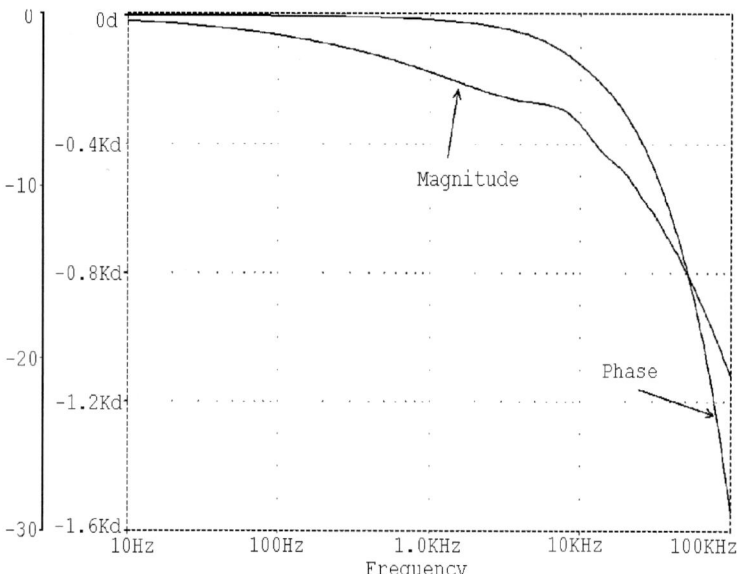

Fig. 4.9. Bode plot of a 10 km transmission line

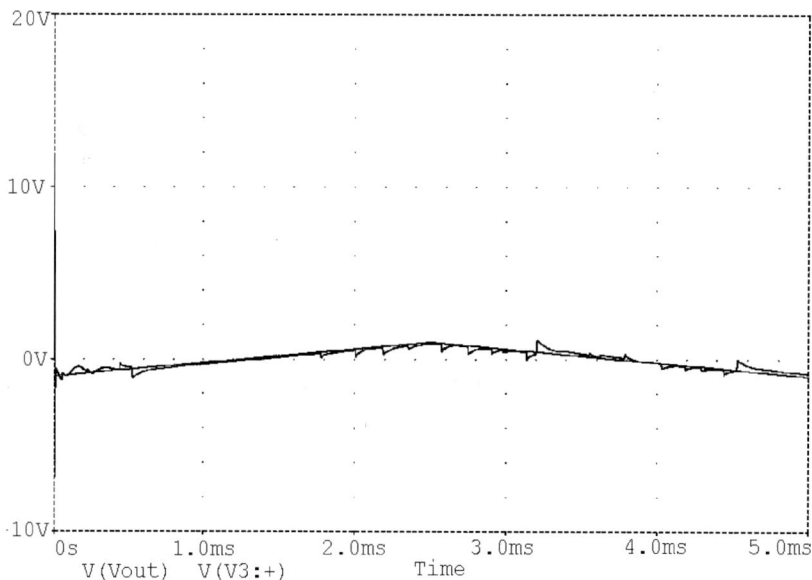

Fig. 4.10. Transmitted and decoded signal for a 1 km line

4.2 Experimental Signal Transmission Using Synchronized SC-CNN   89

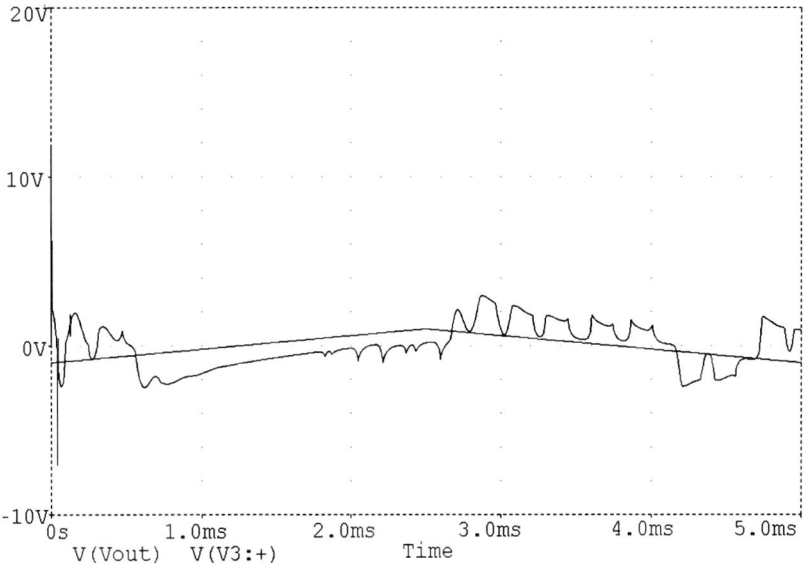

Fig. 4.11. Transmitted and decoded signal for a 10 km line

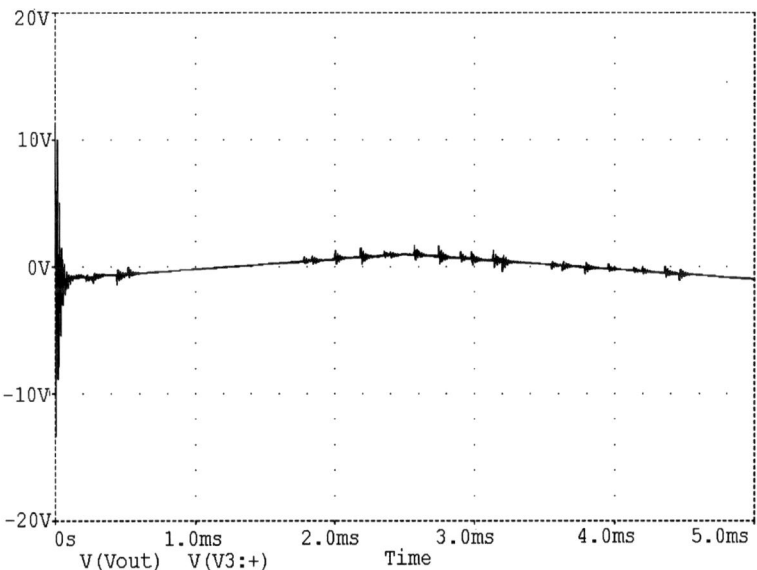

Fig. 4.12. Transmitted and decoded signal for a 1 km line and non-adapted channel

90   4. Synchronization

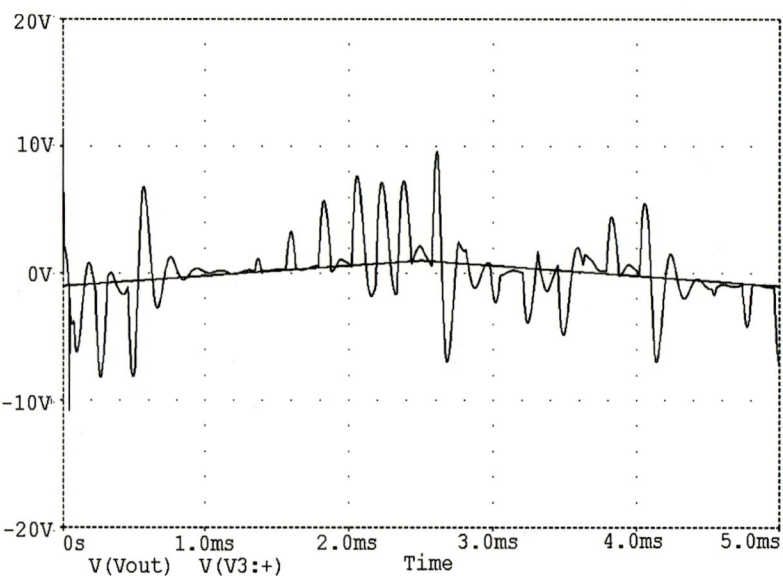

Fig. 4.13. Transmitted and decoded signal for a 10 km line and non-adapted channel

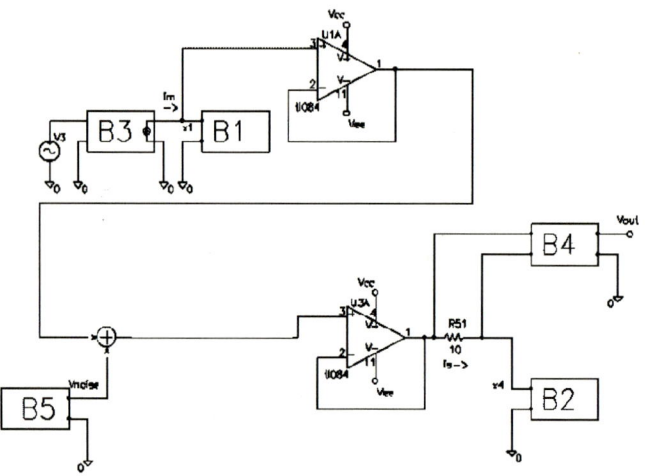

Fig. 4.14. Block scheme for the analysis of the effects of noise

4.2 Experimental Signal Transmission Using Synchronized SC-CNN   91

Three different situations have been considered. The first one deals with *white noise*, whose power spectrum for 1 Vpk (rms), is depicted in Fig. 4.15. In the second case a *pink noise* was added and its spectrum for 1 Vpk (rms), is depicted in Fig. 4.16.

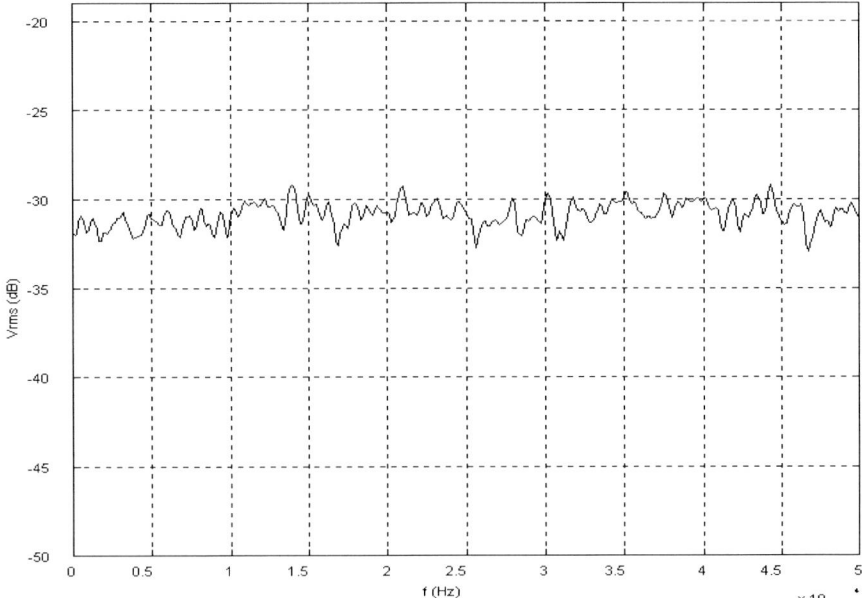

Fig. 4.15. Power spectrum of the white noise

In the third case some sinusoidal disturbances of various frequency and amplitude were added. In order to evaluate the quality of the synchronization the cross-correlation between the state variable x_2 (master) and x_5 (slave) has been considered:

$$R_{x_2,x_5}(\tau) = \lim_{t \to \infty} \frac{1}{T} \int_T x_2(t) x_5(t+\tau) \mathrm{d}t. \tag{4.9}$$

Of course the cross-correlation decreases when the slave is badly synchronized. Moreover, in order to take into account the increased energy supplied to the slave due to the noise and/or disturbance, the cross-correlation functions have been normalized with respect to their maximum values at the origin.

4. Synchronization

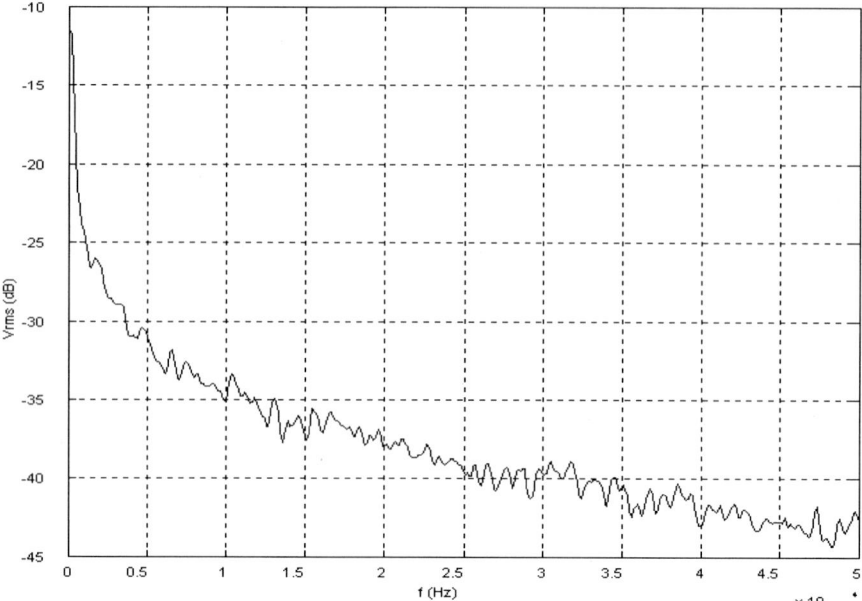

Fig. 4.16. Power spectrum of the pink noise

So the normalized cross-correlation will be defined as:

$$\tilde{R}_{x_2,x_5} = \frac{R_{x_2,x_5}(\tau)}{R_{x_2,x_5}(0)}. \tag{4.10}$$

The results are summarized in Table 4.1; in particular the energy $\epsilon(\tilde{R}_{x_2,x_5})$ of this normalized cross-correlation is reported for the various cases.

The reference values for the following comparisons are those reported in the first row of the table, in which an undisturbed transmission is considered. The next three rows consider the case of additive white noise with increasing power.

As expected, the energy of the normalized cross-correlation decreases as the noise power increases. Analogously, a similar decrease is observed in the case of pink noise (fifth and sixth rows).

The last five rows illustrate the effect of sinusoidal disturbance. It is interesting to note that a tone around 1 kHz can be more harmful, with respect to the synchronization, than a similar tone at a different frequency. This can be understood by observing the spectrum of x_1; in fact, the 1 kHz tone is located around the frequency range in which the coupling signal has the majority of its power. Besides, it can be conjectured that such an external signal could drive the system to a bifurcation point.

It is worth noting, from the table, that the synchronization remains relatively insensible to the additive noise.

Table 4.1. Measurements in presence of additive noise and sinusoidal disturbance

Additive signal	$V_{pk}(\text{rms})/f$ (Hz)	$\epsilon(\tilde{R}_{x_2,x_5})$ (Joule)
No noise	0	49.3387
White noise	2	43.4927
White noise	3	41.0844
White noise	5	32.5768
Pink noise	2	45.7839
Pink noise	3	39.6112
Sine	1/1K	31.8274
Sine	1/10K	36.0531
Sine	1/100	45.0874
Sine	2/1K	31.1263
Sine	1/560	34.1867

4.3 Chaotic System Identification

In this section we describe the new approach to chaotic system parameter identification. It is based on the Pecora–Carroll cascaded synchronization approach. The novel identification procedure is fairly general and it has been applied to the Chua oscillator as an example.

Using the new approach, the chaotic circuit whose parameters have to be estimated, is considered as a master system and one (at least) of its state variables is used as a driving signal for an identical circuit used as a slave system. The slave system responds to the driving signal following a state trajectory that depends on this input signal and on its parameters; these, in general, are different from the unknown master parameters. The distance between the master's state variable, used to drive the slave, and its corresponding slave's state variable is used to define a performance index.

When the master and the slave have the same parameter set then they synchronize, so their corresponding state variables are asymptotically equal. This means that, in this case, the performance index reaches its global minimum. The optimization of the performance index is performed by *Genetic Algorithms (GA)* [116, 117].

4.3.1 Description of the Algorithm

Let us consider an autonomous[6] nonlinear circuit or system whose mathematical model is known, whereas its parameters are unknown and have to be estimated.

---

[6] This hypothesis is not restrictive, because it is well known that a non-autonomous system can be described by an autonomous model augmenting its original model with suitable additional variables and equations as seen in Appendix A.

4. Synchronization

A state representation of this system is considered:

$$\dot{x} = f(x), \tag{4.11}$$

with $x \in \mathbb{R}^n$. As explained in Sect. 4.1.1, this system can be partitioned into two subsystems $x = (v, w)$:

$$\begin{aligned}\dot{v} &= g(v, w), \\ \dot{w} &= h(v, w),\end{aligned} \tag{4.12}$$

where $v = [x_1, .., x_m]'$, $g = [f_1(x), .., f_m(x)]'$, $w = [x_{m+1}, .., x_n]'$, $h = [f_{m+1}(x), .., f_n(x)]'$. This partitioned system can be used to realize a cascaded synchronization scheme as above discussed.

Equations (4.12) represent the master system while the slave $x'' = (v'', w')$, driven by v, is:

$$\begin{aligned}\dot{v}'' &= g(v'', w'), \\ \dot{w}' &= h(v, w').\end{aligned} \tag{4.13}$$

This setup is shown in the block diagram of Fig. 4.17.

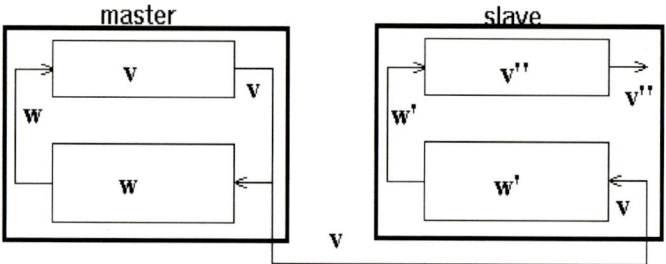

Fig. 4.17. The synchronization scheme adopted

In the slave system, the r unknown parameters $p = [p_1, .., p_r]'$ can initially be arbitrarily assigned.

Let us suppose that v are the only variables available from the master system and that:

$$\{\hat{v}(k)\}, k = 0, 1, .., M \tag{4.14}$$

is its corresponding sampled time series (δ being the sampling time). Analogously $\{\hat{v}''(k)\}$ is the time series corresponding to v''. Let us introduce the following index.

Definition 4.3.1. *Let us define the distance between $\hat{v}$ and $\hat{v}''$ as:*

$$I(p) = \sqrt{\sum_{k=0}^{M} \left\{ (\hat{v}_1 - \hat{v}_1''(p))^2 + .. + (\hat{v}_m - \hat{v}_m''(p))^2 \right\}}. \tag{4.15}$$

It is clear that this definition is independent from the nature of the time series, so their possible chaotic features are not a problem.

Our identification problem can therefore be formulated as an optimization problem. In fact, the master and the slave will synchronize if they have identical parameters and, in this case, as from Def. 4.1.1, the index (4.15) has the global minimum.

Therefore the parameters $\boldsymbol{p}$ must be changed according to this goal; GAs (genetic algorithms) have been used to achieve it. In particular a GA in a standard form has been adopted [116]. Besides reproduction, one point crossover and mutation, the elitist strategy has also been utilized.

The time series (4.14) is used to drive a simulated slave system whose parameters are chosen by a GA-based program. It is worth noting that the time discretization, which is necessary for the computer simulation of the slave, inevitably introduces a numerical error that is an increasing function of the discretization step size. This aspect has to be taken into account in order to evaluate the quality of the results obtained.

4.3.2 Identification of the Chua Oscillator

The above procedure is now applied to the case of the Chua oscillator.

The driving signal can be chosen from among x, y and z. The proper choice depends on the conditional Lyapunov exponents and hence on the considered systems parameters. This means that, in order to obtain the synchronization, some parameter choices require the x variable as driving signal while others require y or z. In the following, the case in which x is used as driving signal is discussed. However, the procedure is quite general and an example in which the z variable is used to drive the slave systems is presented in the next section. The state equations (3.8) of Chap. 3 can be partitioned into two subsystems: the first one, composed of x only, and the second one, composed of y and z. So, using the above terminology, $\boldsymbol{v} = x$ while $\boldsymbol{w} = (y, z)$; then, the slave (driven by x) is described as

$$\begin{aligned} \dot{x}'' &= \alpha(y' - h(x'')), \\ \dot{y}' &= x - y' + z', \\ \dot{z}' &= -\beta y' - \gamma z', \end{aligned} \quad (4.16)$$

where the $r = 5$ parameters are $\boldsymbol{p} = (\alpha, \beta, \gamma, m_0, m_1)$ and the objective index is

$$I(\boldsymbol{p}) = \sqrt{\sum_{k=0}^{M} (\hat{x} - \hat{x}''(\boldsymbol{p}))^2}. \quad (4.17)$$

The slave system has been simulated by using a fixed step-size fourth-order Runge–Kutta algorithm; this step-size has been chosen to be equal to the sampling time δ of the driving time series $\hat{x}$.

4.3.3 Examples

In this section three different examples of the application of the new method to Chua oscillator case are reported; the parameters of the Chua oscillator used as master circuit have been fixed to known values. The identification procedure has been applied in order to recover these values just from the given time series.

Table 4.2. Parameters of the double-scroll case

Master	Slave
$\alpha=9$	$\alpha=9.77$
$\beta=14.286$	$\beta=14,619$
$\gamma=0$	$\gamma=0.157$
$m_0=-0.142857$	$m_0=-0.122$
$m_1=0.285714$	$m_1=0.277$
$I(p)$	1.5721

GA parameters	
popsize	100
generations	100
prob. cross.	0.6
prob. mut.	0.003

In the first case the parameters have been chosen to obtain a double scroll attractor, that is: $\alpha=9$, $\beta=14.286$, $\gamma=0$, $m_0=-0.142857$, $m_1=0.285714$. The sampling time has been chosen as $\delta=0.1$ and $M=2000$ samples have been considered. x has been used as synchronization signal. The results of the identification are shown in Table 4.2 together with the parameters of the chosen genetic algorithm.

In order to evaluate the results obtained, the attractor generated by the master system and the one obtained with the estimated parameters, when they are disconnected, have been superposed in Fig. 4.18; Fig. 4.19 shows the synchronization signal x and x'' together with the synchronization error $e = x - x''$.

In the second example the parameters have been chosen to obtain the attractor shown in Fig. 4.20, that is: $\alpha=-4,08685$, $\beta=-2$, $\gamma=0$, $m_0=-0.142857$, $m_1=0.2857143$. Let us refer to this attractor as No. 2.

In this case the sampling time has been chosen as $\delta = 0.1$ with $M = 2000$ samples and the results are shown in Table 4.3.

In this second example the synchronization can be only accomplished by using the z variable as the driving signal. Again the original attractor and the one corresponding to the estimated parameter set have been superposed as shown in Fig. 4.21; while Fig. 4.22 shows the synchronization signals z and z'' together with the synchronization error $e = z - z''$.

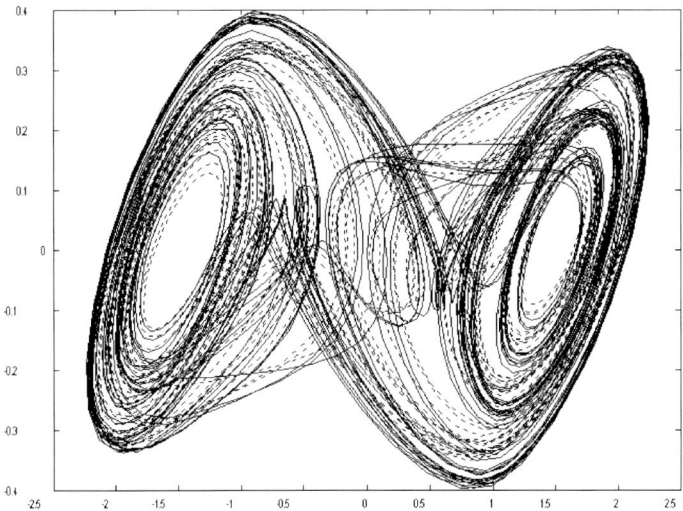

Fig. 4.18. Superposed attractors of the separate master and slave systems in the $x - y$ plane. The dotted line refers to the master while the solid one is the slave

Table 4.3. Parameters of the second example

Master	Slave
$\alpha = -4.088685$	$\alpha = -4.113$
$\beta = -2$	$\beta = -2.157$
$\gamma = 0$	$\gamma = -0.013$
$m_0 = -0.142857$	$m_0 = -0.121$
$m_1 = 0.285714$	$m_1 = 0.267$
$I(\boldsymbol{p})$	1.4922
GA parameters	
popsize	100
generations	140
prob. cross.	0.6
prob. mut	0.006

In the last case the parameters have been chosen to obtain the attractor No. 3, shown in Fig. 4.23, that is: $\alpha=6.579$, $\beta=10.898$, $\gamma=-0.0447$, $m_0=-0.18197$, $m_1=0.3477$.

The sampling time has been chosen as $\delta=0.01$ and $M=6000$ samples have been considered. x has been used as synchronization signal. The results of the identification are shown in Table 4.4 together with the parameters of the adopted genetic algorithm.

In order to evaluate the results obtained, the attractor generated by the master system and the one obtained with the estimated parameters, when they are disconnected, have been superposed in Fig. 4.24; while Fig. 4.25

98   4. Synchronization

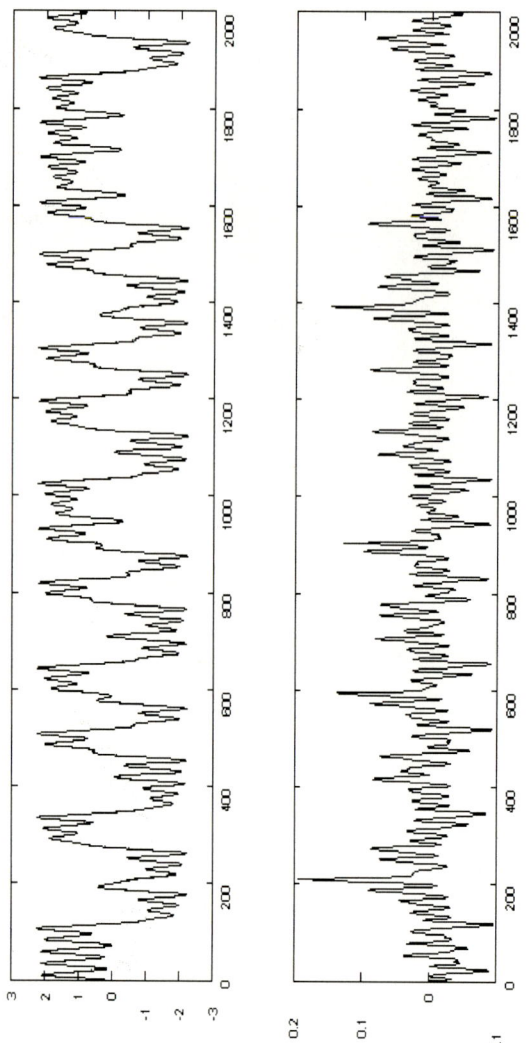

Fig. 4.19. Upper trace: Superposed variables x and x'' of the master and slave systems for the double scroll. The dotted line refers to the master while the solid one is the slave. Lower trace: Synchronization error $e = x - x''$

4.3 Chaotic System Identification   99

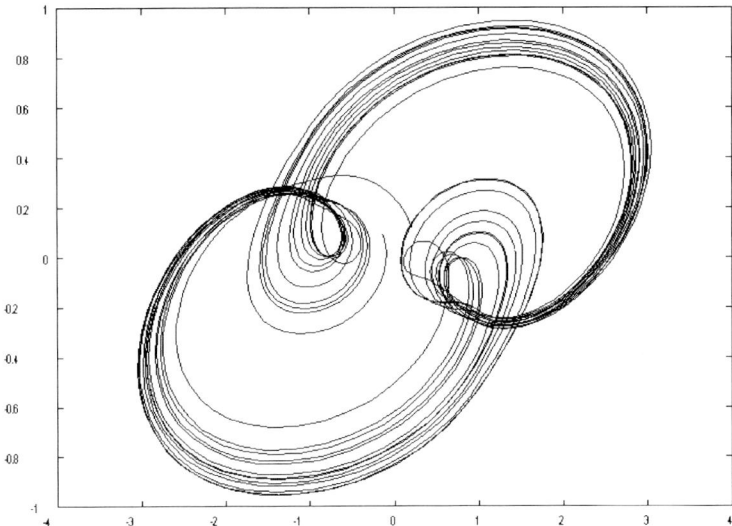

Fig. 4.20. Phase portrait of attractor No. 2 in the $x - y$ plane

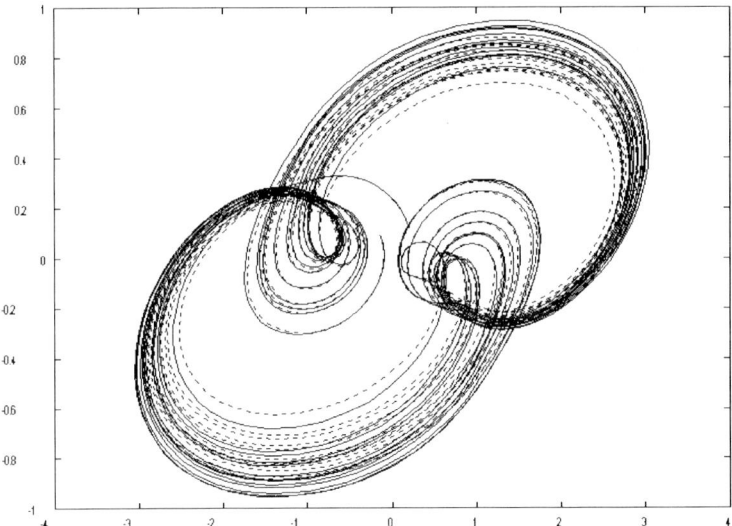

Fig. 4.21. Superposed attractors of the separate master and slave systems in the $x - y$ plane for the attractor No. 2. The dotted line refers to the master while the solid one is the slave

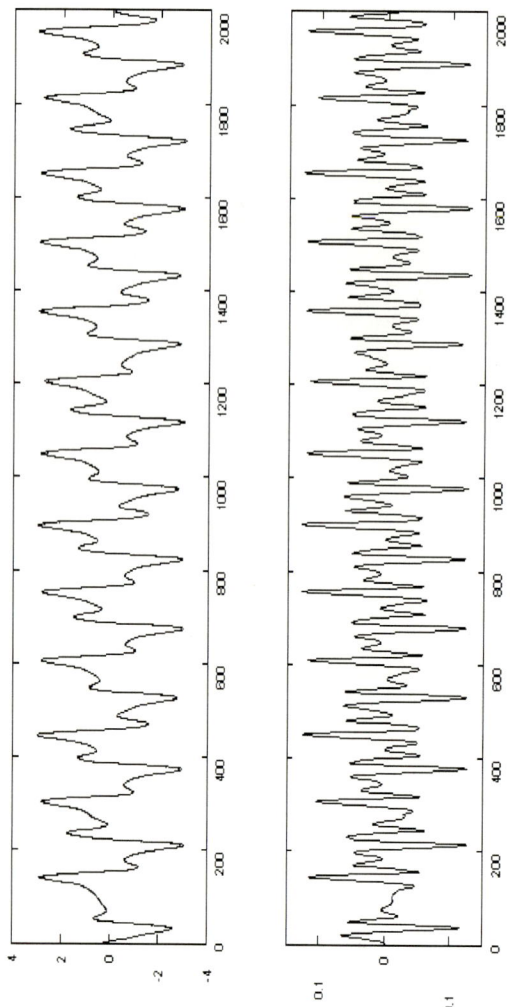

Fig. 4.22. Upper trace: Superposed variables z and z'' of the master and slave systems for the attractor No. 2. The dotted line refers to the master while the solid one is the slave. Lower trace: Synchronization error $e = z - z''$

shows the synchronization signals x and x'' together with the synchronization error $e = x - x''$.

The accuracy of the estimation can be increased if more samples, smaller step-sizes and more generations are used. Of course this implies increased computational costs so a trade-off is necessary. Moreover, some dynamics are more difficult to estimate, that is, the parameters of the master must be very close to those of the slave in order to obtain the correct synchronization.

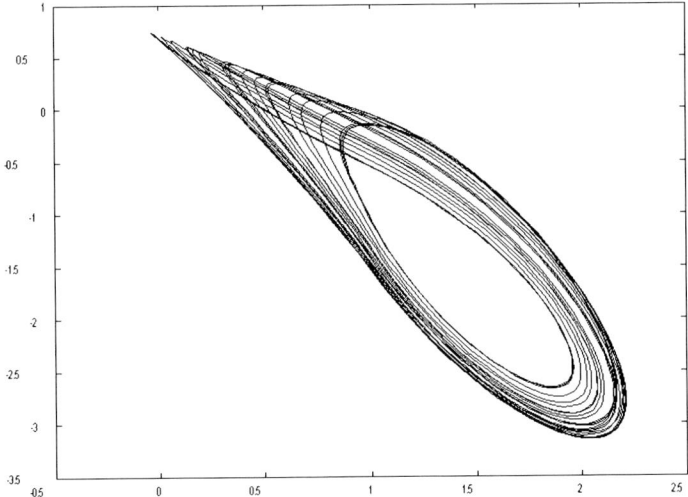

Fig. 4.23. Phase portrait of attractor No. 3 in the $x - z$ plane

Table 4.4. Parameters of the third case

Master	Slave
$\alpha=6.579$	$\alpha=6.195$
$\beta=10.898$	$\beta=10.695$
$\gamma=-0.0447$	$\gamma=-0.041$
$m_0=-0.18197$	$m_0=-0.179$
$m_1=0.3477$	$m_1=0.35$
$I(\boldsymbol{p})$	2.0539
GA parameters	
popsize	80
generations	200
prob. cross.	0.6
prob. mut.	0.001

This feature is related to the structural stability[7] of the considered dynamics, namely to the system's ability to retain its qualitative properties under small perturbations of the parameters or of the model.

4.4 Summary and Conclusions

Section 4.2 of this chapter described the experimental study of the synchronization of chaotic SC-CNNs. The inverse system approach was applied in order to realize a chaotic based transceiver.

---

[7] See Sect. A.8 in Appendix A.

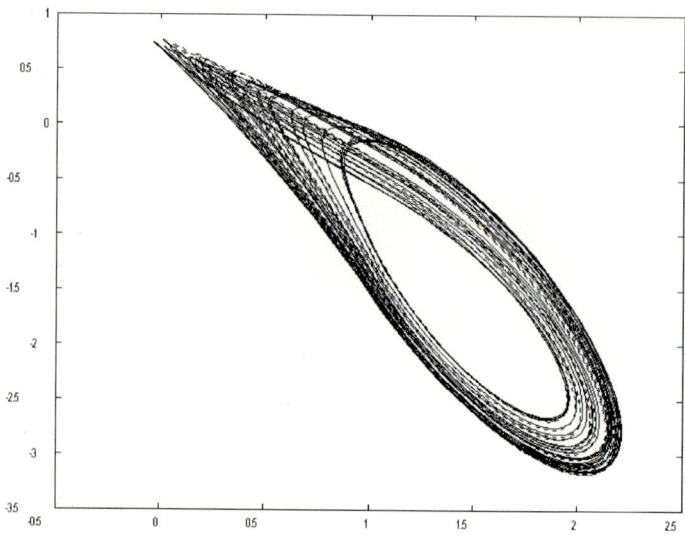

Fig. 4.24. Superposed attractors of the separated master and slave systems in x-y plane for the attractor No. 3. The dotted line refers to the master while the solid one is the slave

A experimental analysis of the effects of a non-ideal channel on the synchronization set-up was presented. Particular attention was devoted to the use of coaxial line as a coupling media. Moreover the effects of additive noise and sinusoidal disturbances were evaluated.

From these last measures it emerges that a tone with a particular frequency can be disastrous. The collected data represent the starting point for the design of an optimal communication channel with respect to the length and the equalization of the line.

To this end it can easily be argued that for short distances (less than few kilometers), as in the case of a local communication system (e.g. in a building), there may be no need for equalization or even line impedance matching.

However this is mandatory for longer communication channels.

Section 4.3 presented a new method to identify the parameters of nonlinear circuits. The procedure has been formulated as a global optimization problem and it has been tackled using a genetic algorithm.

It has been applied to estimate the five dimensionless parameters of the chaotic Chus oscillator and three experimental examples have been reported.

In addition, the accuracy of the method was discussed. With the proposed approach a circuit model for a chaotic behavior could be obtained. In fact, many different attractors have been observed in nonlinear circuits and the strategy introduced represents a useful tool to determine the parameters of a circuit model that best fit a chaotic time-series. Moreover, it should be

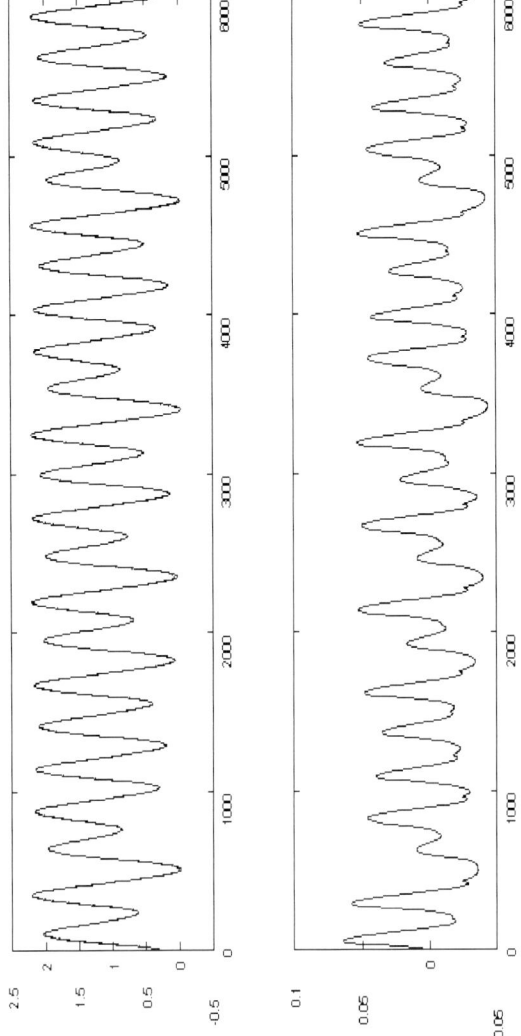

Fig. 4.25. Upper trace: Superposed variables x and x'' of the master and slave systems for the attractor No. 3. The dotted line refers to the master while the solid one is the slave. Lower trace: Synchronization error $e = x - x''$

noticed that the above method could potentially be used to eavesdrop on communication coded in chaos. In fact, it could be used to estimate the parameters of the nonlinear circuit used as modulator in a chaotic carrier cryptography system.

5. Spatio-temporal Phenomena

In this chapter some spatio-temporal phenomena arising in arrays of coupled nonlinear circuits are considered. In particular it is shown that many well-known phenomena, already observed in arrays of coupled Chua oscillators, can be similarly obtained in a two-layer Chua and Yang CNN model, which has a much simpler circuit topology.

5.1 Analysis of the Cell

In order to be able to consider the complex spatio-temporal behaviour described in the following sections it is necessary to develop a thorough analysis of the single cell.

Zou and Nossek [73, 74, 118] made extensive studies of the following two-cell Chua and Yang CNN model:

$$\begin{aligned} \dot{x}_1 &= -x_1 + (1+\mu)y_1 - sy_2, \\ \dot{x}_2 &= -x_2 + sy_2 + (1+\mu)y_2, \end{aligned} \quad (5.1)$$

which, for some choice of the parameters μ, $s \in \mathbb{R}$ originates a stable limit cycle, symmetric with respect to the axis in the phase plane.

However, studies of the different active media in which the phenomena exist, have shown that systems with non-symmetric oscillations, bi-stable and slow–fast regimes are appropriate candidates [14, 82, 119]. Therefore it is necessary to modify the above model.

The nonlinear state equations of the cell in which we are interested are of the form:

$$\dot{x} = f(x), \quad (5.2)$$

obtained from (5.1) by introducing two non-zero constant bias terms $i_1, i_2 \in \mathbb{R}$ [120, 121]:

$$\begin{aligned} \dot{x}_1 &= -x_1 + (1+\mu)y_1 - sy_2 + i_1, \\ \dot{x}_2 &= -x_2 + sy_2 + (1+\mu)y_2 + i_2, \end{aligned} \quad (5.3a)$$

with

$$y_i = \frac{1}{2}(|x_i + 1| - |x_i - 1|). \tag{5.3b}$$

For this model, some of the results derived by Zou and Nossek still hold. However many others cannot be valid anymore, because the introduction of the bias terms leads to the disappearance of some fixed points and to a substantial modification of the limit cycle.

Hence, the study of the planar vector field (5.3a) must be repeated almost from scratch.

In order to analyze the dynamic behavior of (5.3a) it is useful to partition the phase plane into certain regions, in which (5.3a) can be locally studied as a linear[1] vector field. In these regions, the Hartman–Großman theorem[2] gives us thorough information on the system trajectories.

Definition 5.1.1. *The phase plane $\mathbb{R}^2$ can be partitioned into the following nine regions:*

1. *Linear region:*

$$D_0 \doteq \{x \in \mathbb{R}^2 \; : \; |x_i| < 1, \; i = 1, 2\}. \tag{5.4}$$

2. *Saturation regions:*

$$\begin{aligned}
D_s^{++} &\doteq \{x \in \mathbb{R}^2 \; : \; x_1 \geq 1, \; x_2 \geq 1\}, \\
D_s^{+-} &\doteq \{x \in \mathbb{R}^2 \; : \; x_1 \geq 1, \; x_2 \leq -1\}, \\
D_s^{--} &\doteq \{x \in \mathbb{R}^2 \; : \; x_1 \leq -1, \; x_2 \leq -1\}, \\
D_s^{-+} &\doteq \{x \in \mathbb{R}^2 \; : \; x_1 \leq -1, \; x_2 \geq 1\}.
\end{aligned} \tag{5.5}$$

3. *Partial saturation regions:*

$$\begin{aligned}
D_p^{+,l} &\doteq \{x \in \mathbb{R}^2 \; : \; x_1 \geq 1, \; |x_2| < 1\}, \\
D_p^{l,-} &\doteq \{x \in \mathbb{R}^2 \; : \; |x_1| < 1, \; x_2 \leq -1\}, \\
D_p^{-,l} &\doteq \{x \in \mathbb{R}^2 \; : \; x_1 \leq -1, \; |x_2| < 1\}, \\
D_p^{l,+} &\doteq \{x \in \mathbb{R}^2 \; : \; |x_1| < 1, \; x_2 \geq 1\}.
\end{aligned} \tag{5.6}$$

It is obvious that all of them are disjointed and their union is $\mathbb{R}^2$.

For convenience it is also useful to define:

$$\begin{aligned}
D_s &\doteq D_s^{++} \cup D_s^{+-} \cup D_s^{--} \cup D_s^{-+}, \\
D_p &\doteq D_p^{+,l} \cup D_p^{l,-} \cup D_p^{-,l} \cup D_p^{l,+}.
\end{aligned} \tag{5.7}$$

---

[1] In fact as an affine, not exactly linear.
[2] See Theorem A.7.1 in Appendix A.

5.1.1 Fixed Points

The fixed points are obtained by solving the system of nonlinear algebraic equations:

$$f(x) = 0. \tag{5.8}$$

Depending on the choice of the parameters $s, \mu, i_1, i_2 \in \mathbb{R}^n$, some or all of the candidate solutions of (5.8) can be real fixed points or virtual fixed points [90]. Nonetheless, virtual fixed points give as much information on the dynamics as the real equilibria.

Theorem 5.1.1. *In any of the regions defined in Def. 5.1.1 there can be, at most, one equilibrium point.*

Proof. The proof is straightforward if we consider that system (5.8) is locally affine within any of the above defined regions (5.4–5.6). Therefore it can admit one, none or infinite solutions. However, the latter case happens iff the two equations composing the system geometrically correspond to the same straight line. This case is not structurally stable and so it is not considered here. □

Specifically, the following candidate solutions are obtained:

$$\boldsymbol{P}_0^{l,l} = \left[-\frac{i_1}{\mu} + \frac{s}{\mu} \cdot \left(\frac{si_1 + \mu i_2}{\mu^2 - s^2} \right), \frac{si_1 + \mu i_2}{\mu^2 - s^2} \right]', \tag{5.9a}$$

which is the candidate to belong to D_0;

$$\begin{aligned}
\boldsymbol{P}_s^{++} &= [1 + \mu - s + i_1, \ 1 + \mu + s + i_2]', \\
\boldsymbol{P}_s^{+-} &= [1 + \mu + s + i_1, \ -1 - \mu + s + i_2]', \\
\boldsymbol{P}_s^{--} &= [-1 - \mu + s + i_1, \ -1 - \mu - s + i_2]', \\
\boldsymbol{P}_s^{-+} &= [-1 - \mu - s + i_1, \ 1 + \mu + s + i_2]',
\end{aligned} \tag{5.9b}$$

which are the candidates to belong to D_s^{++}, D_s^{+-}, D_s^{--}, D_s^{-+} respectively, and

$$\begin{aligned}
\boldsymbol{P}_p^{l,+} &= \left[\frac{s - i_1}{\mu}, \ 1 + \mu + \frac{s}{\mu} \cdot (s - i_1) + i_2 \right]', \\
\boldsymbol{P}_p^{+,l} &= \left[1 + \mu + \frac{s}{\mu} \cdot (s + i_2) + i_1, \ -\frac{s + i_2}{\mu} \right]', \\
\boldsymbol{P}_p^{l,-} &= \left[-\frac{s + i_1}{\mu}, \ -1 - \mu - \frac{s}{\mu} \cdot (s + i_1) + i_2 \right]', \\
\boldsymbol{P}_p^{-,l} &= \left[-1 - \mu - \frac{s}{\mu} \cdot (s - i_2) + i_1, \ \frac{s - i_2}{\mu} \right]',
\end{aligned} \tag{5.9c}$$

which are the candidates to belong to $D_p^{l,+}$, $D_p^{+,l}$, $D_p^{l,-1}$, $D_p^{-1,l}$ respectively.

108   5. Spatio-temporal Phenomena

If the candidate belongs to the corresponding region then it represents a real fixed point. Otherwise it is a virtual fixed point conditioning the behavior of the vector field in the corresponding region.

For the sake of clarity let us consider a simple example. If $\boldsymbol{P}_0^{l,l} \notin D_0$ then it is a virtual equilibrium. Moreover let us assume that, from the extension of the linearization (in D_0) to $\mathbb{R}^2$, it emerges that $\boldsymbol{P}_0^{l,l}$ would be a stable equilibria. All the trajectories in D_0 represent the restriction of the ones obtained from the extension in $\mathbb{R}^2$. But this means that all the trajectories entering D_0 will eventually go out attracted by the virtual fixed point $\boldsymbol{P}_0^{l,l}$.

Similar conclusions can be formulated for the other cases.

The output nonlinearity is piecewise linearly (PWL) continuous and so differentiable almost everywhere:

$$\frac{\partial y_i}{\partial x_i} = \begin{cases} 0 & \text{if } |x_i| > 1, \\ 1 & \text{if } |x_i| < 1. \end{cases} \qquad \frac{\partial y_i}{\partial x_j} = 0, \quad i \neq j. \tag{5.10}$$

Hence we can consider the Jacobian of (5.3a):

$$J = D\boldsymbol{f} = \begin{pmatrix} -1 + (1+\mu)\frac{\partial y_1}{\partial x_1} & -s\frac{\partial y_2}{\partial x_2} \\ s\frac{\partial y_1}{\partial x_1} & -1 + (1+\mu)\frac{\partial y_2}{\partial x_2} \end{pmatrix} \tag{5.11}$$

and the corresponding characteristic equation:

$$|\lambda I - J| = \lambda^2 + \lambda \left[2 - (1+\mu)\left(\frac{\partial y_1}{\partial x_1} + \frac{\partial y_2}{\partial x_2}\right) \right]$$
$$+ \left\{ \frac{\partial y_1}{\partial x_1}\frac{\partial y_2}{\partial x_2}\left[s^2 + (1+\mu)^2\right] - (1+\mu)\left(\frac{\partial y_1}{\partial x_1} + \frac{\partial y_2}{\partial x_2}\right) + 1 \right\}, \tag{5.12}$$

which provides us with the eigenvalues of the linearized systems corresponding to the candidate fixed points (5.9).

The corresponding results are summarized in Table 5.1.

Table 5.1. Summary of the stability of the (real/virtual) fixed points

	$\boldsymbol{P}_0^{l,l}$	$\boldsymbol{P}_s^{++}, \boldsymbol{P}_s^{+-}, \boldsymbol{P}_s^{-+}, \boldsymbol{P}_s^{--}$	$\boldsymbol{P}_p^{l,+}, \boldsymbol{P}_p^{+,l}, \boldsymbol{P}_p^{l,-}, \boldsymbol{P}_p^{-,l}$
	$\lambda_{1,2} = \mu \pm js$	$\lambda_{1,2} = -1$	$\lambda_1 = \mu, \lambda_2 = -1$
$\mu > 0$	unstable focus	stable node	saddle node
$\mu < 0$	stable focus	stable node	stable node
$\mu = 0$	center	stable node	marginally stable node

From this table it is clearly seen that for $\mu = 0$ non–hyperbolic points do appear (so the Hartman–Großman theorem does not hold). This value delimits the boundary between two qualitatively different behaviors of the candidate fixed points. Indeed a bifurcation is observed.

In contrast, for $\mu < 0$, if at least one of the nine candidates is a real fixed point, then the system is completely stable. This is certainly true if $|i_1| < 1$ and $|i_2| < 1$ because, in this case, we would have at least $\boldsymbol{P}_0^{l,l} \in D_0$. Moreover, analogously to what was done in [74], it can be proved by the Bendixon's Criterion[3] that a cycle limit cannot exist.

However, the case in which we are interested is $\mu > 0$. In this case a *Hopf-like* bifurcation cause the appearance of a stable limit cycle.

5.1.2 Limit Cycle and Bifurcations

Henceforth, let us assume $\mu > 0$, $|i_1| < 1$ and $|i_2| < 1$. In other words there is always, at least one fixed point (i.e. $\boldsymbol{P}_0^{l,l} \in D_0$).

Let us investigate the local bifurcation for $\mu = 0$.

The Hopf theorem[4] cannot be directly applied because $\boldsymbol{P}_0$ is a center for $\mu = 0$ and moreover $f \notin C^k$, $k \geq 4$. Nevertheless, looking at Table 5.1, it can be seen that one has all the conditions for the the birth of a limit cycle.

In fact the following theorem holds.

Theorem 5.1.2. *Let us consider the following real numbers:*

$$s^{++} = \max\{s - i_1, -s - i_2\}, \tag{5.13a}$$
$$s^{+-} = \max\{-s - i_1, s + i_2\}, \tag{5.13b}$$
$$s^{--} = \max\{s + i_1, i_2 - s\}, \tag{5.13c}$$
$$s^{-+} = \max\{i_1 - s, -s - i_2\}, \tag{5.13d}$$
$$s^* = \min\{s^{++}, s^{+-}, s^{--}, s^{-+}\}, \tag{5.13e}$$

and the dynamic system (5.3a) with $|i_1| < 1$ and $|i_2| < 1$. Then $\mu = 0$ is a bifurcation point and, for $\mu \in (0, s^*)$, $\boldsymbol{P}_0^{l,l} \in D_0$ is an unstable focus surrounded by a stable limit cycle Γ.

Proof. From Table 5.1 it is known that for $\mu > 0$ the only candidate stable fixed points are $\boldsymbol{P}_s^{++}, \boldsymbol{P}_s^{+-}, \boldsymbol{P}_s^{-+}, \boldsymbol{P}_s^{--}$. If all of these are virtual fixed points then there will be no stable fixed point in the phase plane. Moreover if we draw a ring-like region Ψ with radius $R \gg 1 + \mu + s + \max\{i_1, i_2\}$ and not including the remaining unstable fixed points, then the vector field (5.3a) can be approximated by

$$\begin{aligned}\dot{x}_1 &\approx -x_1, \\ \dot{x}_2 &\approx -x_2\end{aligned} \tag{5.14}$$

near the outer boundary of the simply-connected region. This means that all the trajectories cross the outer boundary of Ψ from the outside into the inside.

---
[3] See Theorem A.10.2 in Appendix A.
[4] See Theorem A.10.4 in Appendix A.

110   5. Spatio-temporal Phenomena

On the other hand, because of the instability of $P_0^{l,l}$, all the trajectories will also cross the inner boundary of Ψ from the outside to the inside. Therefore, according to the Poincaré–Bendixon theorem[5] there exists a limit cycle Γ inside Ψ.

It only remains to prove that $P_s^{++}, P_s^{+-}, P_s^{-+}, P_s^{--}$ are all virtual equilibria. For this to be true, all of them must be located outside their corresponding saturation regions $D_s^{++}, D_s^{+-}, D_s^{-+}, D_s^{--}$. Let us consider $P_s^{++} = [1 + \mu - s + i_1, \ 1 + \mu + s + i_2]'$. If either its first component or its second component are smaller than 1, then it is located outside D_s^{++}. It is easily verified that this is true if $0 < \mu < s^{++}$. Repeating the same reasoning for the other three points, similar inequalities are derived using s^{+-}, s^{--} and s^{-+}. Hence, to have all of them simultaneously outside their respective regions we must have $0 < \mu < s^*$.

This proves the thesis. □

Corollary 5.1.1. *If $\mu > s^*$ then system (5.3a) is completely stable and so $\mu = s^*$ is another bifurcation value.*

Proof. In fact, in this hypothesis, at least one of $P_s^{++}, P_s^{+-}, P_s^{-+}, P_s^{--}$ will belong to a saturation region and will be a stable node. It will attract any trajectory entering the saturation region and so Γ vanishes. □

The above results reduce to those proved in [74] when $i_1 = i_2 = 0$, the latter just being particular cases of the above.

5.1.3 Slow–Fast Dynamics

The limit cycle considered in [74] for $\mu < s$ is symmetric. Moreover, when $\mu = s > 0$, the stable equilibria of the saturation regions coincide with the unstable equilibria of the partial saturation regions. Therefore the trajectories inside the saturation regions are attracted to these fixed points, while they are repelled from them inside the partial saturation regions. Furthermore, the respective manifolds intersect in such a way as to create a symmetric heteroclinic orbit connecting the four equilibria. This is the result of a suitable degeneration of the stable symmetric limit cycle.

However, when $\mu > s > 0$, the structurally unstable heteroclinic orbit disappears, all of the 4 fixed points of the saturation region appear simultaneously, and the system becomes completely stable.

In our case, thanks to the introduction of the non-zero biases i_1 and i_2, it is possible to have:

1. all of $P_s^{++}, P_s^{+-}, P_s^{-+}, P_s^{--}$ outside the respective saturation regions but
2. only one of them (say P_s^{--} for example) close to the boundary with the corresponding saturation region (D_s^{--} in the example considered).

---
[5] See Theorem A.10.1 in Appendix A.

5.1 Analysis of the Cell

Let us consider this case for $\boldsymbol{P}_s^{--} \notin D_s^{--}$, $\boldsymbol{P}_s^{--} \in D_s^{l,-}$ but close to the boundary between D_s^{--} and $D_s^{l,-}$. Under this condition, the asymmetric limit cycle Γ is distorted close to the virtual equilibria $(\boldsymbol{P}_s^{--})$. In fact, in the saturation region D_s^{--} the trajectories behave as if attracted by the virtual point $\boldsymbol{P}_s^{--}$. Moreover, in D_s^{--} the flow tends towards this point exponentially with time:

$$\boldsymbol{x}(t) = e^{-t}\boldsymbol{x}_0 + (1-e^{-t})\boldsymbol{P}_s^{--}, \qquad \boldsymbol{x}(t), \boldsymbol{x}_0 \in D_s^{--}. \tag{5.15}$$

Conversely, no stable equilibria are present in the partial saturation region, since once they have entered $D_s^{l,-}$, the trajectories are accelerated so as to to leave the region.

This means that the flow undergoes a sudden variation in the rate of change of the state variables close to the virtual fixed point. In fact, let $x_{10} = -1 - \mu + s + i_1$ and $x_{20} = -1 - \mu - s + i_2$ be the two coordinates of $\boldsymbol{P}_s^{--} = [x_{10}, x_{20}]'$, then:

$$\lim_{x_1 \to x_{10},\, x_1 \in D_p^{l,-} \cap \Gamma} y_1(t) = x_1(t), \qquad \lim_{x_1 \to x_{10},\, x_1 \in D_s^{--} \cap \Gamma} y_1(t) = -1, \tag{5.16}$$

with $y_2(t) = -1$. Hence the change of rate is clearly shown by the following limits:

$$\begin{aligned}
\lim_{x \to \boldsymbol{P}_s^{--},\, x_1 \in D_s^{--} \cap \Gamma} \dot{x}_1(t) &= 0, \\
\lim_{x \to \boldsymbol{P}_s^{--},\, x_1 \in D_s^{l,-} \cap \Gamma} \dot{x}_1(t) &= (1+\mu)(s + i_1 - \mu), \\
\lim_{x \to \boldsymbol{P}_s^{--},\, x_1 \in D_s^{--} \cap \Gamma} \dot{x}_2(t) &= 0, \\
\lim_{x \to \boldsymbol{P}_s^{--},\, x_1 \in D_s^{l,-} \cap \Gamma} \dot{x}_2(t) &= s(s + i_1 - \mu).
\end{aligned} \tag{5.17}$$

If similar situations do not arise in the neighbors of the other regions, the remaining part of Γ is covered without significant variations in speed.

This implicitly proves that:

Proposition 5.1.1. *The vector field (5.3a) can have a slow–fast dynamics for an appropriate choice of its parameters.*

In practice, if desired, and if

1. only one of the four stable virtual equilibria $\boldsymbol{P}_s^{++}, \boldsymbol{P}_s^{+-}, \boldsymbol{P}_s^{-+}, \boldsymbol{P}_s^{--}$ is placed on the boundary of its saturation region and
2. the equilibrium corresponding to the adjacent partial saturation region is placed on it creating a single saddle point,

then a homoclinic orbit can be obtained.

5.1.4 Some Simulation Results

After the general analysis carried out in the previous sections we have all the results to enable us to choose the parameters for the desired dynamics.

In particular, the following parameters will be considered: $\mu = 0.7$, $s = 1$, $i_1 = -0.3$ and $i_2 = 0.3$. Figure 5.1 shows the modification of $x_1(t)$, as the bias terms are introduced, showing the slow–fast dynamics discussed earlier.

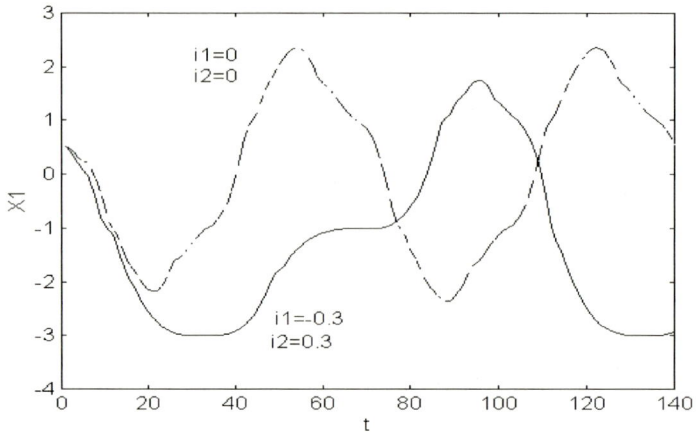

Fig. 5.1. Modification in state variable trend with the addition of bias

Figure 5.2 shows the corresponding limit cycle modification. From this figure it can be seen directly how, while without any bias term the limit cycle is covered at almost constant speed, with the addition of the biases, the limit cycle is no longer centered at the origin, but at the new equilibrium point, and the speed of running through it is faster in some parts than in others.

In particular, in Fig. 5.3 the modified limit cycle is reported together with the trends of $f_1(x) = 0$ and $f_2(x) = 0$. It can be seen that the speed variation of the state variables, while running onto the limit cycle, is accomplished in the neighborhood of the two points $Q_1(-3,1)$ and $Q_2(-1,-2.4)$, which correspond, for our choice of the parameters, to $\boldsymbol{P}_p^{-,l}$ and $\boldsymbol{P}_s^{--}$, respectively.

The time evolution of the state variables x_1 and x_2 is depicted in Fig. 5.4.

5.2 The Two-Layer CNN

Many nonlinear *reaction–diffusion* partial differential equations (PDEs) have shown self–organizing patterns [14, 119] and in [21] the concept of reaction–diffusion CNNs has been formalized in order to reproduce similar behavior

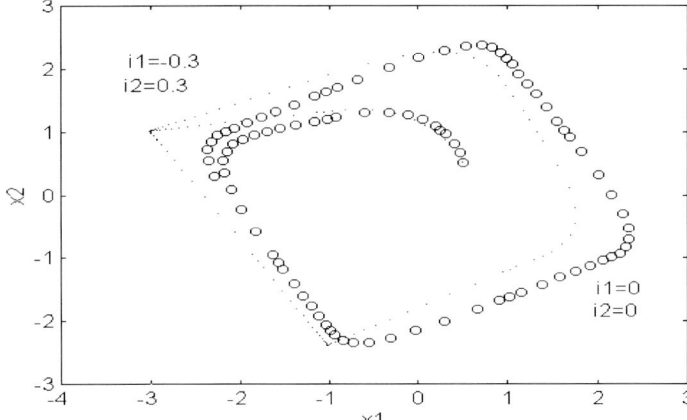

Fig. 5.2. Limit cycle modification with the addition of bias

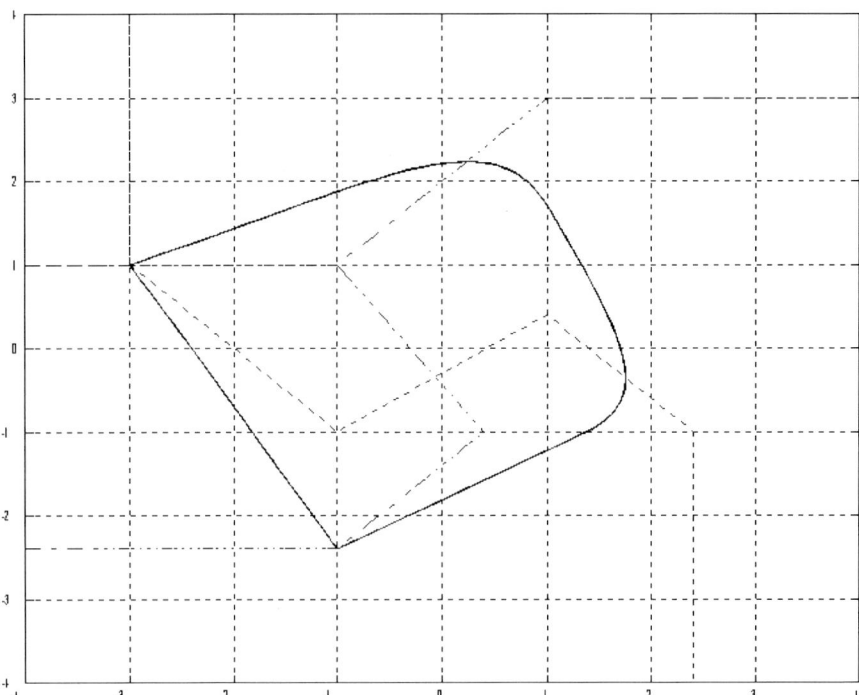

Fig. 5.3. The limit cycle (*solid line*) and the contours of the PWL equations $f_1(\boldsymbol{x}) = 0$ and $f_2(\boldsymbol{x}) = 0$ (*dashed lines*). (horizontal x_1, vertical x_2)

114   5. Spatio-temporal Phenomena

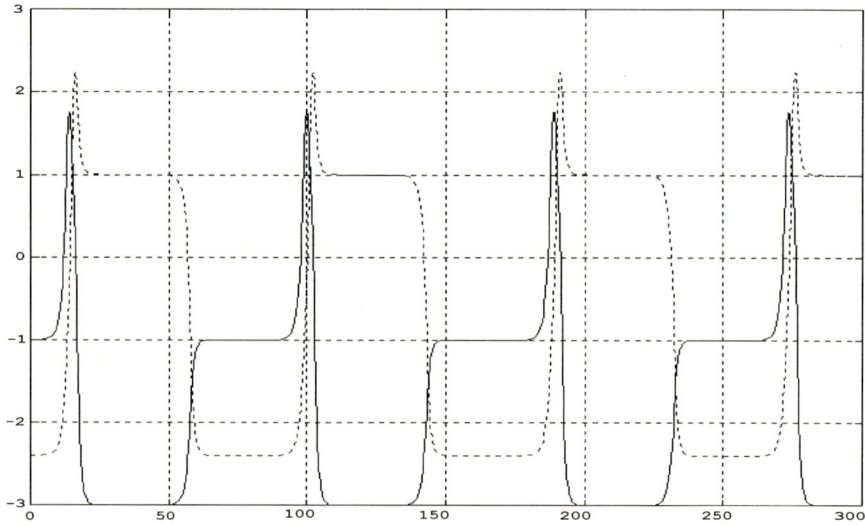

Fig. 5.4. Time evolution of x_1 (*solid line*) and x_2 (*dashed line*)

in CNNs. Here, a new CNN with constant templates, with circuit topology simpler than the CNNs reported in literature[6] [14] is introduced by suitably coupling the second–order cells introduced above.

Let us consider the following state equations:

$$\begin{aligned}
\dot{x}_{1;i,j} &= -x_{1;i,j} + (1+\mu)y_{1;i,j} - sy_{2;i,j} + \\
&\quad + D_1 \cdot (y_{1;i+1,j} + y_{1;i-1,j} + y_{1;i,j-1} + y_{1;i,j+1} - 4y_{1;i,j}) + i_1 \,, \\
\dot{x}_{2;i,j} &= -x_{2;i,j} + sy_{1;i,j} + (1+\mu)y_{2;i,j} + \\
&\quad + D_2 \cdot (y_{2;i+1,j} + y_{2;i-1,j} + y_{2;i,j-1} + y_{2;i,j+1} - 4y_{2;i,j}) + i_2 \,, \\
&\quad 1 \leq i \leq M\,, \quad 1 \leq j \leq N\,.
\end{aligned}$$

(5.18)

It can be seen that while the two layers interact within each single cell generating the oscillatory, slow–fast behavior described above, conversely, the interaction with the neighboring cells is obtained separately by means of two *circulant diffusion templates* with diffusion coefficients D_1 (for the first layer) and D_2 (for the second one) [14].

There is no direct interaction between layer 1 of a cell $C(i,j)$ and layer 2 of its neighbors and vice versa. Moreover, it can be noted that while in the examples reported in [21] the Laplacian templates weighted the *state variables* of the neighboring cells, conversely, in (5.18), consistent with the

---

[6] Until now all the spatio–temporal phenomena that we will study have been reported exclusively for CNNs composed of coupled Chua oscillators.

multi–layer CNN definition in Chap. 1, the Laplacian template is a *feedback template* that weighs the *outputs*.

Hence, with little modification compared to (5.18), the new CNN is formally defined as follows:

Definition 5.2.1. *The state model of the new two-layer CNN with constant templates is:*

$$\dot{x}_{ij} = -x_{ij} + A * y_{ij} + B * u_{ij} + I \,, \tag{5.19a}$$

where $x_{ij} = [x_{1;i,j} x_{2;i,j}]'$, $y_{ij} = [y_{1;i,j} y_{2;i,j}]'$ *and* $u_{ij} = [u_{1;i,j} u_{2;i,j}]'$ *are the state, the output and the input of the CNN respectively while A, B and I are the feedback, control and bias templates respectively. The cloning templates are*

$$A = \begin{pmatrix} A_{11} & A_{12} \\ A_{21} & A_{22} \end{pmatrix}, \ B = 0, \ I = \begin{pmatrix} i_1 \\ i_2 \end{pmatrix}, \tag{5.19b}$$

where

$$A_{11} = \begin{pmatrix} 0.5D_1 & D_1 & 0.5D_1 \\ D_1 & -5D_1 + \mu + 1 & D_1 \\ 0.5D_1 & D_1 & 0.5D_1 \end{pmatrix}, \ A_{12} = \begin{pmatrix} 0 & 0 & 0 \\ 0 & -s & 0 \\ 0 & 0 & 0 \end{pmatrix},$$

$$A_{21} = \begin{pmatrix} 0 & 0 & 0 \\ 0 & s & 0 \\ 0 & 0 & 0 \end{pmatrix}, \ A_{22} = \begin{pmatrix} 0.5D_2 & D_2 & 0.5D_2 \\ D_2 & -5D_2 + \mu + 1 & D_2 \\ 0.5D_2 & D_2 & 0.5D_2 \end{pmatrix}. \tag{5.19c}$$

Zero–flux (Neumann) boundary conditions pertain.

5.3 Traveling Wavefronts

Reaction–diffusion systems can be considered as an ensemble of a large number of identical subsystems, coupled to each other by diffusion. These systems of coupled cells can often be met in living structures where transport processes take place, such as living neural tissues, physiological systems, eco–systems, as well as in chemical reactions or in combustion [122, 123]. Traditionally, the local subsystems are defined through a set of nonlinear differential equations. In this context, CNNs represent a powerful tool for their modeling and real–time simulation.

In this section, the CNN defined in Sect. 5.2.1 will be used to generate spatio–temporal phenomena with the following choice of parameters:

$$\mu = 0.7, \quad s = 1, \quad i_1 = -0.3, \quad i_2 = 0.3, \quad D_1 = D_2 = 0.1. \tag{5.20}$$

5.3.1 Autowaves

The term *autowave* was first introduced by R. V. Khorhlov to indicate "autonomous waves" [119, 123]. They represent a particular class of nonlinear waves, which propagate without forcing functions, in strongly nonlinear active media [124, 125]. Their propagation takes place at the expense of energy stored into the active medium; such energy is used to trigger the process into adjacent regions.

This phenomenon is often encountered in combustion waves or in chemical reactions, as well as in many biological processes, such as propagation in nerve fibers or heart excitation.

Autowaves posses some typical characteristics that are fundamentally different from those of classical waves in conservative systems. Their shape remains constant during propagation, reflection and interference do not take place, while diffraction is a property shared by classical and autowaves.

In Fig. 5.5 the formation of two propagating fronts of an autowave can be observed. A 44 × 44 array has been considered. The grey levels in the figures represent the output values: the black colour represents +1 while the white one represents −1.

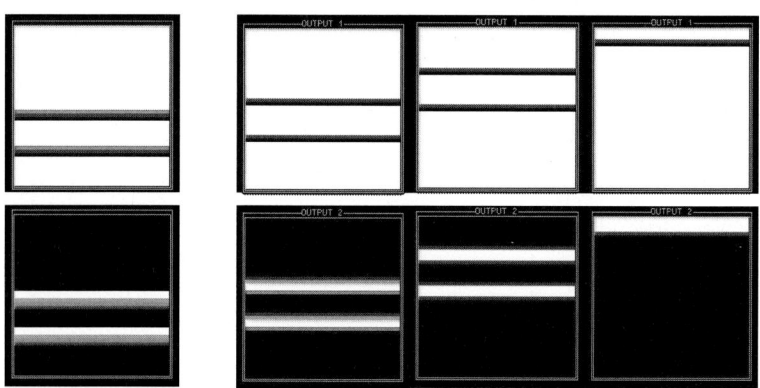

Fig. 5.5. Generation of two autowave fronts

The snapshots in the upper part of the figure represent the outputs of the first layer of the above described CNN at particular time periods. The corresponding snapshots for the outputs of the second layer are reported in the lower part of Fig. 5.5. The initial conditions for the two layers are reported in the first two snapshots at the left-hand side of the figure.

From this figure the main properties of autowaves can be easily observed: the wavefront shape remains unchanged during propagation and no reflection takes place.

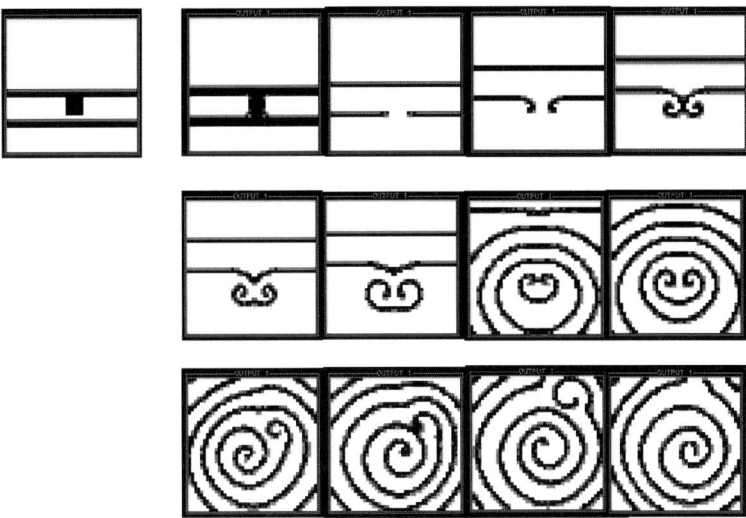

Fig. 5.6. Generation of a reverberator from an autowave break

Figure 5.6 shows the mechanism of initiation of the most important type of autowave source: the *reverberator*. It consists of a rotating vortex similar to an Archimedean spiral. It spontaneously takes place in inhomogeneous media, in which autowaves can break while propagating.

The snapshots in Fig. 5.6 represent the evolution of the first layer in the CNN array. The second layer outputs will be not reported, being simply the opposite of the corresponding first layer outputs. The first snapshot on the left-hand side represents the initial condition on the first layer. It simulates the tail of the autowave front which, due to the medium's inhomogeneity, is somewhat longer in the middle of the picture. The second wavefront cannot propagate and two wave-breaks occur. The broken waves propagate more slowly than the preceding wavefront: from the consecutive wavefront positions it is observed that the wave-breaks begin to curl and rotate, each one forming two *spiral waves*.

Since more than one spiral wave exists in the medium, the wavefronts emitted by each spiral annihilate when colliding, rather than penetrating one another; therefore no interference takes place. All of the characteristics of autowaves have therefore been verified.

Moreover, since the wave-breaking does not take place exactly in the middle of the array (as can be seen from the initial condition), one spiral will rotate at a higher frequency. This one, eventually, will suppress all of the other spirals, as can seen in the last snapshot of Fig. 5.6.

118   5. Spatio-temporal Phenomena

Slightly different initial conditions lead to the formation of two spiral waves in the second wavefront, and in the onset of a circular wavefront in the first one, as can be seen in Fig. 5.7.

Fig. 5.7. The reverberator annihilates a circular wave, as seen from experiments in various active media

Just like all the experiments carried out in chemical active media [119], the reverberator suppresses the concentric wave. Depending on the level of inhomogeneity in the medium, the wave emitted by a rotating reverberator can produce new reverberators, demonstrating their ability to reproduce themselves. This property, usually met in experiments, has been successfully simulated in Fig. 5.8.

The simulations described in this section show the main properties of spirals in active media: their ability to occur in inhomogeneities during propagation, to suppress other wave sources and to reproduce themselves. Such considerations are very important in order to gain further insight into the basic mechanisms of ventricular fibrillation and also of a dangerous type of cardiac arrhythmia, in which the onset of reverberators suppresses the normal heart pacemaker, causing a dramatic increase in the cardiac rate, as seen in experiments on animals [14].

5.3.2 Labyrinths

In this section, we describe an experiment similar to that of Perez-Munuzuri et al. [82], but accomplished with lower-order circuits. Traveling wavefronts go all over a labyrinth. This example is potentially interesting for engineer-

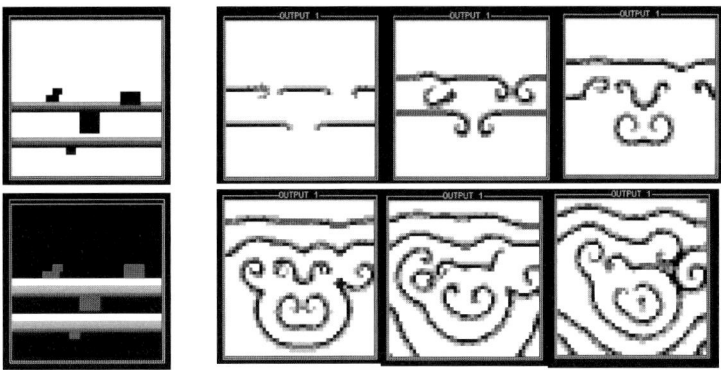

Fig. 5.8. Onset of a Reverberator from medium inhomogeneity and its reproduction

ing purposes, since it represents a suitable application of autowaves for autonomous robot path planning, printed circuit board routing, and so on.

Inputs have been used to define the labyrinth forcing both layers. The control template is not zero anymore:

$$B = \begin{pmatrix} B_{11} & B_{12} \\ B_{21} & B_{22} \end{pmatrix}, \ B_{11} = \begin{pmatrix} 0 & 0 & 0 \\ 0 & 1 & 0 \\ 0 & 0 & 0 \end{pmatrix}, \ B_{22} = \begin{pmatrix} 0 & 0 & 0 \\ 0 & 1 & 0 \\ 0 & 0 & 0 \end{pmatrix}, \quad (5.21)$$

$$B_{21} = B_{12} = 0.$$

The input values of the two layers are the same and they are reported in Fig. 5.9 where, as usual, the black pixel represents +1 while the others are all zero.

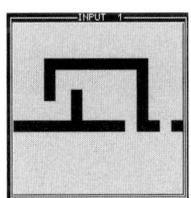

Fig. 5.9. Inputs of the two layers for the labyrinth experiment

In Figs. 5.10 and 5.11 the outputs of the two layers are shown. In this case the output of the second layer is more interesting than in other cases presented because, at the end of the whole diffusion process, it will reproduce the entire labyrinth that has been traversed by the wavefronts of layer 1. As in the experiments reported in [82] the wavefronts of layer 1 propagate

120   5. Spatio-temporal Phenomena

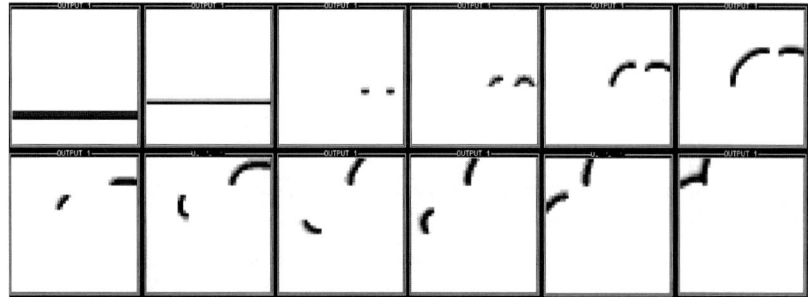

Fig. 5.10. Outputs of layer 1 for the labyrinth experiment

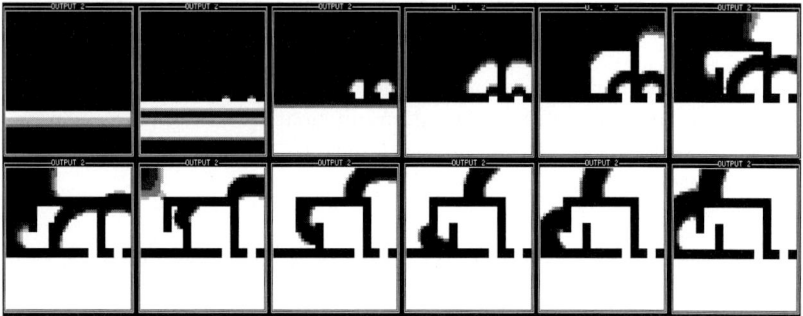

Fig. 5.11. Outputs of layer 2 for the labyrinth experiment

throughout the medium with a constant speed, breaking into different arms at each fork. These features could be used to find the shortest path between two points in the labyrinth.

5.4 Pattern Formation

Autocatalytic chemical reactions coupled with molecular diffusion can generate patterns in biological, chemical and biochemical systems, following Alan Turing's model of morphogenesis [122].

Such phenomena arise from the interaction between different chemicals (also called *morphogens*), which react with each other and spatially diffuse in the chemical medium, until a steady-state spatial concentration pattern has completely developed.

A typical reaction–diffusion model displays the so called *activator–inhibitor mechanism* suggested by Gierer and Meinhardt [122]:

$$\frac{\partial A}{\partial t} = F_1(A, I) + D_A \nabla^2 A,$$
$$\frac{\partial I}{\partial t} = F_2(A, I) + D_I \nabla^2 I. \quad (5.22)$$

Here A and I are the chemical concentrations of the activator and inhibitor, respectively, $F_1(A, I)$ and $F_2(A, I)$ nonlinear functions and D_A and D_I the diffusion coefficients.

The diffusion phenomenon takes place spatially: in particular, the activator is responsible for the initial instability in the medium, and the pattern formation starts. Once this phase is completed, the inhibitor supplies stability. A necessary condition for such a phenomenon to take place is that $D_A \ll D_I$.

In the following it will be shown that, with a slight modification of the above two-layer CNN, pattern structures based on a mechanism similar to the reaction–diffusion phenomenon studied by Turing develop. Therefore, patterns obtained with these methods will be referred to as *Turing patterns*.

5.4.1 Condition for the Existence of Turing Patterns in Arrays of Coupled Circuits

Goraş et al. [126–128] dealt with the problem of obtaining Turing Patterns in arrays of coupled circuits. In particular, in order for such a generalized CNN to generate Turing patterns, the following conditions have to be satisfied:

1. the cells of the two-dimensional grid CNN should be at least of second order (the aim is to simulate the interaction between at least two chemical species);
2. the isolated cells have a stable equilibrium point;
3. when the cells are coupled, the equilibrium point, which corresponds to a homogeneous pattern, becomes unstable, and a non-uniform pattern corresponding to another equilibrium point emerges.

The single cell has the following state model:

$$\dot{\boldsymbol{x}} = \boldsymbol{g}(\boldsymbol{x}), \quad (5.23)$$

where $\boldsymbol{x} \in \mathbb{R}^2$, $\boldsymbol{g} \colon \mathbb{R}^2 \to \mathbb{R}^2$, and the Jacobian matrix

$$D\boldsymbol{g} = \begin{pmatrix} g_{11} & g_{12} \\ g_{21} & g_{22} \end{pmatrix}. \quad (5.24)$$

Without loss of generality let us assume that (5.23) has a fixed point at the origin. For this equilibrium to be stable the following condition on the Jacobian elements in a neighbor of the origin must hold:

$$g_{11} + g_{22} < 0,$$
$$g_{11} g_{22} - g_{12} g_{21} > 0. \quad (5.25)$$

Circuit (5.23) is then coupled to its neighbors. The coupling is obtained by augmenting (5.23) with Laplacian templates. The equations of the array are then

$$\dot{\boldsymbol{x}}_{ij} = g(\boldsymbol{x}_{ij}) + D\nabla^2 \boldsymbol{x}_{ij}, \tag{5.26}$$

where $\boldsymbol{x}_{ij} \in \mathbb{R}^2$ is the state of the cell $C(i,j)$, $D = \begin{pmatrix} D_1 & 0 \\ 0 & D_2 \end{pmatrix}$ is the diagonal matrix of the diffusion coefficients and ∇^2 is the two-dimensional discretized approximation of the Laplacian operator [21]:

$$\nabla^2 \boldsymbol{x}_{ij} \doteq \begin{pmatrix} 0 & 1 & 0 \\ 1 & -4 & 1 \\ 0 & 1 & 0 \end{pmatrix} * \boldsymbol{x}_{ij}. \tag{5.27}$$

Let us consider again the origin of the single cell state and its linearization within a neighbor of the the origin in this new situation. A necessary condition for the origin to become unstable upon coupling is [127]

$$\begin{aligned} D_2 g_{11} + D_1 g_{22} &> 0, \\ (D_2 g_{11} - D_1 g_{22})^2 + 4 D_1 D_2 g_{12} g_{21} &> 0. \end{aligned} \tag{5.28}$$

This condition is necessary to have a *band of unstable spatial modes* in the array [127]. Indeed the complete meaning of this second condition is more complex than that mentioned. A complete discussion can be found in the cited bibliography and it is beyond the scope of this section.

The domain in the parameter space for which the conditions for a Turing instability are fulfilled will be called *the Turing space* of the CNN.

Besides (5.25) and (5.28), the following conditions should be satisfied too [128]:

1. inside the band of unstable modes at least one spatial mode should exist;
2. the initial conditions should be such that at least one of the unstable modes is activated, and
3. the nonlinearity should be such that the final pattern is bounded and stationary.

5.4.2 Turing Patterns in the Two-Layer CNN

In order to obtain Turing patterns from the two-layer CNN some modifications must be made to the cell's equation [129, 130]. First of all the bias coefficients are reset to zero. Secondly, the parameter μ, defining the self-feedback of the cell, has to be perturbed by adding a new parameter ϵ.

This is discussed in the following proposition.

Proposition 5.4.1. *The $M \times N$ two-layer CNN with equation*

$$\begin{aligned}\dot{x}_{1;ij} &= -x_{1;ij} + (1+\mu+\epsilon)y_{1;ij} - sy_{2;ij} + D_1\nabla^2 y_{1;ij}, \\ \dot{x}_{2;ij} &= -x_{2;ij} + sy_{1;ij} + (1+\mu-\epsilon)y_{2;ij} + D_2\nabla^2 y_{2;ij},\end{aligned} \quad (5.29)$$

and zero-flux boundary conditions satisfies the necessary conditions to admit Turing patterns for an appropriate choice of its parameters.

Proof. First of all it is trivially verified that (5.29) admit the origin as fixed point, regardless of the parameters values.

In order to check condition (5.25), let us consider (5.29) in a neighbor of the origin:

$$\begin{aligned}\dot{x}_{1;ij} &= (\mu+\epsilon)x_{1;ij} - sx_{2;ij} + D_1\nabla^2 x_{1;ij}, \\ \dot{x}_{2;ij} &= sx_{1;ij} + (\mu-\epsilon)x_{2;ij} + D_2\nabla^2 x_{2;ij},\end{aligned} \quad (5.30)$$

and its Jacobian matrix

$$D\mathbf{g} = \begin{pmatrix} g_{11} & g_{12} \\ g_{21} & g_{22} \end{pmatrix} = \begin{pmatrix} \mu+\epsilon & -s \\ s & \mu-\epsilon \end{pmatrix}. \quad (5.31)$$

So (5.25) reduces to

$$\begin{aligned} \mu &< 0, \\ \mu^2 + s^2 &> \epsilon^2. \end{aligned} \quad (5.32)$$

In order to have a band of spatially unstable modes, the second necessary condition (5.28) must be met. This reduces to:

$$\begin{aligned}\epsilon &> -\mu\frac{D_2+D_1}{D_2-D_1}, \\ \mu(D_2-D_1) &+ \epsilon(D_2+D_1) > 4s^2 D_1 D_2.\end{aligned} \quad (5.33)$$

Conditions (5.32, 5.33) define the Turing space of (5.29).

The other conditions mentioned above still need to be met. Let us look for a solution of (5.29) by decoupling it into MN-decoupled systems of two first-order linear differential equations and considering the MN orthonormal space-dependent eigenfunctions $\Phi_{MN}(m,n,i,j)$ of the discrete Laplacian operator:

$$\begin{aligned}\nabla^2 \Phi_{MN}(m,n,i,j) &= \Phi_{MN}(m,n,i+1,j) + \Phi_{MN}(m,n,i-1,j) \\ &+ \Phi_{MN}(m,n,i,j+1) + \Phi_{MN}(m,n,i,j-1) \\ &- 4\Phi_{MN}(m,n,i,j) = -k_{mn}^2 \Phi_{MN}(m,n,i,j),\end{aligned} \quad (5.34)$$

where k_{mn}^2 are the corresponding spatial eigenvalues.

In particular, for the zero-flux boundary conditions, the spatial eigenfunctions and eigenvalues assume the following form:

$$\Phi_{MN}(m,n,i,j) = \cos\frac{(2i+1)m\pi}{2M}\cos\frac{(2j+1)n\pi}{2N},$$
$$k_{mn}^2 = 4\sin\left\{\sin^2\frac{m\pi}{M}+\sin^2\frac{n\pi}{N}\right\},$$
(5.35)

with $m = 0,\ldots,M$, $n = 0,\ldots,N$

From the relations which arise between the spatial eigenfunctions and the temporal eigenvalues, it emerges that the connection between diffusing cells can generate patterns if

1. at least one of the temporal eigenvalues has a positive real part (as a function of k_{mn}^2) and
2. inside the band of unstable modes B_u, at least one spatial mode exists.

The bandwidth depends on the cell parameters as well as on the diffusion coefficients. In particular, the limits of B_u are the values k_1^2, k_2^2:

$$k_{1,2}^2 = \frac{1}{2D_1 D_2}\left[(D_2 f_1 + D_1 g_2) \pm \sqrt{(D_2 f_1 + D_1 g_2)^2 + 4D_1 D_2 f_2 g_1}\right]. \tag{5.36}$$

Let us choose the following set of parameters:

$$\mu = -0.1,\ s = 2,\ \epsilon = 2,\ D_2 = 1,\ D_1 = 0.01. \tag{5.37}$$

Using such a set of parameters, the band B_u of unstable modes turns out to be the following: $k_1^2 \leq B_u \leq k_2^2$, with $k_1^2 = 0.0564$ and $k_2^2 = 187.33$.

Because of the *zero–flux* boundary conditions and the initial conditions (5.35), and restricting the analysis to the case $M = N = 2$ (with no loss of generality), the following four spatial eigenvalues can be found from relations (5.35):

$$k_{00}^2 = 0,\ k_{01}^2 = k_{10}^2 = 2,\ k_{11}^2 = 4. \tag{5.38}$$

Therefore, for the above parameter choice, three spatial eigenvalues are contained in the band of unstable modes. Thus it is possible to obtain three pattern configurations, each one with two available polarities: in total a set of six different patterns.

This proves the thesis. □

5.4.3 Simulation Results

In the following simulations, particular care must be taken concerning the accuracy of the numerical integration. Besides the boundary conditions, which have been set to *zero–flux*, the integration time step used was $\delta = 0.01$. Inaccurate integration can cause the appearance of spurious steady-state conditions.

The $M = N = 2$ Case. Let us consider the case $M = N = 2$. Starting from zero initial conditions, excite some of the four cells belonging to the first layer with a constant input randomly selected between −0.1 and 0.1 while the second layer has zero initial conditions. As it can be seen from Fig. 5.12, six different patterns have been obtained at the first layer output, as analytically derived in the previous section.

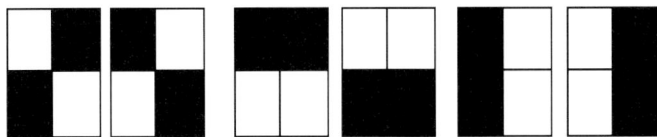

Fig. 5.12. Patterns arising at the first layer output when $M = N = 2$

The $M, N > 2$ Case. Hereby the zero initial conditions are represented in white, while black represents an initial state set to one. Conversely for steady state conditions black and white represent +1 and −1 respectively.

Fig. 5.13 shows the pattern formation phenomenon.

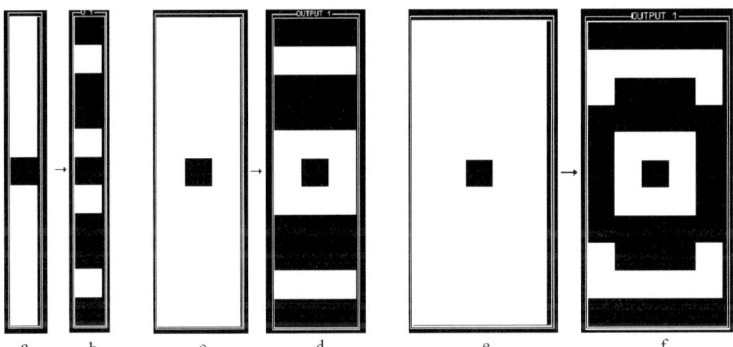

Fig. 5.13a–f. Patterns arising from different initial conditions at the first layer output for various M and N values

The initial conditions are depicted in Figs. 5.13a,c,e, while the final patterns are shown in Figs. 5.13b,d,f, respectively. Figs. 5.13a,b show the the case of a 11×1 CNN array; Figs. 5.13c,d a 11×3 CNN, and Figs. 5.13e,f a 11×5 CNN. It is seen that the number of cells available for propagation determines the final pattern geometry.

This is also seen from Fig. 5.14. In particular, the pictures on the left hand side show the initial conditions, while those on the right side show the corresponding final configuration. Figs. 5.14a and b show the case of a 11×11 array. In Figs. 5.14c and d the size is increased to 21×21 both with

126   5. Spatio-temporal Phenomena

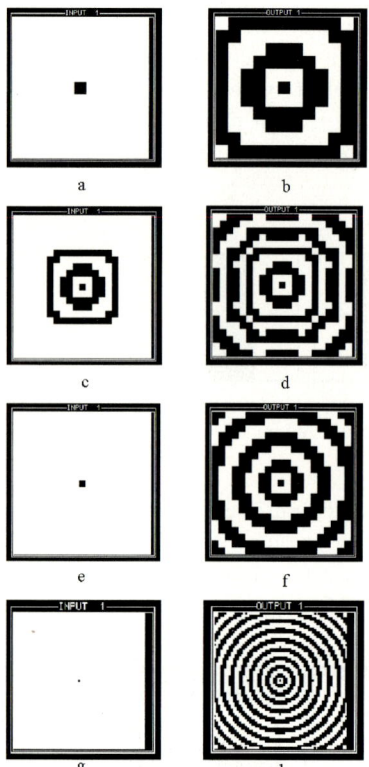

Fig. 5.14a–h. Another set of patterns arising from various initial conditions (first layer output) when $M = N > 2$

a modification on the initial condition. Instead, on Figs. 5.14e,f and 5.14g,h the same kind of initial condition used in Fig. 5.14a is considered while the corresponding sizes are (again) 21×21 and 44×44 respectively.

Figure 5.15 shows the propagation of a unique horizontal stripe (initial condition) towards a steady state condition in which the strip propagates through the available 11×10 array (Figs. 5.15a,b), or a 21×10 array (Figs. 5.15a,b).

A slightly different initial condition (Fig. 5.15e) in the 21×21 leads to the formation of a new pattern (Fig. 5.15f).

The final pattern configuration is shown to depend, once again, on the initially available cell dimension and on the initial conditions.

It is interesting to note how the results of these simulations resemble the rearrangement of skin stripes and spots observed in many animals during growth. As a matter of fact, mathematical models based on reaction–diffusion PDEs have been reported in literature [122, 131]. The same initial condition

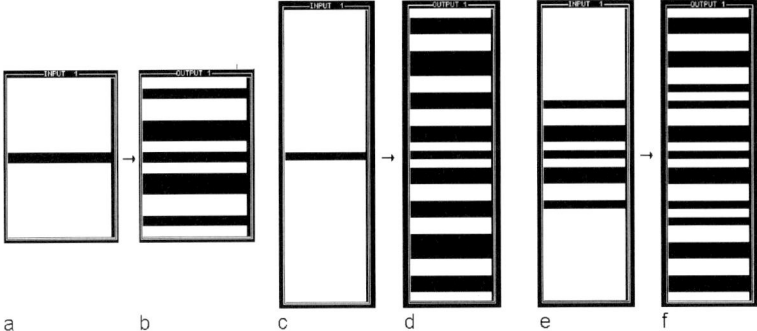

Fig. 5.15a–f. Pattern arising from the propagation of a horizontal strip (initial condition)

can give stripe or spot formation according to the different part of the animal skin.

5.5 Sensitivity to Parametric Uncertainties and Noise

Let us now consider how the behavior of the two-layer CNN is affected by parametric uncertainties and noise. The case of a spiral wave obtained with the CNN discussed in Sect. 5.2 is considered first. Thereafter, the case of Turing patterns will be considered.

5.5.1 Spiral Wave: Parametric Uncertainty

Consider the case in which all the pairs of capacitors $C_{1;i,j}$ and $C_{2;i,j}$ associated to the two layers of each cell $C(i,j)$ of the CNN array are affected by uncertainty, namely:

$$C_{1;i,j} = C_0 + \delta C_{1;i,j}, \quad C_{2;i,j} = C_0 + \delta C_{2;i,j}, \tag{5.39}$$

where C_0 is the nominal value (always implicitly assumed to be equal to unity) of the capacitors while $\delta C_{1;i,j}$ and $\delta C_{2;i,j}$ are their uncertainties. In this condition, the cells' capacitors have different values randomly distributed within a bounded range. This can be considered as a model for an inhomogeneous active medium.

In Fig. 5.16 the snapshots corresponding to the case in which $\delta C_{1;i,j}$ and $\delta C_{2;i,j}$ vary in the range $[-0.2, +0.2]$ (20 percent of the nominal value) are shown. It is seen that, in spite of the significant amplitude of the uncertainties, the spiral is only put out of shape.

128   5. Spatio-temporal Phenomena

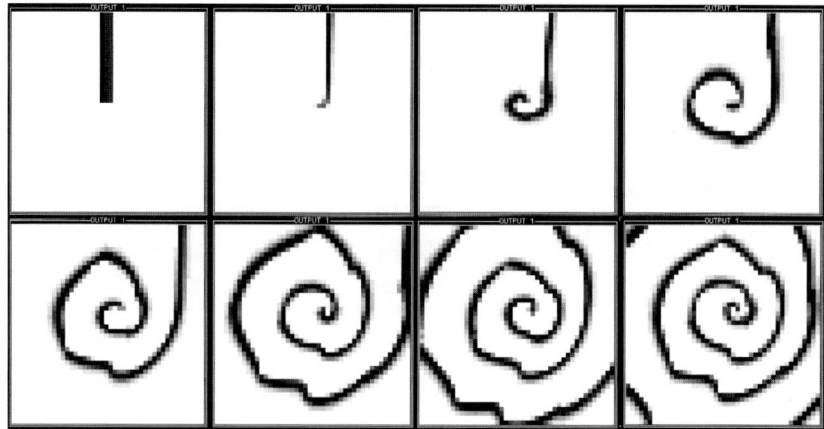

Fig. 5.16. Formation of a spiral in the presence of uncertainties in the capacitor values

Fig. 5.17. Formation of a spiral in the presence of uncertainty concerning the feedback template

Let us now take into account the case in which only the cloning templates are affected by uncertainty. Namely, all the elements of the feedback template matrices reported in Def. 5.2.1 have been randomly perturbed within bounded ranges. The template coefficients obtained in this way are then used for all the cells of the array.

The results obtained for an uncertainty of 1% of the nominal values is depicted in Fig. 5.17. It can be seen that the uncertainty causes the generation of undesired wavefronts. These collide with one another and with the rising spiral annihilating themselves and the outer part of the spiral. Nevertheless, observing the subsequent snapshots it is seen that, due to the fact that the inner part of the spiral is protected by its external wave fronts, the spiral is able to develop (even though it takes longer).

5.5.2 Spiral Waves: Presence of Noise in the Initial Conditions

As discussed above, the initial conditions of the CNN are in the range $[-1, +1]$, and they determine the formation of spirals, patterns and so on. Here, noise has been added to the matrices of initial conditions that lead to the formation of a spiral. The noise considered is uniformly distributed in the range $[-\delta, +\delta]$ along the entire CNN lattice.

In the following examples it is shown that even with significant values of δ the noise is rapidly filtered by the diffusion effect and it does not change the qualitative behavior of the spiral.

Figure 5.18 shows the case in which $\delta = 0.4$. It is worth noting that due to the limited number of colors used to represent the output values, the noise peaks are not immediately visible in all the pictures. However, their effects on the spiral can be observed in the snapshots. As anticipated the noise has been rapidly filtered and the qualitative behavior is maintained.

Analogously, in Fig. 5.19 the case with $\delta = 0.8$ is shown. In this case, due to the fact that the noise peaks are comparable with the amplitude of the noise-free initial conditions, some spurious wave fronts can be generated. In the case reported this is annihilated by the growing spiral and the last snapshot shows a fully developed spiral. However, in general, some undesired spirals can be generated by the noise and they cannot necessarily be annihilated.

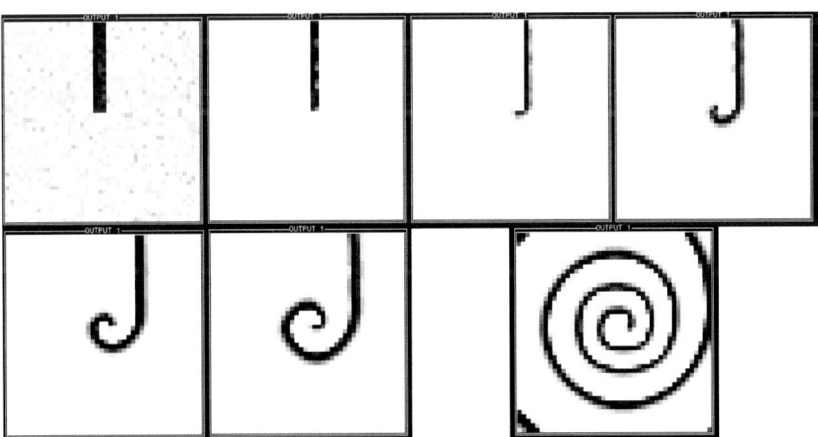

Fig. 5.18. Formation of a spiral in the presence of noise in the initial conditions ($\delta = 0.4$)

130   5. Spatio-temporal Phenomena

Fig. 5.19. Formation of a spiral in the presence of noise in the initial conditions ($\delta = 0.8$)

5.5.3 Patterns: Parametric Uncertainties

According to the theory, the presence of noise in Turing patterns affects only the geometry of the final configuration. Therefore it will be not considered. What is interesting is to examine the extent to which an uncertainty in the CNN parameters will affect the stability of the final configuration.

Again, let us first consider the case of uncertainties on the capacitors. Figure 5.20a shows the initial conditions, Fig. 5.20b depicts the final pattern in the disturbance-free case; Figs. 5.20c,d,e,f report the final patterns when the cell capacitors have an uncertainty of 10%, 20%, 40%, 50% on the nominal value, respectively. Namely $\delta C_{1;i,j}$ and $\delta C_{2;i,j}$ vary in the range $[-0.1, +0.1]$, $[-0.2, +0.2]$ and so on. Higher uncertainty levels prevent the steady state condition from being reached. It can be noted that uncertainty, in this case, plays the role of varying the geometry of the final pattern.

Let us now consider the case of uncertainties on the feedback templates. Figure 5.21 shows the final patterns when the template values have been perturbed by 0%, 10%, 20%, 40% of their nominal value, respectively.

Figure 5.22 shows the results obtained when the capacitors have been perturbed in the range 10% to 50%, and, at the same time, the templates have been perturbed in the range 10% to 60%. From such figures it can be observed that these phenomena are fairly robust with respect to disturbances.

5.5 Sensitivity to Parametric Uncertainties and Noise

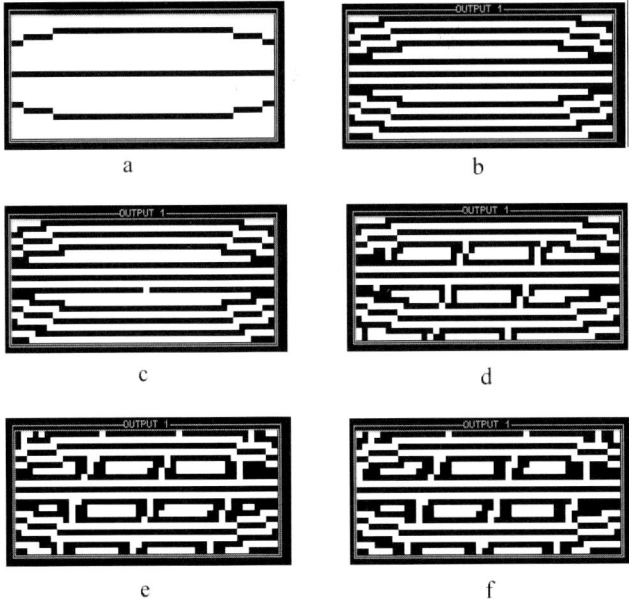

Fig. 5.20a–f. Pattern arising from particular initial conditions (upper left corner picture) with uncertainties in the capacitors

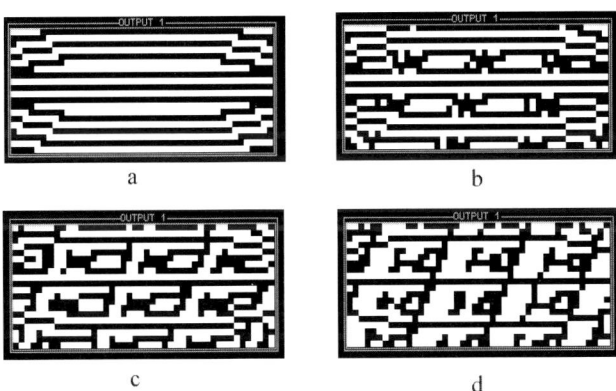

Fig. 5.21a–d. Pattern evolution with uncertainties in the template coefficients

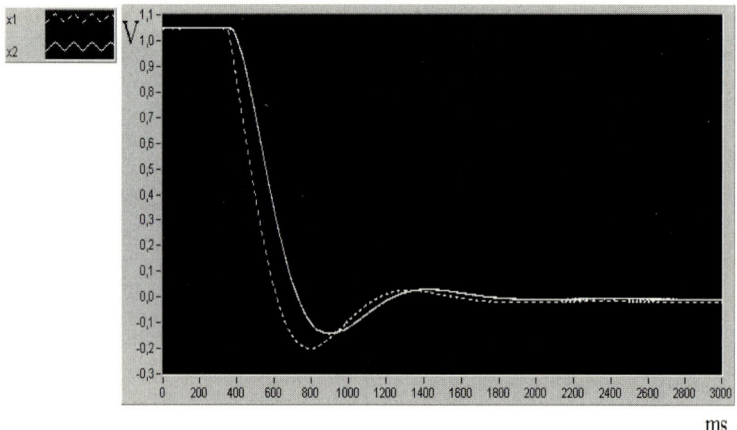

Fig. 6.6. Measured state variables x_1, x_2 for the circuit of Fig. 6.4

6.1.3 Realization of the Laplacian Couplings and Boundary Conditions

The implementation of the laplacian couplings has been simply performed by using some operational amplifiers in summing configuration. A schematic diagram is shown in Fig. 6.7. The output of such a circuit is then connected to the non-inverting input of the operational amplifier in the cells previously discussed by means of suitable resistors whose values will represent the diffusion coefficients D_1, D_2 and varying the offset resistor R_n accordingly. The *zero–flux* boundary conditions can be easily realized. In fact considering

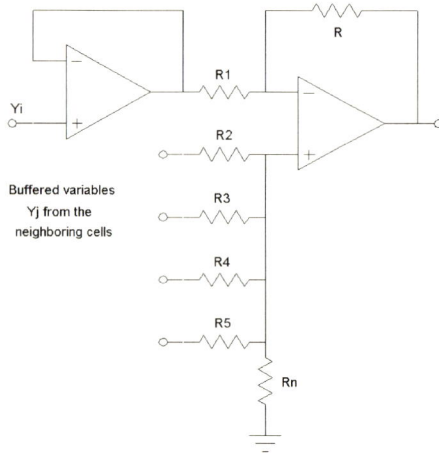

Fig. 6.7. Circuit realization of the laplacian couplings

(5.18) the influence of the boundary cells is realized by the discrete laplacian. When a cell $C_{i,j}$ belongs, for example, to the right boundary edge of the array, the dynamics of the missing cell, say $C_{i,j+1}$, is assumed to match the $C_{i,j}$ dynamics. Therefore in this case the discrete laplacian simply modifies to: $D_k \cdot (y_{k;i+1,j} + y_{k;i-1,j} + y_{k;i,j-1} - 3y_{k;i,j})$, $k = 1, 2$. Similar arguments also hold for cells belonging to a corner position, for example to the upper right one for which one has: $D_k \cdot (y_{k;i+1,j} + y_{k;i,j-1} - 2y_{k;i,j})$, $k = 1, 2$. Such considerations, in terms of the circuits, translate into a suitable modulation of the ratio R/R_1 (Fig. 6.7) in the realization of the laplacian couplings for the boundary cells.

6.1.4 Realization of the Main Board

All the circuits previously discussed realize a CNN array capable of displaying self-organization phenomena. To this end a main board has been designed and built, able to realize all the necessary connections between the circuits, which can easily be located on to the board with suitable slots. Moreover in the whole frame it has to be possible to easily modify or change the resistors realizing the couplings between circuits. A particular section in the board has therefore been implemented, in which all the resistive couplings can take place. The board has also been provided with a section to perform all measurements on the state and output variables. Figure 6.8 is a schematic diagram showing the main board. Moreover, for the direct visualization of the phenomena, a LED matrix of dimension 5×5 has been built. In this way each diode represents a cell in the CNN array. The LEDs used are able to change

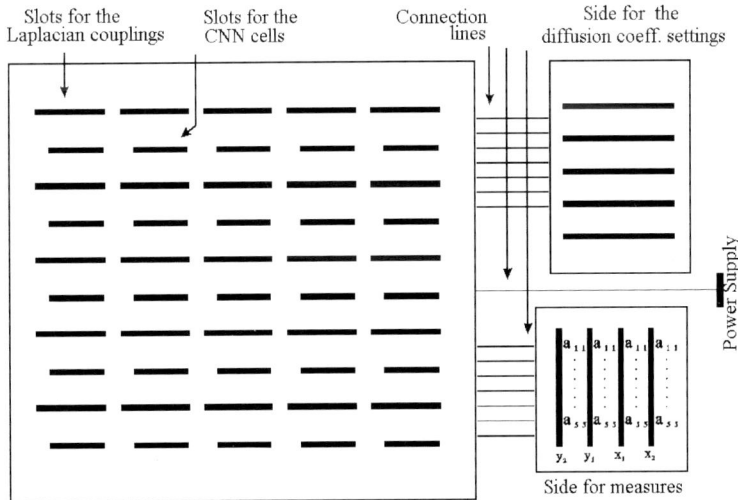

Fig. 6.8. Main board scheme

their emitted light color from green to red as the applied voltage changes from -1 V to $+1$ V. In such a way it is possible to quickly gain enough information about the CNN system state. In order to report the results of the experiments in the clearest way, a data acquisition system, based on a Windows PC and a data acquisition board National AT-MIO-16E-10 have been used, together with the `LabView` program for data management and visualization. The experiments on autowaves will be presented by means of figures drawn from a virtual instrument panel (built using the `LabView` program), showing the temporal trends of certain variables useful for understanding the experiment, together with a table in which the index of the monitored cells is reported, together with their representation (in solid or dotted line). A matrix scheme of the cells, standing for the LED matrix, is also reported to better explain the experiment. The virtual instrument panel for Turing pattern experiments presents the temporal trends of certain variables belonging to the CNN cells. Moreover patterns will be also presented as they appear in the first layer output of the CNN matrix. In particular, the white color represents $+1$ V, black -1 V, while the shaded cells are set to 0 V. The cells will be identified as belonging to a matrix. Therefore, for example, A_{12} will represent the cell positioned in the first row and second column in the CNN matrix.

6.1.5 Autowave Experiments

The previously discussed setup has been adopted to perform experiments on autonomous phenomena in wave propagation in active media. The results are presented with particular emphasis on the circuit implementation. Neither the tolerance usually met in commercially available discrete components nor the presence of noise seem to affect the results of the experiments: The whole structure is therefore suitable for hardware implementation. Indeed, as outlined previously, these phenomena already revealed their robustness in a number of simulations (see Chap. 5) but measures on a real circuit frame acquire more importance. The first experiment performed regards the propagation of an autowave through the CNN matrix. Initially all cells were tuned in order to show the same dynamics, namely a stable limit cycle with slow–fast regime, compatible with the component tolerance (10% for capacitors and 1% for resistors). The initial conditions for this experiment consisted in setting each cell of the first row of the matrix to the following voltages: $x_{10} = +1$ V, $x_{20} = -1$ V. If such initial conditions hold, all of the other cells, with approximately zero initial conditions, will soon synchronize with the propagation of the wave generated by the first row of the matrix. Of course, if no other initial conditions are imposed on the matrix, at the end of the propagation through the fifth row, the autowave vanishes and a competition between waves induced by each cell starts. Such competing wavefronts mutually annihilate, leading to the onset of an autowave propagating in a particular direction, which depends on the positions of some "dominant" circuits which bias the whole phenomenon. In our case, the favored steady state direction is the one along the matrix diagonal. In order to re-obtain the autowave propagation along the matrix rows (i.e. the desired direction), the initial conditions have to be set to those of the cells of the first row at the end of each propagation through the CNN matrix. Figure 6.9 shows the LabView acquired image of the time evolution of the first layer output of the cells belonging to the first and fourth row of the CNN matrix, as is also outlined by the schematic picture representing the CNN matrix reported in the lower right corner of Fig. 6.9. In Fig. 6.10 the time evolution of the first and fourth row of the matrix are also reported in a 3D graphic plot, to outline the dynamics of the measured phenomenon. Taking into consideration also the table reported at the top right of Fig. 6.9, one can conclude that the cells of the two rows depicted (and therefore all the cells belonging to each row) proceed well synchronized in time, with unchanged amplitude and shape, as is typical of autowaves. The slight imbalance of the rising and falling fronts is mainly due to the circuit component tolerance. It can also be concluded that the time required for the autowave to propagate through the CNN matrix corresponds to about 0.66 circuits per second.

Figure 6.11 shows the propagation of an autowave from the central cell of the CNN matrix. In this experiment the initial conditions $x_{10} = +1\,\mathrm{V}$ and $x_{20} = -1\,\mathrm{V}$ were locked for the cell A_{22}, which, therefore, acts as a source of concentric waves. The wavefront propagates, synchronizing the cells po-

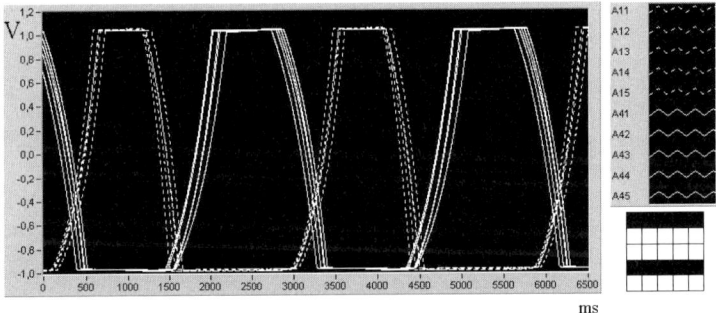

Fig. 6.9. Autowave propagating through the CNN matrix rows

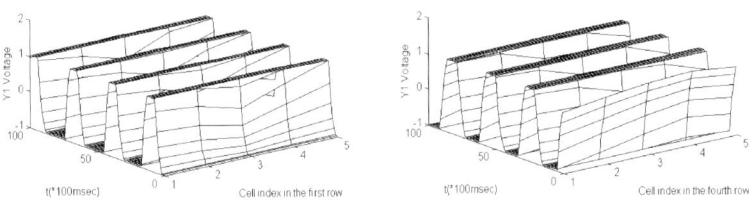

Fig. 6.10. Time evolution of the first and fourth rows of the CNN matrix during the propagation of an autowave

sitioned at the same distance from the "pace maker cell" A_{22}. Taking into consideration the simulation results presented in the previous chapter as well as the experiments explained above, it emerges that, in spite of the influences of the circuit component tolerance, the underlying reaction–diffusion phenomenon takes place successfully and the fronts generated show typical autowave behaviour in this case too.

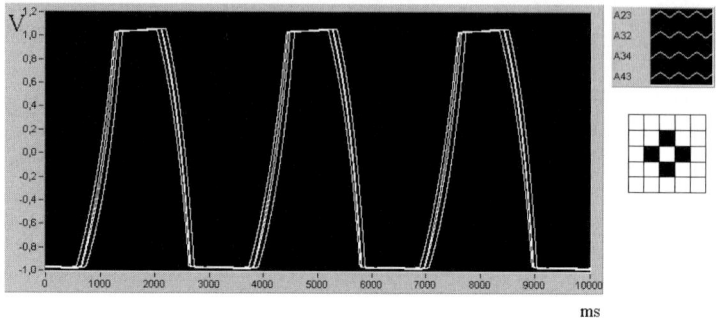

Fig. 6.11. Concentric wave from the CNN matrix central cell

6.2 Pattern Formation and Propagation

In this second section the reaction–diffusion dynamics showing *patterning* processes, very common in biology [122] will be reproduced on a real CNN based circuit which realizes this kind of reaction–diffusion rules.

At the beginning of the experiments, all the cells were tuned in such a way as to realize circuits where the origin should be a stable equilibrium point, according to the parameter set presented in the previous chapter. Initial conditions on all cells are crucial in this case in order to successfully perform the experiment. Actually pattern formation, in the simulation results shown in the literature, is generally studied in large circuit arrays. Indeed, when a circuit realization is considered, the possibility of obtaining patterns reflecting a certain geometry is greatly affected by noise. Therefore this task can be better achieved in small circuit arrays, or in circuits where some cells are added after the initial group of cells has reached an equilibrium condition, for example, to simulate animal coat development during growth, in which the Turing patterning process is fully realized. In the case at hand, it has been observed that, given an initial condition in a certain set of cells, during the earliest part of the spatio–temporal evolution, the cells lying at a certain distance from the set of cells chosen for setting the initial conditions, are affected by a spurious, noise-induced, diffusion-driven instability. Such

6.2 Pattern Formation and Propagation

stimuli precede the actual diffusion induced by the genuine signal propagation. Therefore, even if a steady state configuration is found, the "geometry saving" propagation, typical of Turing patterns can be affected. This set of experiments was therefore carried out in a 3×3 CNN, and a comparison between calculations and experimental results will be reported.

Larger CNN structures can be considered while studying the propagation phenomenon in growing structures, simulating living cell reproduction. Such an example will be presented at the end of the section.

Calculations similar to those ones reported in the previous chapter show that, in the case $M = N = 3$, eight possible pattern configurations are allowed, according to the extension of B_u [133]. In fact, from (5.35, 5.36), the limits of B_u are the values k_1^2, k_2^2 [127]:

$$k_{1,2}^2 = \frac{1}{2D_1 D_2}$$

$$\times \left[[D_2(\mu + \epsilon) + D_1(\mu - \epsilon)] \pm \sqrt{[D_2(\mu + \epsilon) - D_1(\mu - \epsilon)]^2 - 4 D_1 D_2 s_1 s_2} \right]$$

Taking into consideration the chosen CNN cell parameters and the diffusion coefficients, one obtains: $k_1^2 = 1.96$; $k_2^2 = 109$. From (5.35), in the case $M = N = 3$, the following unstable spatial eigenvalues can be found:

$$k_{mn}^2 = \begin{pmatrix} -1.14 & 2.5063 & 2.5063 \\ 2.5063 & 5.6326 & 5.6326 \\ 5.6326 & 5.6326 & 5.6326 \end{pmatrix}. \tag{6.5}$$

The temporal eigenvalues associated with the spatial modes already determined can be obtained using the following relation [127]:

$$\mathrm{Re}\left\{\lambda(k_{mn}^2)\right\}$$

$$= \mathrm{Re}\left\{ \mu - k_{mn}^2 \frac{D_1 + D_2}{2} + \sqrt{(-\epsilon + k_{mn}^2 \frac{D_1 - D_2}{2})^2 - s_1 s_2} \right\}.$$

In our case they take on the following values:

$$\lambda_{mn}^2 = \begin{pmatrix} 0.0000 & 0.6463 & 0.6463 \\ 0.6463 & 0.7726 & 0.7726 \\ 0.6463 & 0.7726 & 0.7726 \end{pmatrix}. \tag{6.6}$$

Therefore, for our parameter choice, eight spatial modes are contained in B_u, and thus it is possible to obtain eight possible pattern configurations, as seen from Fig. 6.12, which depicts all patterns arising from the spatial eigenfunctions of the type (5.35) related to their eigenvalues k_{mn}^2.

Since each pattern presents two available polarities, the total number of different pattern configurations is sixteen. Of course, the configuration

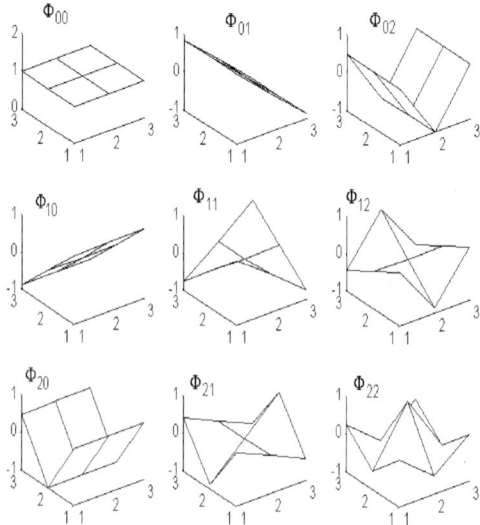

Fig. 6.12. Turing pattern configurations for $M = N = 3$

corresponding to Φ_{00} arising from k_{00}^2 cannot occur, since k_{00}^2 does not lie inside B_u.

When more than one pattern is unstable, a kind of competition between modes arises. It has to be mentioned that the conditions for the onset of Turing patterns are based on a linearization of the CNN state equations. In particular, an analysis of the cells in the linear region around the origin has been performed. Therefore, once the diffusion process starts, the system, sooner or later, will become nonlinear, and the final pattern geometry will be influenced by the nonlinear interaction among other temporal modes. Therefore, one can predict only how patterns will begin to evolve.

It is tempting to think that the spatial mode, and therefore its associated eigenfunction, corresponding to the temporal eigenvalue with the most positive real part, will most probably appear; but initial conditions, as well as the unavoidable imbalance among the circuit parameters play a crucial role in the pattern formation phenomenon in the circuit realization, as will be seen in the experiments reported below. This part strictly reflects what commonly happens in living structures.

Of course, among such patterns, the most favored ones are those possessing the largest positive temporal associated eigenvalues, (such as Φ_{12} through Φ_{22} in Fig. 6.12). Moreover, such conditions are strictly dependent on the set of parameters employed. In fact, all the unstable temporal eigenvalues are of the same order of magnitude, and therefore no clear supremacy exists among them. However, such considerations can be made only by referring to simulations and calculations based on a linearization of the CNN state

6.2 Pattern Formation and Propagation 145

equations. In the real case, each cell realization has its own intrinsic noise that biases its behavior and therefore some particular pattern configurations are more favored than others. This can easily be seen in the following experiment, shown in Fig. 6.13, in which the CNN cells are allowed to evolve from $near-zero\ initial\ conditions$.

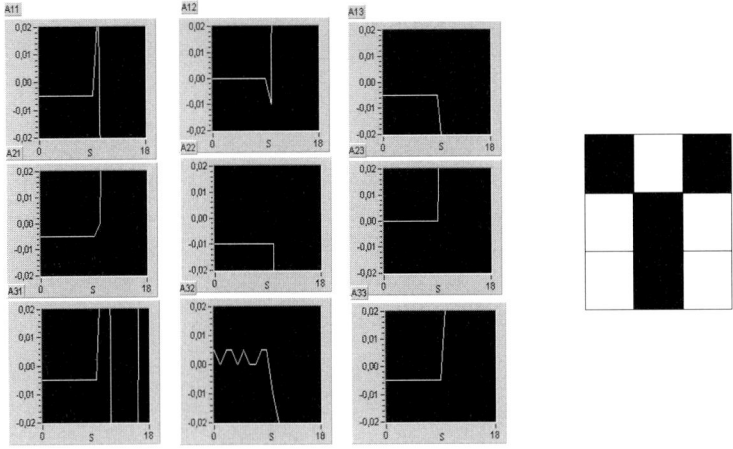

Fig. 6.13. Turing Pattern from "near zero" initial conditions

In fact, as can be observed from the plots in Fig. 6.13, each circuit starts from a condition $near\ zero$, say $\bar{V}_0$. The final pattern configuration therefore is $biased$ by the $disturbances$ inside each cell and by the circuit tolerance, and only incidentally matches one of the patterns with the largest positive temporal mode (Φ_{12} in Fig. 6.12). Of course, changing the cell positions inside the main board, the final pattern configuration will change accordingly, as can be seen from Fig. 6.14. Moreover, some pattern configurations arise though they are not among the ones predicted by linear theory, since $nonlinear\ competition$ between unstable modes takes place when nonlinearities set in. These latter patterns can be considered as "spurious", or as defects in the CNN structure. Therefore it is possible to drive the structure itself to some desired geometry avoiding the "defects" by introducing suitable initial conditions so as to excite the modes responsible for the desired final pattern geometry. Initial conditions have to assume values sufficiently larger than the voltages $\bar{V}_0$, whose greatest absolute value is around 0.01 V (see Fig. 6.13), otherwise other spurious configurations can arise. In Fig. 6.15 the desired checkerboard pattern is obtained by imposing an initial condition $V_0 = -0.15$ V on the central cell A_{22} and leaving the other cells with "near zero" initial conditions. Smaller initial conditions for the cell A_{22} lead to the onset of spurious patterns. The time evolution of the cell outputs allows one

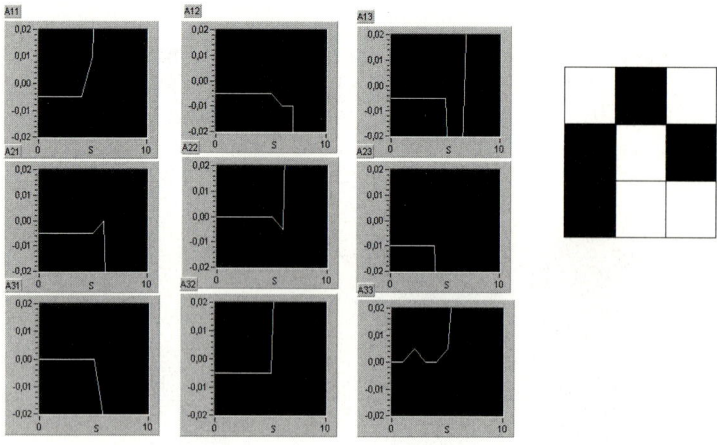

Fig. 6.14. Turing Pattern from "near zero" initial conditions

to appreciate the role of the diffusion-driven instability which leads to the formation of the stable checkerboard pattern depicted on the right-hand side of the figure. It should be mentioned that the pattern control is realized only with suitable initial conditions, and not with the addition of forcing actions. Therefore the structure of the CNN is not further complicated.

The possibility of controlling the final pattern configuration is further emphasized in the following experiments. Figure 6.16 depicts the initial con-

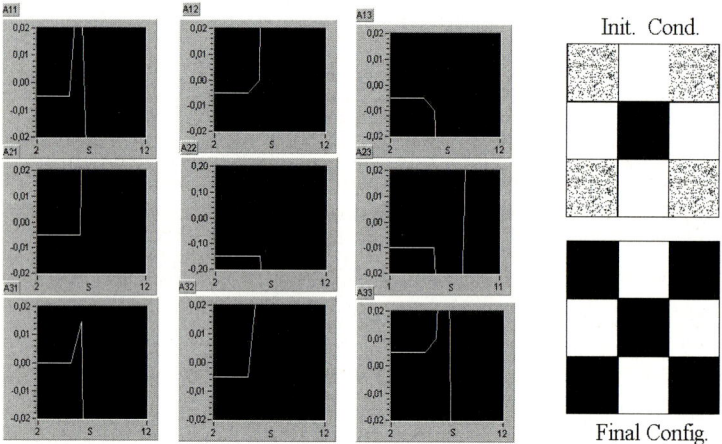

Fig. 6.15. Checkerboard Turing pattern

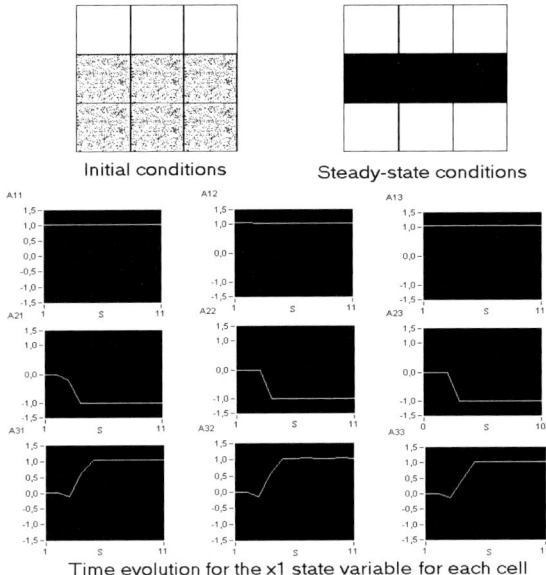

Fig. 6.16. Formation of a stripe pattern

ditions leading to the formation of a stripe, the corresponding time evolution of the first layer outputs, as well as the final pattern configuration.

Such an experiment can also be viewed as the propagation of a stripe from the first to the third row of the CNN matrix.

The next experiment refers to a phenomenon which can easily be observed in the animal coat, looking at the particular parts linking the body with legs or tail. In these zones the coat growth takes place in two different directions and cannot be recast into a unique (radial or linear) direction, as it commonly can in the other parts of the animal body: There are two main directions of growth, which are mutually orthogonal. Therefore the propagation of a stripe, which proceeds in one direction, near such *scapular sites*, has to "change" direction. Such a situation is encountered in the next experiment, in which a stripe is forced to propagate through a corner region. Figure 6.17a depicts the initial conditions, which are zero for all of the cells, except for the two at the bottom (namely A_{43} and A_{44} in Fig. 6.17a), set to 1 V (white color). Moreover, the two cells A_{11} and A_{21} are physically disconnected from the other ones: they will become connected subsequently, when simulating the growth phenomenon. Figure 6.17b shows the steady-state conditions: the stripe propagates through the corner cells. In fact the cells now set to 1 V are A_{13}, A_{23} and A_{24}, which represent the "boundary" between horizontal and vertical stripes. All the other cells have been set to -1 V (black color). If A_{11} and A_{21} are now connected to the CNN array with zero initial conditions, simulating the growing process, they both move to 1 V, thus confirming the propagation of the stripe to the left-hand side (Fig. 6.17c). Figure 6.18 shows the time evolution of the first state variable of the cells, during this experiment.

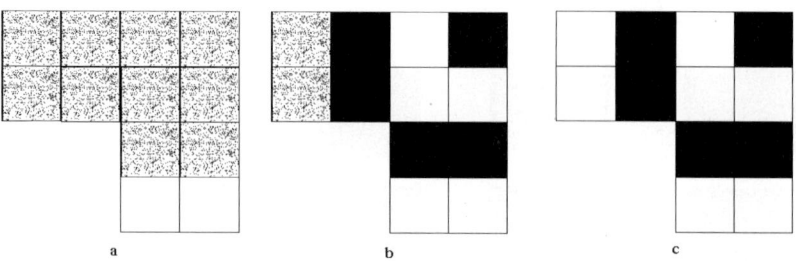

Fig. 6.17a–c. Propagation of a stripe through a corner: matrix view

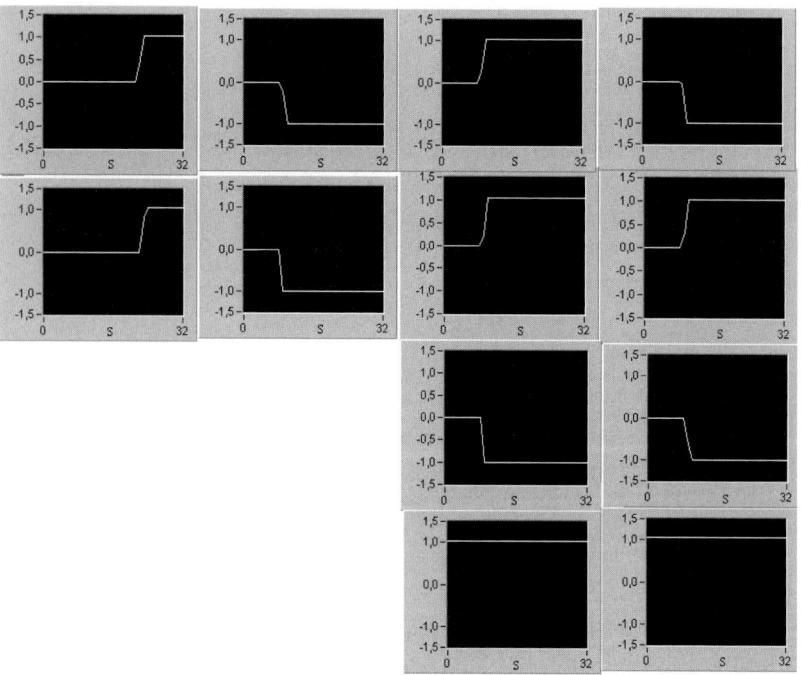

Fig. 6.18. Propagation of a stripe through a corner: time evolution

6.3 CNNs for Generating and Controlling Artificial Locomotion

In this section the reaction–diffusion CNN model, whose hardware implementation has been fully described earlier, is used to generate and control locomotion in some mechatronic devices [134]. The strategy proposed has the aim of simulating, from a functional point of view, that part of the brain devoted to generating and controlling locomotion in animals.

6.3.1 Links to Biological Locomotion

Experiments on simple invertebrates have begun to reveal the principles that lie behind the anatomic and functional structures of the nervous systems and have helped us to build mathematical descriptions. Signal transmission among neurons is a phenomenon which takes place both in space and in time and involves an ensemble of mutually connected neurons. Their dynamics is governed by the laws of nonlinear diffusion. Therefore such phenomena can be efficiently described by particular solutions of nonlinear partial differential equations (PDEs), known as reaction–diffusion equations [122]. These solutions describe wave-like phenomena which are self-sustained oscillations and have all of the properties of autowaves [119]. In the simplest moving animals, including, for example, nematodes, but also in some mollusks, such as squids, some types of locomotion are directly induced by the propagation of an impulse signal (autowave) along the neuron axon. The soft body structure is able to synchronize with the traveling wave and a wave-like locomotion begins. In more developed animals, like insects, from a high level point of view the autowave propagation can be still supposed to generate locomotion, but the neural structure has been improved and is highly organized and much more complex. In such cases, the pure physiological approach seems impracticable, and a high-level, functional type of approach appears more suitable.

Here, the central nervous system (CNS) is responsible for specific patterns of motor neuron impulses during coordinated motion. The central hypothesis is that there is a central pattern generator (CPG), within the CNS, that produces the basic motor programs [132]. Indeed, the CPG's neural organization has been shown to be more complex than needed to merely generate motor oscillations. In fact its capacity also to generate plateau potentials or oscillations is a key issue for rhythm generation, and also for driving the transitions between various types of locomotion (walking, running, swimming) [135]. More specifically, rhythmic movements that drive locomotion effectors (muscles) are triggered by a group of neurons, here called *local pattern generating neurons* (LPGN). They in turn are controlled by a higher level neural center, called *Center of Command Neurons* (CCN), which fixes a particular locomotion scheme based either on specific signals coming from the CNS, or on feedback deriving from sensory inputs. In such a way the output of the pattern generator can be modified so as to adapt locomotion the the environment [132, 136].

There are cases in some invertebrates in which a single command neuron is able to excite a whole coordinated behavior, and measures show that such neuron is active for the whole duration of that behavior [137], though normally a small collection of command interneurons may be activated during a natural behaviour.

In the simple animal *Aplysia*, an interesting property of some command neurons is their ability to produce plateau potentials, while in general rhythmic motor systems, neurons with intrinsic oscillatory dynamics, chemically or hormonally modulated, are commonly found [138].

From a behavioral point of view, the whole locomotion system (CPG) therefore appears to be a complex activator–inhibitor system characterized by a hierarchical organization in which a group of neurons (the CCN), due to sensory or central excitations, activates other groups of neurons (the LPGN) that generate the appropriate timing signals for the type of locomotion induced by the command neurons. Such a schematic representation is shown in Fig. 6.19. It is important to note that what is stated above are ideas used by

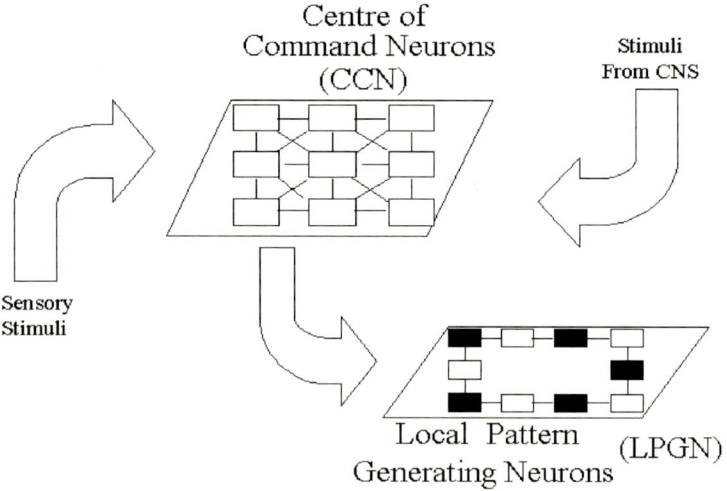

Fig. 6.19. Schematic representation of a central pattern generator (CPG)

neurobiologists as working hypotheses, and, when applied to any specific motor system, may have to be modified to provide a more accurate theory concerning the function of that specific motor system. From the point of view of the neural signal processing, nonlinear reaction–diffusion PDEs are currently employed in mathematical biology to model the excitation and propagation of impulses in nerve membranes. Moreover, motion in some invertebrate species

6.3 CNNs for Generating and Controlling Artificial Locomotion

is driven by the firing of a unique giant neuron provided with a long axon: the propagation of an impulse train takes place in the longitudinal direction of the animal, causing locomotion. Such considerations have led to the idea of using autowaves for locomotion generation and control purposes. In fact, since autowaves efficiently describe propagation in nerve tissues and since they are solutions of nonlinear PDEs that can be represented by reaction–diffusion CNNs, such structures can be used to generate locomotion (which is a classical spatio-temporal phenomenon) by directly driving the actuators of walking robots. The first prototype is a ring-worm-like structure, in which locomotion is accomplished through an autowave propagating from the back to the front part of the body. Another example describes a novel kind of "conveyor belt", able to propagate autowaves on itself. In fact the belt does not move in the direction of the object destination: each element of the structure moves in such a way as to generate a wave which induces the objects to move towards their destination. Moreover, since autowaves do not suffer from reflection phenomena, due to boundaries and also inhomogeneities in the medium, a two-dimensional structure can be imagined in which any two objects can be moved in any direction or to any destination point, without changing anything in the actuation system. It is sufficient to fix the desired trajectory for each of the objects and fix the corresponding mask at the CNN input. Such a versatile structure could lead to a dramatic improvement in industry automation. Finally artificial locomotion is investigated through an Hexapode Walking Robot, so as to reproduce walking in insects. This case seems to be quite interesting, not only from the point of view of the locomotion generation, but, above all, from the aspects of the coordination and control between various types of locomotion (walking, swimming,...). In this case the CNN is composed of two main parts: one is devoted to generating autowaves, so as to simulate an artificial LPGN, and the other is responsible for the particular locomotion type obtained with a stable pattern configuration. To this steady state condition corresponds a particular leg combination of the robot moving with the rhythm of the same wavefront. Therefore this part of the CNN can be called an artificial center of command neurons (CCN). A stable condition of the CCN corresponds to a particular locomotion type. Particular conditions under which the robot has to change locomotion, translate directly into an induction of instability into the CCN state, which has to migrate towards another stable pattern. This enables the robot to assume a locomotion type more suitable to the environmental stimuli. The transition between stable pattern configurations via a diffusion driven instability caused by external stimuli, is then used to control artificial locomotion. Moreover, as outlined earlier, a unique CNN cell can generate Turing patterns or autowaves by a suitable parameter variation [133]; therefore, the whole task is accomplished on a unique CNN architecture, a special purpose *CNN universal machine*, able to generate and control the locomotion phenomenon.

Based on the previous considerations, the following subsections describe several examples of mechatronic devices. In order to explain the whole strategy of motion generation and control, the schematic diagram of Fig. 6.19 will be employed.

6.3.2 WORMBOT: A Ring-Worm-like Walking Robot

In neurobiology nematodes are the subject of intensive study; in particular, one of them, the *Caenorhabditis Elegans*, has been morphologically mapped, enhancing the key role of the generation of traveling muscle activation waves [139]. Moreover, undulatory locomotion is not uncommon among terrestrial and marine animals. In some simple cases, for example in the lamprey, the undulatory swimming rhythm generator has been mathematically described as a chain of coupled nonlinear oscillators [140], and an attempt made to find a suitable coupling among the oscillators so as to emulate the whole behavior. If such couplings are seen as realizing diffusion effects, it is quite natural to model the pattern generation networks by a reaction–diffusion CNN. Using the experimental CNN frame described above, this first experiment regards the propagation of an autowave front through the boundary of the CNN matrix, in such a way as to have a kind of fiber where the autowave is able to propagate. The experimental setup is shown in Fig. 6.20.

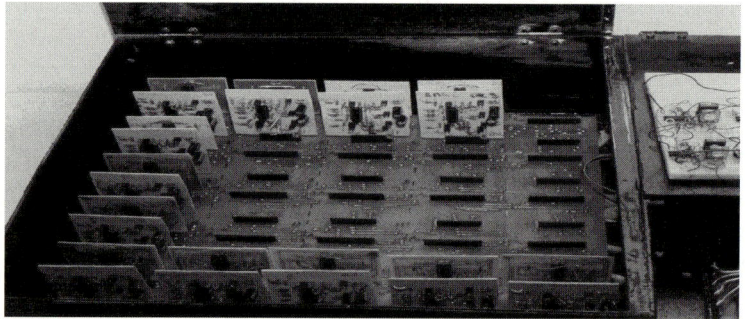

Fig. 6.20. The reaction–diffusion CNN board

The experimental results are reported in Fig. 6.21, which shows the autowave propagation for the first output of one of the cells of the fiber. Such signals are used to directly drive, in an analog way, the position of some servo motors used to realize the locomotion. The mechanical structure built and the snapshots showing the locomotion phenomenon are shown in Figs. 6.22–6.24.

They depict a mechatronic device emulating the structure of a ring worm with ten legs. The autowaves are allowed to drive the servomotors, each one moving a pair of legs, in a direction which goes from the back to the front part of the body. The propagation acts so as to push the body with feedforward,

6.3 CNNs for Generating and Controlling Artificial Locomotion

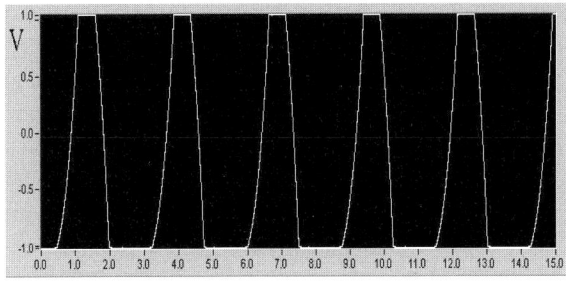

Fig. 6.21. Autowave propagation

Fig. 6.22. Ring worm robot locomotion: first step

Fig. 6.23. Ring worm robot locomotion: second step

Fig. 6.24. Ring worm robot locomotion: last step

initiating the locomotion phenomenon, in just the way in which it actually occurs in nematodes, from both a macroscopic and behavioral point of view. This is the first and the simplest experiment performed to study aspects of *locomotion control*.

In fact, referring to Fig. 6.19, in this case only the LPGNs have been modeled. However, even in this simple implementation, it is possible to control the direction of locomotion as well. In fact, autowaves are by definition autonomous waves, therefore the CNN input has not yet been used. If the simple mechanical structure is built so that each leg is driven by the output of a cell, and the autowave is allowed to propagate in a matrix having two rows, the CNN inputs can be used to make the autowave propagate along the two legs of each single ring of the wormbot, causing a change of direction during locomotion. In fact the inputs prevent the autowave from propagating in some cells, which in this way are frozen. The corresponding legs are prevented from moving, while the opposite ones move regularly, realizing the direction change. The flexibility of the approach and the possibility of controlling the locomotion direction, even in this simple example, make the approach particularly appealing.

6.3.3 REXABOT: An Hexapode Reaction–Diffusion Walking Robot

Insects are among the biological creatures most studied by neurophysiologists and neurobiologists for their robustness and adaptability. Their nervous system is somewhat complicated, but locomotion modeling is possible, at least from a functional point of view. Here we continue to consider autowaves as responsible for the locomotion pattern generation, but we investigate them on the CCN which, as previously said, supplies the right stimuli for the actuation of a specific pattern generation to the muscle system (Fig. 6.19). These command neurons can be modeled as a collection of neurons which generate a "locomotion scheme", which in turn is mapped onto the LPGN. The command neurons can be modeled by a reaction–diffusion CNN, whose output are plateau signals, therefore steady state commands, which reflect the scheme for the right connection between the pattern generation (autowave) and the actuation system robot legs in order to realize a particular locomotion type. Moreover the command neurons must possess a discrete quantity of steady state configurations, in order to realize a number of different locomotion types. Finally, the particular steady state condition corresponding to a particular locomotion type has to be directly controlled by different types of sensory inputs, and possibly in parallel. For example, the speed with which an insect varies its kind of locomotion, from slow to fast, depends on the number of different sensory inputs that concurrently excite the command neurons. With such considerations, and recalling the basic properties of Turing patterns, their application to modeling the command neuron dynamics appears

6.3 CNNs for Generating and Controlling Artificial Locomotion

very natural for the following characteristics of Turing patterns obtained by CNN's:

- even in a CNN composed of a small number of cells, it is possible to obtain a certain number of different steady state patterns. For example, as previously shown, in a reaction–diffusion CNN 3×3, eight different stable pattern configurations are available;
- it is easy to drive the network towards the desired pattern by setting suitable boundary and initial conditions;
- a particular steady state configuration attained can be changed to approach another desired one by forcing the state values of some cells;
- the forcing actions that make the Turing Pattern configuration migrate towards another steady state pattern can act in parallel and contribute to speed up the transient towards the new pattern. This characteristic derives from the fact that a certain pattern can be obtained from different configurations of initial conditions on each of the cells in the array.

Fig. 6.25. The hexapod prototype

The structure depicted represents a distributed neural network able to reach a finite number of steady state patterns as a function of some control actions to the network cells. Therefore it appears to be a natural way to model the CCN. Moreover, taking into consideration that the same cell can generate autowaves or Turing patterns with only a parameter change, the same structure can realize the whole scheme of Fig. 6.19, both for generation and fine control of this type of artificial locomotion.

The strategy proposed has been implemented in a walking robotic structure able to emulate the walking of insects. In fact, the structure is characterized by six legs, and it is shown in Fig. 6.25. From the figure it can be seen that each leg has two degrees of freedom and is moved by two servomotors: one drives the vertical position of the foot, while the other one drives the rotation of the leg to realize the locomotion steps. The kinematic has been designed so that both of the motions in each leg can be driven by the same output of one CNN cell, strictly resembling some biological cases where the same command signal triggers an ensemble of muscular fibers which lead to a coordinated motion.

Fig. 6.26. Representation of the insect "fast gait"

Implementation of the control strategy. The classical fast gait of insects (for example the alternating tripod walking of cockroaches) is depicted in Fig. 6.26. Here each horizontal bar represents the time that each leg spends detached from ground (white part), or on ground (black part) [141]. Referring to Fig. 6.19 the LPGN is mapped on the CNN, continuously generating autowave fronts, while the CCN implements a particular Turing pattern configuration. In this case we have used a CNN 3×3 for Turing pattern generation, whose steady state configuration for the fast gait is depicted in Fig. 6.27. Here one sees the direct correspondence between the CNN cells which form the pattern configuration and the hexapod legs.

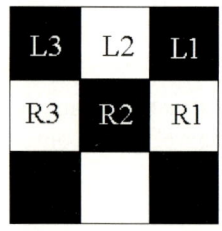

Fig. 6.27. Representation of the insect "fast gait" via a Turing pattern

Referring to Fig. 6.26, and to Fig. 6.15, each cell in the first two rows of the 3×3 CNN matrix directly corresponds to a leg. Therefore such cells have been labeled with the corresponding leg name. The cells belonging to the last row will acquire explicit meaning in the following. Indeed the pattern depicted in Fig. 6.27 corresponds to the classical checkerboard (Fig. 6.15), where, in this case, the initial condition is assumed to be imposed by either external stimuli or by specific commands from the CNS. This pattern configuration directly modulates the pattern generation centers for the generation of the specific locomotion: "fast gait": this is accomplished making the same autowave front drive the legs corresponding to the cells with black color in the Turing pattern configuration. In this case the legs: $L3$, $R2$, and $L1$, will move in accordance with the same wave flow, while the remaining ones are connected to a cell which turns out to be opposite in phase to the cell used to drive the first leg tripod combination. In such a way this type of locomotion can be easily mapped onto a reaction–diffusion CNN. The power of the approach proposed derives from the fact that the same Turing pattern configuration can be reached from different stimuli concurrently. For example, the checkerboard pattern configuration can be obtained by forcing the cell A_{22} to be "high" alone, or simultaneously with the cells A_{11} and/or A_{13}, or some other combinations. Therefore a particular locomotion type can be achieved by the coexistence of certain sensory or central stimuli, strictly reflecting the biological case.

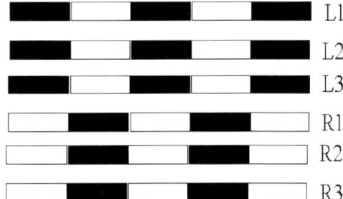

Fig. 6.28. Representation of the insect "fast gait"

Another classical type of locomotion type is "swim", which, for several insect species, can be represented as in Fig. 6.28. Here the legs $R1$, $R2$, and $R3$ are moved synchronously and in antiphase to the legs of the other side of the animal body. This type of pattern can be easily obtained with the framework proposed, as is shown in Fig. 6.29, which corresponds to Fig. 6.16, in which the dashed cells indicate initial conditions which are around zero. Just to emphasize the strength of the approach we can attain the transition between the forms of locomotion "swim" and "fast gait", which is accomplished for instance when an insect leaves water, and begins to move on earth. Some sensory inputs, visual as well as tactile, indicate absence of water and fast gait onset. The transition between these two types of locomotion is depicted in Fig. 6.30, which shows the effect of a perturbation on an existing steady state condition.

Given some initial conditions which lead to the formation of a strip pattern (swim locomotion type), if this steady state condition is perturbed setting

158   6. Experimental CNN Setup and Applications to Motion Control

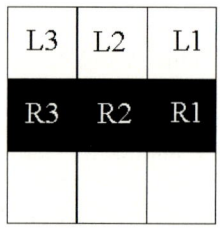

Fig. 6.29. Representation of the insect "swim" via a Turing pattern

the central cell x_1 variable from -1 V to $+1$ V (see the time trend of A_{22} in Fig. 6.30), another state transition takes place, and a new pattern geometry, a checkerboard one (fast gait locomotion type) is reached. Assuming that the sensory input directly controls the state variable x_1 of the A_{22}, the transition between the two types of locomotion – from swimming (Fig. 6.30b), to the fast gait (Fig. 6.30c) – can be realized following an analog signal flow from sensors to the network of command neurons, just as described in Fig. 6.19.

Since now the last row of the CNN 3×3 matrix has not been considered. Indeed the pattern configurations already used can also be obtained by using a 2×3 CNN. Indeed, some more complex pattern configurations, currently under investigation, corresponding to the slow and medium speed gait are stable only if they are realized on a 3×3 CNN. In this case the presence of the last row is also useful because these last mentioned kinds of locomotion

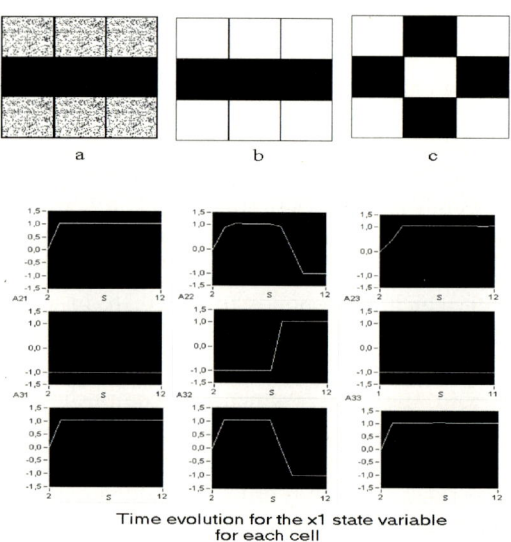

Fig. 6.30. Pattern formation and further evolution after perturbing the previously attained steady state conditions of the central cell

6.3 CNNs for Generating and Controlling Artificial Locomotion 159

require a phase shift between the legs being moved by a unique wavefront. Therefore other cells of the autowave generator have to be used and a suitable means connection can be decided by the configuration of the last row of the 3×3 CNN generating the Turing pattern configurations. Several experiments are in progress in our laboratory to stress and optimize such a strategy.

6.3.4 READIBELT:
Reaction Diffusion Conveyor Belt Autowave Driven

This experiment can be very useful for motion control in automatic production chains: A "conveyor belt" has been built. In this novel kind of transportation system the belt does not move in the direction of the destination of the object: each servomotor is moved up and down directly following the autowave front. The propagation effect, due to the diffusion process, has the effect of "pushing" the object towards its destination. In the simple example reported, a computer mouse ball can be moved on the autowave front and in the direction of the auto waveitself, as it can be directly appreciated from Figs. 6.31–6.33.   This novel application of autowaves is quite interesting. In fact, autowaves proceed with unchanged amplitude and shape along the array and with no reflection, and the direction of propagation can be determined by suitable initial conditions [119]. Therefore, suitably choosing the actuators, a two-dimensional conveyor belt, driven by reaction–diffusion CNNs showing autowaves, makes it possible to realize arbitrary motion of objects from any

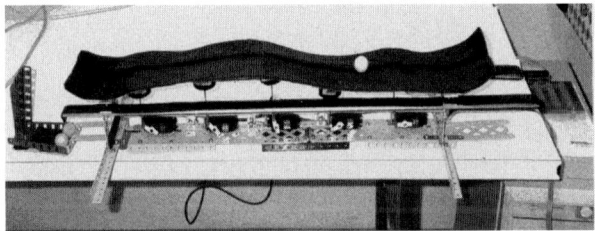

Fig. 6.31. Conveyor belt motion

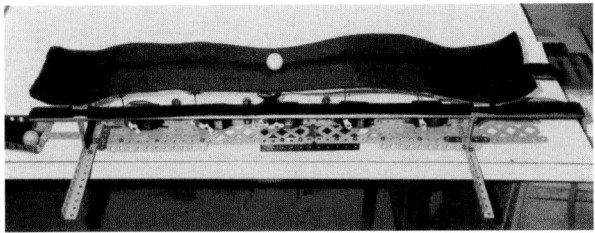

Fig. 6.32. Conveyor belt motion

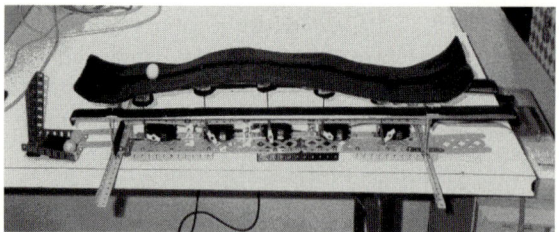

Fig. 6.33. Conveyor belt motion

given starting point to any other point, by suitably modulating the CNN control templates that fix the desired trajectories for the different objects. The example of the labyrinths, reported in the previous chapter, could be useful to clarify this concept. In this case the figure at the CNN input (Fig. 5.9) fixes the trajectory that has to be followed by the objects and the autowave fronts represent the "motor" parts which cyclically "push" the objects towards the target fixed for each of them.

It should be noted that no modification of the mechanical frame is required when a modulation of the trajectories is required. Of course, a study of suitable actuators for each application is necessary to make the approach really advantageous. The control of such mechatronic device via a *CNN universal machine* allows one to obtain a very flexible and robust structure which could lead to a major improvement in the industrial automation field.

6.4 Conclusion

In this chapter a reaction–diffusion CNN scheme has been presented to model, generate, and control artificial locomotion in some mechatronic devices. The strategy outlined has some differences to and advantages over the classical ways to generate robotic and automation locomotion:

- Artificial locomotion is commonly implemented via digital processors. The approach proposed allows one to realize a novel, analog way of generating locomotion, strictly resembling biological guidelines, and based on an analog flowing, reaction–diffusion phenomenon which links local oscillations to global self–organization, as commonly occurs from the bottom to the top of biological evolution.
- The original way in which the biologically inspired walking robots are able to adapt to the environment is powerful. In fact the locomotion change depends on the feedback from external stimuli, which can arrive one at a time or in parallel. The pattern configuration change assures great flexibility and adaptation to the environment.
- The analog implementation of the autowave generation guarantees that the locomotion actuation can reach high speeds.

- The high programmability of the CNN frame allows one to create structures, such as the one discussed in this chapter, able to meet the complexity requirements needed for the Turing pattern formation center.
- The capability of CNNs to process and classify signals [19] allows one to implement in a unified fashion the part devoted to the external stimuli and the frame presented as a whole.

The strategy introduced represents a new way to realize motion control algorithms; in fact the programmable paradigm allows one to implement real-time adaptive trajectory configurations for a wide range of motion applications.

Moreover it has been proved that very simple CNNs with few cells could be the core of new motion control algorithms in biologically inspired walking robots.

Part II

Implementation and Design

Introduction to Silicon Implementations

The remaining chapters describe the implementation and design of an integrated CNN. However, before discussing this, in order to have a glance at the problems involved, it is useful to highlight some of the real limitations imposed by current silicon technology.

The CNN paradigm has been introduced with a view to its suitability for VLSI implementation. Unlike traditional neural networks, in which the full connectivity is one of the major obstacles, the local cell-to-cell interaction allows the realization of CNN arrays often by abutting the single cells.

As was shown at the beginning of Chap. 1, ideally, the simplest cell would be essentially composed by a capacitor and a resistor in parallel fed by a set of voltage controlled current sources (VCCS) whose transconductance gains constitute the templates.

This leads to an initial classification of CNN implementations into *programmable* and *fixed template* CNNs respectively. In fixed template CNNs the values of the cloning templates are set up by the designer and cannot be changed once the chip has been fabricated. Conversely the programmability consists in the possibility, for the user, to set the template values within an allowed range. As expected, fixed template cells generally have simpler implementation and smaller area.

Although fully digital CNN implementations have been presented in the literature [142], most of the CNN chips are either fully analog or mixed–signal architectures [10, 11, 13]. Among these, it is possible to distinguish between digitally programmable and analog–programmable CNNs. The former class includes all of those chips for which the desired template is chosen by a digital word[1] [143–146]. In the latter class, however, both the template and weighted signal vary continuously in a range [147–150].

In order to implement the programmable templates some choices are possible. Among these, the most common and appropriate include the use of *operational transconductance amplifier (OTAs)* [147, 148, 151], *tunable resistors* [152, 153] and *current mirrors* [154], *analog multipliers* [155–157]. Again a design trade-off can be expected. In fact, the more flexible synapse building blocks (such as the multipliers and the linearized OTAs) are the ones that require more area and power.

In Sect. 1.2.4 it was pointed out that the input/output between an integrated CNN array and the external world represents a bottleneck for the performance of the chip. Increasing the silicon area, the number of cells in a two-dimensional CNN grows with the square law, while the number of available pins can only grow linearly. As a consequence, the time spent in the communication of data and results rapidly become significant in relation to the actual processing time.

---

[1] Within a discrete finite set of values.

But although larger available areas allow the integration of wider arrays, this can raise severe problems of yield and reliability. In this regard Ref. [157] gives some figures, based on a typical CNN design: a 50×50 array would have a yield of 70% consuming around 4 W, a 100×100 array would have a yield of only 30% while the yield barely 2% for a 200×200 array. In practice, the maximum silicon area that can be considered for an acceptable yield is limited to about 2 cm^2 or less [158].

Scaling the technology down to smaller size brings the perspective of integrating a larger number of components. On the other hands it forces the designer to deal with reduced voltage power supplies. This is a severe limit in terms of the circuit architectures that can be considered, the signal swing (and hence the signal to noise ratio), in the achievable speed. Indeed, the low voltage/low power design is currently one of the hottest topics in both digital and analog VLSI design [109, 159, 160].

Of course analog designs must inevitably have larger transistors than digital designs to achieve satisfactory component matching. Moreover, smaller channel lengths lead to devices with more severe channel length modulation effects and hence to circuits with reduced output impedance. Adding to this the reduced voltage supply problem, many well-established design techniques (e.g. cascode) become hardly applicable.

These are just a few of the actual limitations that a designer will face in the realization of a real CNN. Therefore, for instance, if programmability and accuracy are the features of interest, the immediate limit will be the area of the cell and thus the number of cells that can be integrated on the chip.

To overcome some of these problems, an alternative is to redefine/modify the CNN model to fit the technological limitations. For instance, an excellent job is done in [158] where a 48×48 cell grid on a 0.5µm standard digital CMOS process with three metal layers is presented. However, the CNN model is different from the one of Chua and Yang[2]. Another example is represented by [147] where first-class designs and very remarkable results are presented for a modified model named *full range* model (again, different from the classical Chua and Yang model). More classical models tend to be restricted to much lower densities [161, 162].

To alleviate the limitation on the number of cells that can be integrated on a single chip different approaches have been proposed. One solution, called *time multiplexing*, consists in partitioning the set of data to be processed (e.g. an image represented by a two-dimensional array of pixels) and processing them block by block [163]. Another alternative, is to design the chips in such a way that all of them can be connected in an array of ICs constituting a larger CNN. This technique is known as the *multi–chip* approach [162]. It is quite clear that in both the approaches the price to pay is in terms of throughput.

---

[2] For instance, the output nonlinearity is substituted by a Heaviside function, the signal range is limited to an interval of positive values only, and the multipliers are substituted by comparators.

Nevertheless the inherent speed offered by the CNN can be enough to yield a quite remarkable overall performance for the desired applications.

In summary, although CNNs are undoubtedly well-suited to IC technology, several design challenges are involved and cannot be underestimated. The larger the number of cells in a CNN, the larger will be the processing power of this inherently parallel architecture. Considering that the total chip area cannot grow beyond certain limits, the density of cells becomes an important parameter in the evaluation. However, this is certainly of little significance unless one also considers the particular CNN model, the range in which it may possibly be programmed, the presence of additional modules (such as memories, digital circuitry and the like), the reliability, the accuracy, the power consumption, the speed and many other fundamental issues. Note, however, that in many applications (some of which have already been discussed in this book) a very reduced number of cells is more than enough.

7. A Four Quadrant S²I Switched-Current Multiplier

Switched-current (SI) circuits [164] represent a feasible alternative to *switched-capacitor* (SC) [102, 165] circuits especially due to their compatibility with digital technology.

Only one switched-current (SI) multiplier has ever been presented until now [166]. The complexity of this circuit required one extra clock phase besides the normal clock phases used in common *2nd generation* SI cells, two explicit capacitors to cope with clock feedthrough problems, and a regulated-cascode architecture to deal with channel modulation effects. The multiplication of any two given currents x and y is accomplished by evaluating their quadratic terms:

$$(x+y)^2 - x^2 - y^2 = 2xy. \tag{7.1}$$

One smart aspect of Leenaert's [166] multiplier is that any quadratic term can be evaluated using the same squarer circuit in different clock phases, avoiding in this way the need for precisely matched squarer circuits as is the case in some continuous-time current multipliers [167].

This chapter presents an alternative implementation [168, 169] to the latter approach by using the S²I switched current technique[1] that has already been proved effective in filtering and converting applications [170–173].

One of the most important features of S²I is that it allows compensation of the analog errors that are mainly due to charge injection from the MOS switches. Essentially, the signal-dependent clock feedthrough is sampled separately and added to the corrupted sampled signal to dramatically reduce the error due to the charge injection. This removes the need for additional "replica circuitry" for compensating the offsets.

Another improvement in the proposed design is that no cascode transistors have been used to minimize errors due to the finite g_m/g_0 ratio. Instead, they were minimized by varying the the current gain of a current mirror. Moreover, thanks to the adoption of the S²I technique, no capacitors have been used, making the implementation more suitable for a standard digital technology.

The multiplier is implemented using a 2 μm n-well CMOS technology. Experimental results are in agreement with the theoretical findings.

---

[1] The name S²I comes from the double sampling of the input.

168   7. A Four Quadrant S²I Switched-Current Multiplier

The following summarizes the most important measurement results:

1. 0.425 million multiplications per second,
2. 1.7% total harmonic distortion for a sinusoidal wave of amplitude 35 mA (50 Hz),
3. 206 kHz of bandwidth,
4. 50 dB signal-to-noise ratio (SNR), and
5. 0.3 mW zero input power consumption for a ±3 V power supply.

A complete set of detailed experimental results is provided at the end of this chapter.

7.1 Detailed Analysis of the S²I Memory Cell

One of the most severe problems affecting SI circuits is the charge injection coming from the MOS switches. This is the well-known problem of *clock feedthrough* [164, 174, 175]. Many different approaches have been developed to cope with this problem including replica circuits [174, 176, 177], dummy switches [147, 164], algorithmic approaches [164, 178], and so on [179]. The S²I technique [159, 160, 170, 171] deals with this problem in a very effective way as we will demonstrate here.

Essentially, the signal-dependent clock feedthrough that corrupts the input signal is sampled and stored in the initial sampling phase of the current copier operation and then algebraically added to the corrupted current to minimize the corresponding error. Likewise, more sophisticated circuit techniques have been applied to this elementary circuit architecture [172, 173] in order to cope with other problems like switch ir-drops and charge injection coming from the drain–gate and drain–substrate capacitances.

We have considered only the basic cell in the design of the multiplier.

The general circuit of a S²I cell and its associated clock phases is shown in Fig. 7.1.

Let us now perform a detailed small signal analysis of the cell. In all cases we refer to Fig. 7.2 which depicts equivalent circuits for the various clock phases. During phase ϕ_{1a} of a given clock period, arbitrarily called period $(n-1)$, the NMOS transistor M_C is diode connected while the PMOS transistor M_F behaves as a current source. This is depicted in Fig. 7.2a. The input current $i_i(n-1)$ is stored in M_C by means of i_{1a}:

$$i_{1a} = \frac{g_{mC}}{g_{mC} + g_{dsF} + g_{dsC}} i_i(n-1). \qquad (7.2)$$

When the switch that realizes the diode-connection of the NMOST is opened, it will inevitably inject an undesired charge into the NMOST due to the clock feedthrough, corrupting in this way the stored current. Unfortunately, this charge is dependent on the input signal i_i itself. Hence, the NMOST is referred

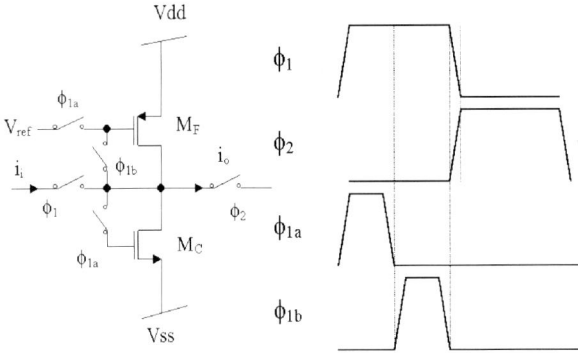

Fig. 7.1. S²I memory cell and clock waveforms

to as the *coarse memory*. The input resistance of the memory cell, during this phase is $r_i = (g_{mC} + g_{dsC} + g_{dsF})^{-1} \cong g_{mC}^{-1}$.

During ϕ_{1b} the current error Δi just introduced is sampled and stored (as the difference of the stored current and the current i_i still present at the input) by diode-connecting the PMOST as depicted in Fig. 7.2b. In this case the PMOST is referred to as the *fine memory*. Here, C_C is the total capacitance between the gate and the source of M_C that stores the current i_{1a} plus the undesired current error. Moreover, due to the drain–gate capacitance C_{dgC} an additional current error is fed back to the gate of M_C.

This is commonly modeled by augmenting the total output conductance g_{0C} of M_C by an additional contribution [164]

$$g_{0C} = g_{dsC} + \frac{C_{dgC}}{C_C + C_{dgC}} g_{mC}. \tag{7.3}$$

So the current i_{1b} stored in the fine memory M_F is

$$i_{1b} = [i_i(n-1) - i_{1a} - \Delta i] \frac{g_{mF}}{g_{0C} + g_{dsF} + g_{mF}}. \tag{7.4}$$

It is worth noting that because $i_i(n-1) \cong i_{1a}$ then, only the coarse memory error is stored in the fine memory. Moreover this also means that most of the input current $i_i(n-1)$ is cancelled by making the actual input impedance smaller than in the previous sub-phase.

In order to evaluate the input resistance r_i shown in the small signal equivalent circuit on the right hand side of Fig. 7.2b, the node equation can be written neglecting the contribution Δi due to the feedthrough:

$$i_i(n-1) - i_{1a} = v(g_{mF} + g_{0C} + g_{dsF}). \tag{7.5}$$

Substituting (7.2) into (7.5) we have

$$i_i(n-1) - \frac{g_{mC}}{g_{mC} + g_{dsF} + g_{dsC}} i_i(n-1) = v(g_{mF} + g_{0C} + g_{dsF}). \tag{7.6}$$

170   7. A Four Quadrant S²I Switched-Current Multiplier

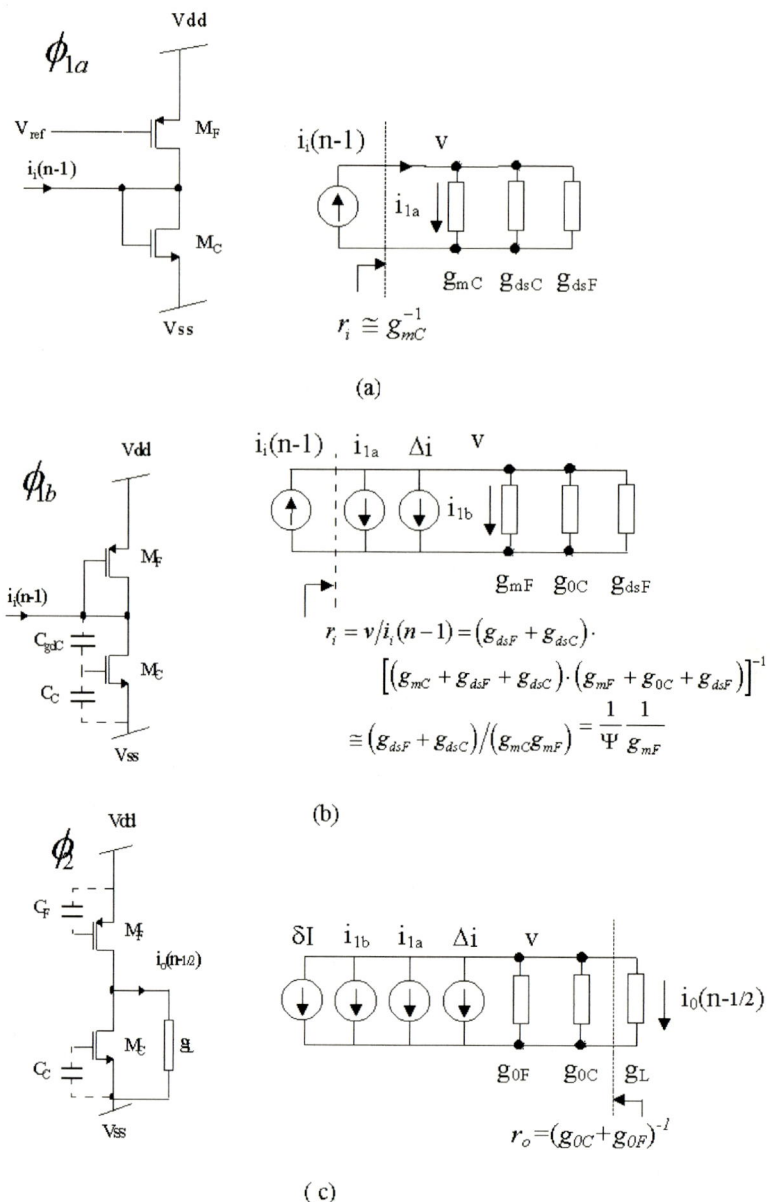

Fig. 7.2. Small signal analysis for the S²I memory cell. Phase: (**a**) ϕ_{1a}; (**b**) ϕ_{1b}; (**c**) ϕ_2

Therefore,

$$\begin{aligned} r_i &= v/i_i(n-1) \\ &= (g_{\text{dsF}} + g_{\text{dsC}}) \cdot [(g_{\text{mC}} + g_{\text{dsF}} + g_{\text{dsC}}) \cdot (g_{\text{mF}} + g_{\text{0C}} + g_{\text{dsF}})]^{-1} \\ &\cong (g_{\text{dsF}} + g_{\text{dsC}})/(g_{\text{mC}} g_{\text{mF}}) = \frac{1}{\Psi} \frac{1}{g_{\text{mF}}}. \end{aligned} \qquad (7.7)$$

It is seen that the dimensionless constant Ψ is the voltage gain of a CMOS inverter, where M_{F} works as a current source while M_{C} serves as driver.

Finally, in the phase ϕ_2, the currents stored in the two transistors are added up at the drain node and supplied to the generic load g_{L}. There is however a fixed output current offset, δI, due to the last signal-independent charge injection but it is of minor concern compared to the compensated signal-dependent error. The final output current $i_o(n-1/2)$ is

$$i_o(n-1/2) = -[\Delta i + i_{1a} + i_{1b} + \delta I] \frac{g_{\text{L}}}{g_{\text{0C}} + g_{\text{0F}} + g_{\text{L}}}, \qquad (7.8)$$

where δI is the offset current due to the signal-independent clock feedthrough in M_{F}, and g_{0F} the total output impedance of M_{F} augmented by the feedback effect due to its drain–gate capacitance. Notice that during this phase the output resistance is $z_o = (g_{\text{0C}} + g_{\text{0F}})^{-1}$.

Substituting relationships (7.2)–(7.4) into (7.8) and grouping common terms, the following expression is obtained for the output current:

$$\begin{aligned} i_o(n-1/2) = &-i_i(n-1)\frac{g_{\text{mF}}}{g_{\text{mF}} + g_{\text{0C}} + g_{\text{dsF}}} \cdot \frac{g_{\text{L}}}{g_{\text{0C}} + g_{\text{0F}} + g_{\text{L}}} \\ &- \Delta i \alpha - \delta I \frac{g_{\text{L}}}{g_{\text{0C}} + g_{\text{0F}} + g_{\text{L}}}, \end{aligned} \qquad (7.9)$$

where

$$\alpha \doteq \left(1 - \frac{g_{\text{mF}}}{g_{\text{mF}} + g_{\text{dsF}} + g_{\text{0C}}}\right) \frac{g_{\text{L}}}{g_{\text{0C}} + g_{\text{0F}} + g_{\text{L}}}. \qquad (7.10)$$

Notice that α is composed of two terms: the first one in brackets is close to 0 while the other one is less than 1. Therefore, in (7.9), the second and third terms can be neglected. Indeed, this proves the clock feedthrough cancellation property of the S^{2}I cell. Let us further simplify (7.9) and make some assumptions. First of all, assume that all the MOST have the same value of small-signal transconductance g_{m}. Two different cases have to be distinguished at this point: if the load consists of a diode connected transistor (e.g. another current mode stage) then $g_{\text{L}} = g_{\text{m}}$. If, instead, the load is another S^{2}I memory cell then ϕ_2 is partitioned into ϕ_{2a} (during which $g_{\text{L}} = g_{\text{m}}$) and ϕ_{2b} (during which $g_{\text{L}} \cong (g_{\text{mC}} g_{\text{mF}})/(g_{\text{dsF}} + g_{\text{dsC}}) = \Psi g_{\text{m}}$).

Let us first consider the case of a diode connected transistor as load. An underestimation of the information retrieved from the memory is accomplished by making $g_{\text{dsF}} = g_{\text{0F}}$ in (7.9). Hence (7.9) can be simplified to

7. A Four Quadrant S²I Switched-Current Multiplier

$$i_o(n-1/2) \cong -i_i(n-1)\left(\frac{g_m}{g_{0C} + g_{0F} + g_m}\right)^2 - \delta I \frac{g_m}{g_{0C} + g_{0F} + g_m}. \quad (7.11)$$

Taking $g_0 = g_{0F} = g_{0C}$ (7.11) can be approximated to

$$\begin{aligned} i_o(n-1/2) &\cong -i_i(n-1)\frac{1}{1+4\frac{g_0}{g_m}} - \delta I \frac{1}{1+2\frac{g_0}{g_m}} \\ &= -i_i(n-1)\frac{1}{1+4\frac{g_0}{g_m}} + i_{\text{offset}}, \end{aligned} \quad (7.12)$$

where i_{offset} represents the output current offset. Given that the ideal transfer function of an S²I cell is $H_i(z) = I_o(z)/I_i(z) = -z^{-1/2}$ we can take the z-transform

$$I_o(z) \cong I_i(z)\frac{H_i(z)}{1+4\frac{g_0}{g_m}} + I_{\text{offset}}(z). \quad (7.13)$$

It is seen that an attenuation due to the finite conductance ratio has to be expected on the retrieved signal. A less detailed analysis taking into consideration only the effect of the channel length modulation of M_F and M_C during the last phase (ϕ_2) will give a perfect signal-dependent clock feedthrough cancellation. The corresponding final expression is

$$I_o(z) \cong I_i(z)\frac{H_i(z)}{1+\frac{g_{0T}}{g_m}} + I_{\text{offset}}(z), \quad (7.14)$$

where $g_{0T} = g_{0F} + g_{0C}$ is the total output conductance of the memory cell.

When, instead, the load of the memory cell is constituted by another S²I cell then the load changes to $g_L \cong \Psi g_m$ during ϕ_{2b}. Therefore g_L is going to dominate over the output impedance of the S²I allowing a complete transfer of the output current $i_o(n-1/2)$ to the next stage. This implies that (7.11) is replaced by

$$i_o(n-1/2) \cong -i_i(n-1)\frac{g_m}{g_{0C} + g_{0F} + g_m} - \delta I \frac{\Psi g_m}{g_{0C} + g_{0F} + \Psi g_m}. \quad (7.15)$$

Notice that the fractional term multiplying δI on the right-hand side of (7.15) is very close to 1. Moreover, taking $g_0 = g_{0F} = g_{0C}$, this can be approximated to

$$i_o(n-1/2) \approx -i_i(n-1)\frac{1}{1+2\frac{g_0}{g_m}} - \delta I. \quad (7.16)$$

In other words, considering that the total output conductance is $2g_0$, then the final expression describing the behavior of the cell is formally equal to (7.14). If, however, as commonly assumed in SI circuit analysis [164], the effect of the device's output impedance (due to the channel modulation effect) is neglected during the sampling phase (namely during ϕ_{1a} and ϕ_{1b}) and considered only during ϕ_2 (analogously to the procedure to obtain (7.14)), then: (a) there is no current attenuation during the sampling phase and (b) the Norton equivalent of the S²I cell (with output conductance equal to $2g_02$) transfers the retrieved current to a load with conductance equal to Ψg_m during f_{2b}. So following final expression is obtained

$$I_0(z) \cong I_i(z) \frac{H_i(z)}{1 + \frac{2g_{0T}}{\Psi g_m}} + I_{\text{off}}(z). \tag{7.17}$$

This analytically proves that the S²I cell not only achieves the feedthrough cancellation but also reduces the error due to the finite g_m/g_0 ratio as compared with common 2nd generation cells [170, 172].

This analytic result is in complete agreement with the simulation [170] and experimental results [172, 173] obtained by Hughes et al.

The sampling frequency in normal 2nd generation cells is limited by the settling time of the cell sampler [164]. In the case of S²I the cell treats the coarse phase ϕ_{1a} settling error as the other errors and so attempts to cancel it during the fine phase ϕ_{1b}. This means that the bandwidth is the same as that of a standard 2nd generation cell.

7.2 The Multiplier Architecture

Starting from the algorithm introduced by Leenaerts et al. [166] an alternative circuit implementation is now presented. The product of two currents x and y fed in at the inputs of the multiplier is accomplished by evaluating the left-hand side of

$$(x + y)^2 - x^2 - y^2 = 2xy. \tag{7.18}$$

A fundamental requirement is that these inputs have to be kept constant during the evaluation process. In order to obtain the square of a current the simple squarer circuit shown in Fig. 7.3 has been considered [180].

For this circuit a relationship including the input/output offset errors gives

$$i_o = I_b + e_1 + \frac{(i_i + e_2)^2}{4I_b} \tag{7.19}$$

where i_o and i_i are the output and input current respectively, I_b is a constant current related to the bias voltage V_{bias}, e_1 is the output offset error, e_2 is the input offset error.

The approach introduced in [166] consists of using the same squarer circuit many times while storing intermediate results in the SI memory cells. In

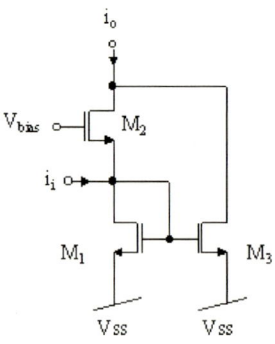

Fig. 7.3. The current squarer

this way the problem of matching many squarer circuits (the problem of the quarter squarer multipliers [167], for example) is avoided.

The block diagram of the whole system realizing the above algorithm is given in Fig. 7.4.

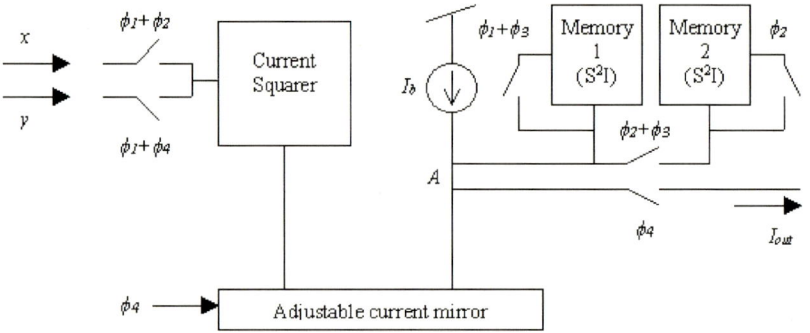

Fig. 7.4. Block diagram of the S^2I multiplier

The algorithm consists of four main steps described by means of four main clock phases: ϕ_1, ϕ_2, ϕ_3, ϕ_4. The circuitry consists of three building blocks: the current squarer, an adjustable current mirror, and two S^2I memory cells. From this figure it is seen that, analogously to [166], a complementary version of the squarer circuit shown in Fig. 7.3 is used. In this way, due to the fact that the adopted technology is a CMOS n-well it is possible to avoid the bulk effect on M_2 by connecting its source to the bulk. Moreover, by using the mirror and the current source I_b, it is possible to remove the DC contribution from the output of the squarer and to store only small signals in the S^2I memory cells.

Let us investigate the algorithm. During ϕ_1 both x and y are added to the input node of the current squarer. The squared result is stored in memory 1. During ϕ_2 only x is fed in; the result of the squaring operation is added to the previous result that was fed (and sign inverted) by memory 1 to node A. This sum is stored in memory 2. During ϕ_3 no input is presented to the squarer that feeds the residual to node A. At the same time the latter partial sum is fed (and sign inverted) to node A by memory 2 and the total current is stored into memory 1. Finally, during ϕ_4 the current y is squared and added to the previous sign-inverted sub-total given by memory 1.

The total current I_{out} is provided at the output. Applying this algorithm to (7.18) by using (7.19) and assuming ideal behavior for the circuits and devices (infinite output/input impedance ratio of the building blocks constituting the circuit) the following first-order relationship is obtained:

$$i_{\text{out}} = \left\{ I_\text{b} + e_1 + \frac{(x+y+e_2)^2}{4I_\text{b}} \right\} - \left\{ I_\text{b} + e_1 + \frac{(x+e_2)^2}{4I_\text{b}} \right\} \\ + \left\{ I_\text{b} + e_1 + \frac{(e_2)^2}{4I_\text{b}} \right\} - \left\{ I_\text{b} + e_1 + \frac{(y+e_2)^2}{4I_\text{b}} \right\} = \frac{xy}{2I_\text{b}}. \quad (7.20)$$

It is seen that thanks to the third term in which no input to the squarer exists, the errors due to the offsets in the squarer are canceled out and the final result is proportional to the product of x and y. Due to the fact that the transfer function of any single S^{2}I memory cell of any kind is of inverted type, the order in which the various terms are evaluated and the intermediate results retrieved is important and can help to avoid the need for additional current-inverter circuits.

As recalled in Sect. 7.1, the output node of a S^{2}I cell can be seen as a virtual ground. This minimizes the channel length modulation effect and so the problem of the finite g_m/g_0 ratio in comparison to other SI cells not utilizing cascode or feedback-based approaches.

However, as seen in Sect. 7.1, an attenuation in the retrieved contents of the memory cells is expected. Among the four steps, the biggest imbalance occurs during ϕ_4 because the squared input signal is directly added to the rest of the previously calculated result without an equal attenuation. Therefore, during this last phase the ratio of the mirror is changed in order to compensate for this imbalance. The mirror has unit gain during all the phases but ϕ_4. Strictly speaking, similar problems are expected also in phases ϕ_2 and ϕ_3. Therefore, for better accuracy a similar ratio-tuning can also be accomplished in those phases with only a small increase in the complexity of the system. However, simulation results have clearly shown that for a satisfying accuracy (nonlinearity of around $\pm 1\%$ FS) this is not necessary. An analysis of this behavior is reported in the next section. The circuit schematic of the complete multiplier is shown in Fig. 7.5.

In this figure it is clearly seen how the current gain of the mirror is slightly changed during ϕ_4 by adding a small-area diode-connected MOST in parallel to the existing one. This, however, will cause an extra bias current emerging from the right branch of the mirror due to the imbalance in I_b. This will add up to the constant offset error present in the output of the S^2I cells and will cause a constant current offset on I_{out}.

This undesired offset can be canceled in two ways. If the multiplier is going to be used alone, an extra current can be added during ϕ_4 by using an additional current source and a current steering switch. This is realized by M_{10} as shown in Fig. 7.5. This for example is what has been done in the fabricated chip whose experimental results are discussed in the following. If, instead, many multipliers are going to be used in the same chip (as in the case of a neural network or in the case of adaptive filters and so on) then a suitable alternative might consist of realizing another multiplier (namely a replica multiplier) without inputs. The output current offset of this multiplier is added to the outputs of all the other multipliers by using current mirrors. In this way the output offset of the other multipliers can be drastically reduced in spite of process variations.

Seven of the nine control signals used to drive the switches are depicted in Fig. 7.6. The other two signals are $\phi_2 + \phi_3 = \overline{\phi_1 + \phi_4}$ and $\phi_3 + \phi_4 = \overline{\phi_1 + \phi_2}$. The nine control signals are obtained as combinations of the various "master" phases. It can be noticed that while the sub-phases used to control the internal switches of the memory cells do not overlap (namely $\phi_{1a} + \phi_{3a}$, $\phi_{1b} + \phi_{3b}$, ϕ_{2a}, ϕ_{2b}), the current steering switches are controlled by signals with overlapping

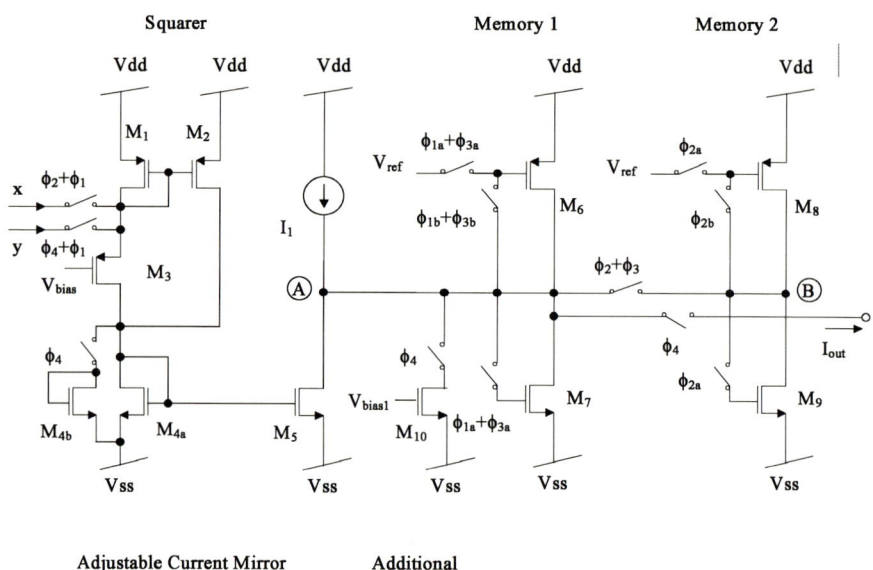

Fig. 7.5. Circuit schematic of the S^2I multiplier

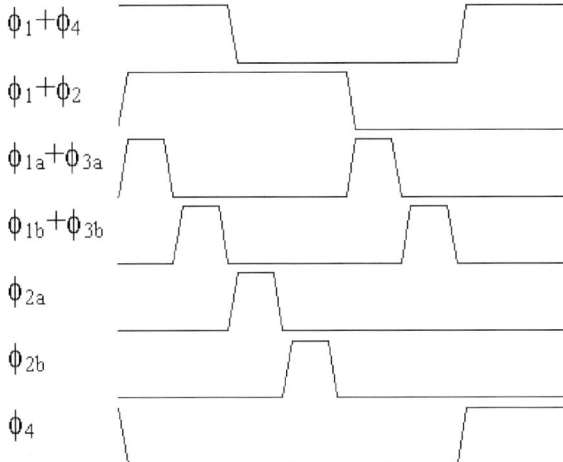

Fig. 7.6. Control signals for the switches

rising and falling edges ($\phi_1 + \phi_4$, $\phi_1 + \phi_2$, ϕ_4). This minimizes the generation of spikes without interfering with the operation of the circuit.

7.3 Analysis and Design of the S²I Multiplier

In this section the behavioral analysis of the S²I multiplier is carried out. The approach presented takes into account the non-idealities of the circuits and devices. Finally, an algorithm for the circuit design of the multiplier is discussed.

7.3.1 Circuit Analysis of the Multiplier

In [170, 172, 173] the behavior and some of the applications of the S²I cell have been discussed. However, in order to analyze the proposed multiplier, the effect of the finite conductance ratio in the memory cells, as well as in the other building blocks, has to be taken into account. These topics are discussed below.

In the following, for the sake of simplicity, it will be assumed that all the g_m's are equal and all the g_{ds}'s are equal.

Strictly speaking, a complete analysis would require one to analyze the behavior of the multiplier in seven phases (ϕ_{1a}, ϕ_{1b}, ϕ_{2a}, ϕ_{2b}, ϕ_{3a}, ϕ_{3b} and ϕ_4). However, taking advantage of the analysis of the S²I memory cell carried out in Sect. 7.1, we can consider its small signal equivalent circuit as follows. During ϕ_1, ϕ_2 and ϕ_3: the memory cell working in the sampling phase (composed by the two sub-phases "a" and "b") has a small signal equivalent circuit consisting of a conductance equal to Ψg_m. The current stored in the cell is the one flowing into this conductance. During the retrieval phase the small signal circuit is composed of an ideal current source supplying the current stored in the previous phase in parallel with a conductance equal to $2g_{ds}$. Notice that, although the cells supply a fixed output offset when the stored current is retrieved, we can nevertheless implicitly consider this offset as part of the squarer offset e_1. During ϕ_4, however, the circuit load is assumed to be the generic conductance g_L.

Therefore the analysis is carried out in the four main phases ϕ_1, ϕ_2, ϕ_3 and ϕ_4.

Let us now refer to the complete circuit schematic shown in Fig. 7.5 and to the small signal equivalents shown in Fig. 7.7.

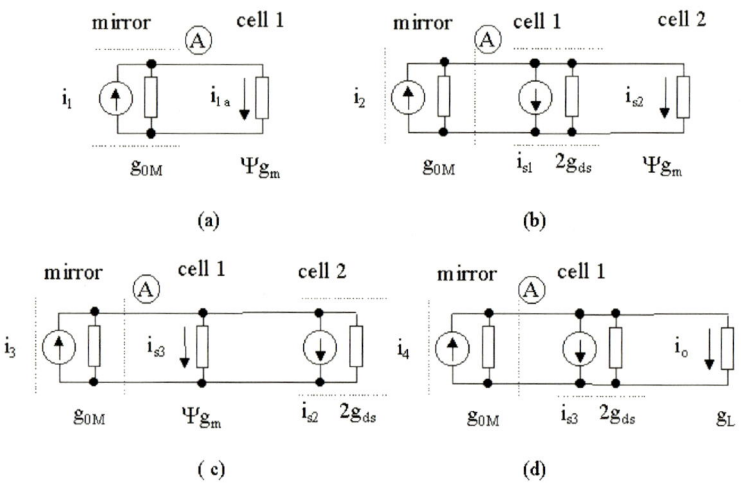

Fig. 7.7. Small signal equivalent circuits for the S^2I multiplier: (a) phase ϕ_1 ; (b) phase ϕ_2; (c) phase ϕ_3; (d) phase ϕ_4

The current source I_1 depicted in Fig. 7.5 is implemented by using a single PMOST supplying a constant current equal to I_b. According to what was discussed in Sect. 7.2, in phase ϕ_1 the mirror $M_{4a} - M_5$ will supply the following current:

$$i_1 = -\left(e_1 + \frac{(x+y+e_2)^2}{4I_b}\right). \tag{7.21}$$

The corresponding small signal equivalent is shown in Fig. 7.7a. Here g_{0M} represents the output conductance of the right-hand side branch of the current mirror. Therefore, following the previous discussion we have $g_{0M} = 2g_{ds}$. Only cell 1 is connected to node A and the input conductance of the memory cell is Ψg_m. So the current stored in it is

$$i_{s1} = \frac{\Psi g_m}{\Psi g_m + 2g_{ds}} i_1. \tag{7.22}$$

In phase ϕ_2 the current i_2 supplied by the mirror is added to i_{s1} retrieved from cell 1 and the result is stored in cell 2. Again, the actual current stored in cell 2 is the one flowing into Ψg_m.

Hence, from the small signal equivalent circuit shown in Fig. 7.7b we have

$$i_2 = -\left(e_1 + \frac{(x+e_2)^2}{4I_b}\right), \quad i_{s2} = \frac{\Psi g_m}{\Psi g_m + 4g_{ds}}(i_2 - i_{s1}). \tag{7.23}$$

In phase ϕ_3 the mirror supplies i_3 which is added to i_{s2} supplied by cell 2 and the result i_{s2} is stored in cell 1. So, considering the small signal equivalent circuit depicted in Fig. 7.7c, we have

$$i_3 = -\left(e_1 + \frac{(e_2)^2}{4I_b}\right), \quad i_{s2} = \frac{\Psi g_m}{\Psi g_m + 4g_{ds}}(i_3 - i_{s2}). \tag{7.24}$$

Finally in phase ϕ_4 the mirror changes its ratio. This is accomplished by inserting M_{4b} in parallel to M_{4a}. The ratio changes from $1:1$ to $1:\beta$ with $\beta < 1$. However, because the bias current on the left branch of the mirror is not changed, the extra current $(1-\beta)I_b$ is supplied by the right-hand branch of the mirror. This represent an output offset that can be canceled as discussed in the previous section.

With regards to the small signal analysis, the mirror supplies i_4, which is added to i_{s3} supplied by cell 1. The result, i_0, flows into the output load g_L. So, from the small signal circuit of Fig. 7.7d we have

$$i_4 = -\left(e_1 + \frac{(y+e_2)^2}{4I_b}\right)\beta, \quad i_{s2} = \frac{g_L}{g_L + 4g_{ds}}(i_4 - i_{s3}). \tag{7.25}$$

Substituting (7.22)–(7.24) into (7.25) the following expression is obtained:

$$i_o = \left\{i_4 - \left[i_3 - \left(i_2 - \frac{\Psi g_m}{\Psi g_m + 2g_{ds}}i_1\right)\frac{\Psi g_m}{\Psi g_m + 4g_{ds}}\right]\frac{\Psi g_m}{\Psi g_m + 4g_{ds}}\right\} \\ \times \frac{g_L}{g_L + 4g_{ds}}. \tag{7.26}$$

Defining:

$$\eta \doteq \frac{\Psi g_m}{\Psi g_m + 4g_{ds}}, \quad \rho \doteq \frac{g_L}{g_L + 4g_{ds}}, \tag{7.27}$$

the relationship (7.26) can be approximated by

$$i_o \cong \left\{i_4 - [i_3 - (i_2 - \eta i_1)\eta]\eta\right\}\rho = \left\{i_4 - i_3\eta + i_2\eta^2 - i_1\eta^3\right\}\rho = i\rho, \tag{7.28}$$

where i is just a current proportional to the actual output current i_0. Let us then analyze i by substituting i_1, i_2, i_3 and i_4 from (7.21)–(7.25) into (7.28). After some algebra the following relationship is obtained:

$$i = e_1(\eta^3 - \eta^2 + \eta - \beta) + e_2^2 \frac{1}{4I_b}(\eta^3 - \eta^2 + \eta - \beta)$$
$$+ \frac{x^2}{4I_b}\eta^2(\eta - 1) + \frac{y^2}{4I_b}(\eta^3 - \beta) \quad (7.29)$$
$$+ \frac{xe_2}{2I_b}\eta^2(\eta - 1) + \frac{ye_2}{2I_b}(\eta^3 - \beta) + \frac{\eta^3}{2I_b}xy.$$

The first two terms of the right-hand side represent an offset, the last term is the desired result while the remaining terms constitute the nonlinear distortion. Note that η is very close to, but less than, 1. Therefore the third and fifth terms, being multiplied by a factor $\eta^2(\eta - 1)$, can be neglected. So the relationship (7.29) can be rewritten as:

$$i = i_{\text{off}} + \frac{y^2}{4I_b}(\eta^3 - \beta) + \frac{ye_2}{2I_b}(\eta^3 - \beta) + \frac{\eta^3}{2I_b}xy,$$
$$i_{\text{off}} = e_1(\eta^3 - \eta^2 + \eta - \beta) + e_2^2 \frac{1}{4I_b}(\eta^3 - \eta^2 + \eta - \beta), \quad (7.30)$$

where i_{off} is the offset current. The nonlinear error is canceled by setting the current mirror ratio as

$$\beta = \beta_0 \doteq \eta^3 = \left(\frac{1}{1 + \frac{4g_{ds}}{\Psi g_m}}\right)^3. \quad (7.31)$$

The final expression is therefore

$$i = e_1(-\eta^2 + \eta) + e_2^2 \frac{1}{4I_b}(-\eta^2 + \eta) + \frac{\eta^3}{2I_b}xy. \quad (7.32)$$

It is worth noting that because $\eta \cong 1$ the offset is almost canceled as well. So finally

$$i_0 \cong \frac{\eta^3 \rho}{2I_b}xy. \quad (7.33)$$

The analysis carried out in this section demonstrates the big advantage of the proposed architecture for the nonlinearity cancellation by selecting the appropriate mirror gain.

7.3.2 Circuit Design

Let us consider the circuit diagram shown in Fig. 7.5 and, for the sake of clarity, assume a symmetric power supply ($V_{\text{dd}} = -V_{\text{ss}}$). Let us consider the two memory cells composed by M_6–M_9 and the corresponding switches. The quiescent drain voltages of the MOSTs are chosen so as to stay at ground. In order to avoid jumping of the drain voltage of these transistors when the switches turn on or off M_6-M_9 are designed for so that

$$|V_{GS}| = |V_{DS}| = |V_{dd}|. \tag{7.34}$$

Therefore $V_{ref} = 0$ V. The bias current I_D can be chosen to have a safe value for the adopted technology. Alternatively, it can be chosen to satisfy other possible requirements on g_m and/or g_{ds} or to yield a desired S/N ratio. From these considerations the aspect ratios are determined as:

$$\left(\frac{W}{L}\right)_{NMOST} = \frac{2I_D}{K'_N(V_{dd} - V_{TN})^2(1 + \lambda_N V_{dd})},$$
$$\left(\frac{W}{L}\right)_{PMOST} = \frac{2I_D}{K'_P(V_{dd} - |V_{TP}|)^2(1 + \lambda_P V_{dd})}, \tag{7.35}$$

where K'_N (K'_P) and V_{TN} (V_{TP}) are the transconductance parameter and the threshold voltage of the NMOST (PMOST), respectively [103]. There is still a degree of freedom to choose either W or L. Two alternatives are possible. If the output impedance of the cell is of major concern then L is fixed and W is determined accordingly. The second alternative regards the area WL. Larger areas minimize the clock feedthrough. On the other hand, smaller areas are necessary for high speed.

The two transistors M_4 and M_5 constitute a current mirror. Hence, to minimize the current error due to the channel modulation effect, the quiescent drain voltage of M_{4a} is chosen to stay at ground. M_1, M_2 and M_3 constitute the squarer circuit and it is assumed that all of them are equal [180]. Hence, from the above considerations it follows that the quiescent drain voltage of M_1 stays at $V_{dd}/2$. The square law holds for $|i_i| < 2I_b$ [180]. Therefore, a minimum value for I_b is fixed because the maximum value of $|i_i|$ is essentially given by $|x_{max}| + |y_{max}|$.

Taking into consideration the above assumptions for the node voltages and the fact that M_1, M_2 and M_3 are biased for a drain current equal to $I_b/2$, while M_{4a} and M_5 are biased for a drain current equal to I_b, the design of these transistors is then straightforward.

In (7.31) the mirror ratio (current gain) has been obtained. Assuming that $(W/L)_5 = (W/L)_{4a}$, the aspect ratio of M_{4b} is obtained as follows:

$$\left(\frac{W}{L}\right)_{4b} = \left(\frac{W}{L}\right)_{4a}\left(\frac{1-\beta}{\beta}\right). \tag{7.36}$$

However, the final design of M_{4b} also depends on the actual voltage drop on the switch in series with M_{4b}. A common measure of the accuracy is given by the difference between the multiplier's actual and ideal output, at full scale, as a percentage of the full scale itself. This is known as the *internal trim* error ϵ [181]. This includes the effects of offset, feedthrough, nonlinearity and scale-factor errors. A plot of ϵ versus β, obtained by HSPICE simulation for a 100 kHz clock frequency and large signals is shown in Fig. 7.8.

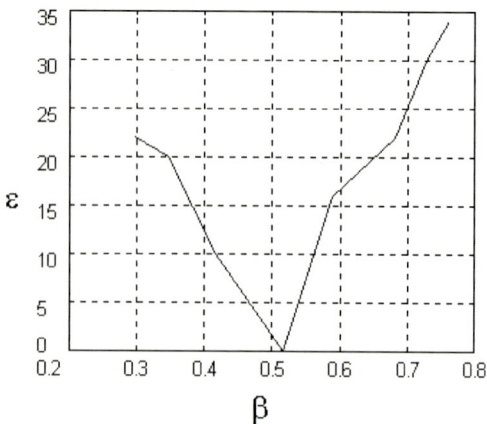

Fig. 7.8. Percentage trim error (ϵ) versus mirror correction factor (β). The total error correction is obtained for the optimal value $\beta_0 = 0.5161$ in the present example

Let us now consider the switches. The control voltages for the switches go from rail to rail while the terminals of the switches stay at a voltage close to ground during the whole operation of the circuit. Therefore the switches can easily be designed by using minimum size either NMOS transistors or CMOS transmission gates. However, for the technology adopted, no appreciable difference has been noticed upon substituting the NMOSTs for the CMOS switches; thus, single NMOSTs have been used in the considered implementation. There is one exception only: Because the voltage at the drain of M_1 is $V_{\mathrm{dd}}/2$ the two switches steering the currents x and y at the input of the squarer are implemented using CMOS switches instead of simple NMOSTs.

A summary of the adopted transistor sizes is reported in Table 7.1.

7.4 Experimental Performance Evaluation

A prototype of the proposed multiplier has been fabricated in MOSIS Orbit N-well 2 µm technology. In this section the experimental results obtained by testing the chip are discussed. A summary of the measured parameters is reported in Table 7.2.

The control signals for the switches have been generated on-chip by means of digital circuitry [182, 183] driven by an external master clock of frequency f_c. The multiplier performs $f_c/4$ multiplications per second since the frequency of the four phases ϕ_1 to ϕ_4 is $f_c/4$. The nominal clock frequency is

Table 7.1. Sizes of the transistors for the circuit shown in Fig. 7.5. All the quantities are in µm.

Squarer:
$M_1 = 36/4$, $M_2 = 36/4$, $M_3 = 36/4$
Adj. Mirror:
$M_{4b} = 4/8$, $M_{4a} = 16/30$, $M_5 = 16/30$
Memory 1:
$M_6 = 28/16$, $M_7 = 16/30$
Memory 2:
$M_8 = 28/16$, $M_9 = 16/30$
Additional Current Source: $M_{10} = 4/18$
Switches: 4/4

$f_c = 400\,\text{kHz}$. Indeed, it has been experimentally verified that the maximum frequency at which the circuit works without appreciable performance degradation (1.5% FS) is $f_c = 1.7\,\text{MHz}$, corresponding to 0.425 million multiplications per second. The input currents have been provided by means of off-chip V-I converters. In particular two Howland circuits have been used [184]. The output current of the multiplier is measured by feeding a $2.2\,\text{k}\Omega$ grounded resistor. The power supply is $\pm 3\,\text{V}$ and the power consumption with zero inputs is $0.3\,\text{mW}$ (essentially constant until $f_c = 1.7\,\text{MHz}$).

The multiplier's ideal transfer characteristic is $I_{\text{out}} = 5000 A^{-1}\, I_x \cdot I_y$. The measured input current ranges from $-35\,\text{mA}$ to $35\,\text{mA}$ with a maximum output current of $\pm 6\,\text{mA}$. It has also been verified that the multiplier still

Table 7.2. Performance of the tested chip

Output function (ideal)	$I_{\text{out}} = 5000 A^{-1}\, I_x \cdot I_y$
Input range	$\pm 35\,\mu\text{A}$
Max output current	$\pm 6\,\mu\text{A}$
Internal trim error ϵ (full scale)	1.0% at $f_c = 400\,\text{kHz}$ 1.5% at $f_c = 1.7\,\text{MHz}$
THD ($I_x = 35\,\mu\text{A}$, $I_y = 35\,\mu\text{A}\,\sin(2\pi 50)$)	1.73%
Accuracy vs. supply	0.2%/%
SNR	50 dB
Output offset	200nA
x-feedthrough	< 200nA
y-feedthrough	200nA
$-3\,\text{dB}$ small-signal bandwidth	200 kHz at $f_c = 1.7\,\text{MHz}$
Full power response	150 kHz at $f_c = 1.7\,\text{MHz}$
Max clock frequency f_c	1.7 MHz
Max throughput	0.420 MOPS at $f_c = 1.7\,\text{MHz}$
Die area	$225 \times 250\,\mu\text{m}^2$
Power consumption	0.3 mW
Power supply	$\pm 3\,\text{V}$

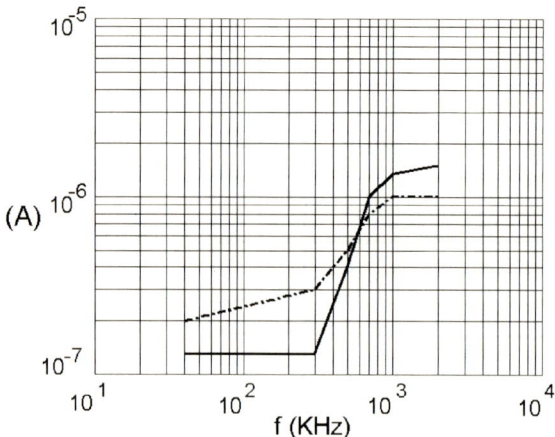

Fig. 7.11. x-feedthrough (dashed line) and y-feedthrough (solid line) versus clock frequency

Finally, the percentage of the variation of the THD at full scale for a variation of the power supply is 0.213%/% (i.e. $-13\,\mathrm{dB}$ %/%) while the *Signal to Noise Ratio* (SNR) is 50 dB (here, as for the measurement of the THD, this is the worst case: the signal is fed to y while x fixes the gain).

Some die micro-photographs of the multiplier prototype are shown in Fig. 7.12. The whole area, including the references for biasing the circuit is $225 \times 250\,\mu\mathrm{m}^2$.

7.5 Summary

The design, analysis and experimental results of a S^2I switched-current multiplier have been presented. A comprehensive analysis to understand the sources of offsets and nonlinearities of our circuit has been performed.

It has been found that by appropriately setting current-mirror gain the nonlinearity can effectively be canceled. An IC prototype was fabricated using MOSIS n-well 2 μm technology. Experimental results are consistent with theoretical findings.

This kind of multiplier is used to realize the programmable synapses of the cellular neural network that will be discussed in the next chapter.

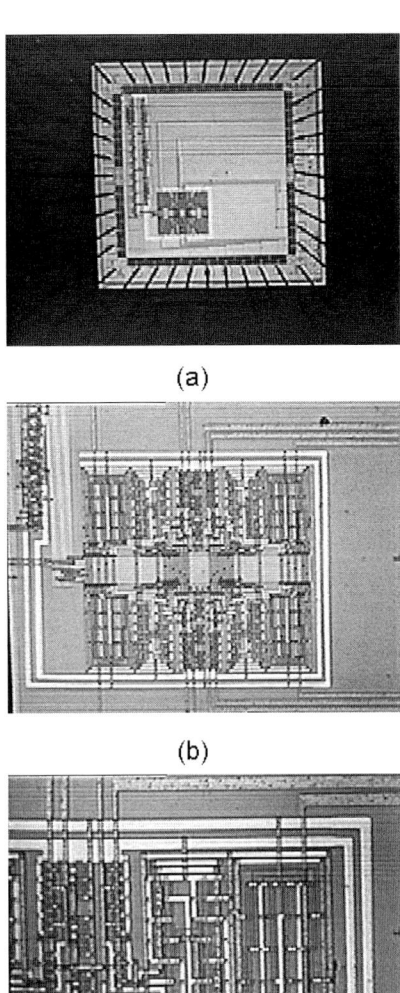

Fig. 7.12. Die micro-photographs of the multiplier prototype. (**a**) The whole chip, (**b**) detail of four identical multipliers, (**c**) one multiplier with references and bias circuitry

8. A One-Dimensional Discrete-Time CNN Chip for Audio Signal Processing

Although most of the CNN applications and corresponding VLSI implementations involve two-dimensional arrays, one-dimensional cellular neural networks have recently attracted increasing attention. For the latter, quite different applications have been reported, ranging from 1-D signal processing [153, 185, 186] to instrumentation and control [187]. In particular a 1-D CNN architecture able to emulate the behavior of FIR filters and to perform the *Daubechies Wavelet Transform* [188] has been thoroughly discussed in [185, 186].

Furthermore, unlike the case of 2-D CNNs for image processing, the applications reported for 1-D CNNs require only a very reduced number of cells [153, 185, 186]. This chapter presents a VLSI implementation of a new 1-D discrete-time programmable CNN suitable for the above mentioned applications [189]. It is based on the well-known S^2I technique.

Moreover, the introduction of a time-multiplexing scheme allows a more efficient use of the hardware.

8.1 System Architecture

A block diagram of the proposed architecture is shown in Fig. 8.1.

The main blocks are an analog shift register (i.e. a tapped delay line) and a set of locally connected cells. A cascade of eight delay elements (sr0–sr7) composes the shift register. Input data enters the shift register on the left (through sr0) and the samples are shifted to the right, passing from sr0 to sr1 and so on. The cells of the CNN receive their inputs from the shift register and from the outputs of the neighbors. The cloning templates are provided externally making the proposed architecture programmable. The state equation describing the behavior of the cell at position c is

$$x_c(n+1) = \sum_{d \in N(c)} A_{c,d} y_d(n) + B_{c,d} u_d(n), \qquad (8.1)$$

where $x_c(n+1)$ is the updated state of the cell, $y_c(n) = f(x_c(n))$ is its output, $u_c(n)$ is the output of the shift register srC (e.g. u_1 is the output of sr1), $A_{c,d}$

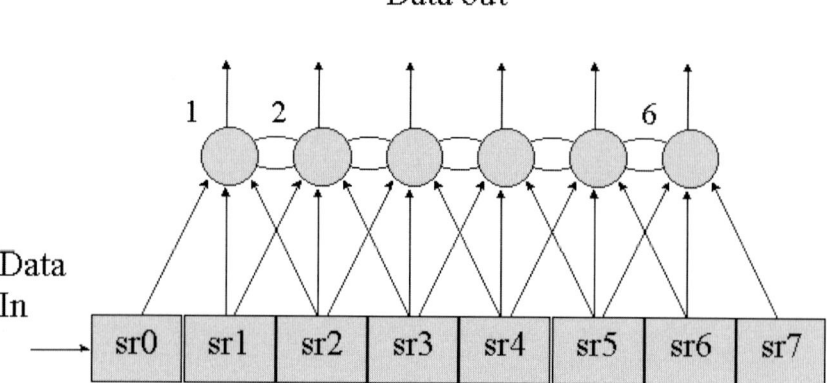

Fig. 8.1. Block diagram of the proposed architecture

are the feedback templates while $B_{c,d}$ are the control templates, $N(c)$ is the neighbor set of cell c defined as follows:

$$N(c) = \{d \ : \ |d - c| \leq 1\} \ . \tag{8.2}$$

As shown in Fig. 8.1, the proposed architecture includes six CNN cells and eight delay cells (the self-feedback is implicit in the cells of Fig. 8.1 and it has not been shown simply to avoid clutter).

In practice, in the real implementation, the first delay stage sr0 is not necessary and the input ("Data In" on Fig. 8.1) can be directly provided in place of it.

8.2 The Tapped Delay Line

A cascade of S^2I delay cells[1] composes the analog shift register. A full delay is obtained by cascading two S^2I half-delay cells whose circuit diagram and corresponding switch control signals can be seen in Fig. 7.1.

A replica of the current at the output of the cell is needed for the next delay cell and for any other circuit requiring it as input. Therefore, a current mirror with multiple output branches is placed in-between every full-delay block. In particular, cascode current mirrors have been used in order to obtain very high output impedance. The circuit of a shift register cell is shown in Fig. 8.2.

An impedance is placed between the output and the input node of the half-delay cells in Fig. 8.2. It represents an NMOST identical to the ones

---

[1] See Sect. 7.1 in Chap. 7.

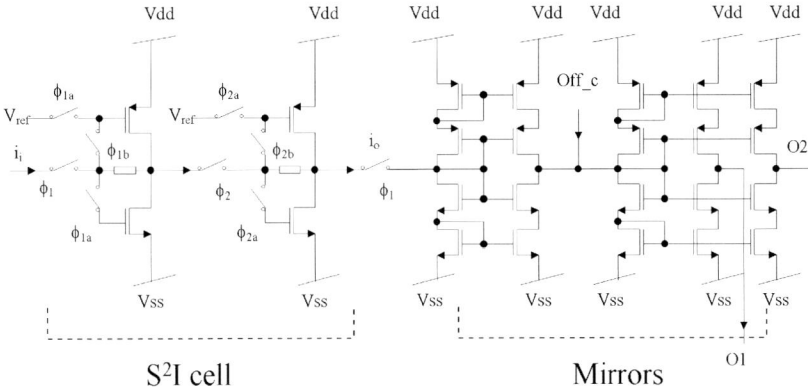

Fig. 8.2. Analog shift register cell

used for the switches with its gate connected to the positive power supply. It compensates for the *ir* drop due to the output switch [172].

In Sect. 7.1 it was proved that the transfer function of a S²I half-delay cell, when loaded by another identical one, is approximately

$$I_o(z) = I_i(z)\frac{H_i(z)}{1+\frac{2g_0}{\Psi g_m}} + I_{\text{off1}}(z), \tag{8.3}$$

where $H_i(z) = -z^{-1/2}$ is the ideal transfer function of a half-delay cell, $I_{\text{off1}}(z)$ represents a constant output offset $\Psi = g_m/2g_0$ is the voltage gain of an inverter in which M_C acts as driver while M_F acts as current source.

However, when the S²I half-delay cell is loaded by a conductance equal to g_m then the transfer function is approximately given by

$$I_o(z) = I_i(z)\frac{H_i(z)}{1+2\frac{g_0}{g_m}} + I_{\text{off2}}(z), \tag{8.4}$$

where $I_{\text{off2}}(z)$ represents a constant output offset very close to $I_{\text{off1}}(z)$. Therefore, due to the sign inversion, when two of these cells are cascaded in order to obtain a full delay, the two offsets practically cancel each other. In our case, the first half-delay cell is loaded by the second one, while the current mirror loads the second one. Therefore the total transfer function of the two blocks in cascade is given by

$$G(z) = \frac{z^{-1}}{1+\frac{2g_0}{g_m}\left(1+\frac{1}{\Psi}\right)+\frac{4g_0^2}{\Psi g_m^2}} \cong \frac{z^{-1}}{1+2\frac{g_0}{g_m}}. \tag{8.5}$$

This implies that the delay cell slightly attenuates the delayed signal. This attenuation could be minimized by making use of cascode transistors in the half-delay cell [190]. However, in the proposed design, it has been preferred to compensate it by choosing a suitable current gain for the current mirror following the delay cell. In fact, from a *Monte Carlo analysis*, the latter solution has shown a better robustness to parameter variation. Finally, in order to compensate a possible residual output offset current, another delay cell with zero input has been created. This cell's output current is the residual offset. This is copied and subtracted from the output of all the other delay cells by means of cascode mirrors (it is added to the intermediate node of the current mirrors using the terminal Off_c as shown in Fig. 8.2).

This is very similar to the well-known replica circuit technique [177].

The aspect ratios of the transistors used for the S^2I half-delay cells are $(W/L)_{\text{NMOS}} = 16\,\mu\text{m}/30\,\mu\text{m}$ and $(W/L)_{\text{PMOS}} = 28\,\mu\text{m}/16\,\mu\text{m}$ for a quiescent current of $20\,\mu\text{A}$. The aspect ratios of the transistors of the first mirror are $(W/L)_{\text{NMOS}} = 5\,\mu\text{m}/4\,\mu\text{m}$ and $(W/L)_{\text{PMOS}} = 23\,\mu\text{m}/4\,\mu\text{m}$ for the first one and $(W/L)_{\text{NMOS}} = 4\,\mu\text{m}/4\,\mu\text{m}$ and $(W/L)_{\text{PMOS}} = 18\,\mu\text{m}/4\,\mu\text{m}$ for the second one. The quiescent current is approximately $5\mu\text{A}$ and imposes the limits of the dynamic range. The final output branch is identical to the output branch of the first current mirror.

All the switches are NMOSTs with aspect ratio $(W/L)_{\text{NMOS}} = 4\,\mu\text{m}/4\,\mu\text{m}$.

The large sizes adopted are imposed by the reliability of the available technology. The layout of the tapped delay line is shown in Fig. 8.3. The linear current-to-voltage converters are shown on the right-hand side; they provide off-chip observation points.

8.3 CNN Cells

The cell's behavior is characterized by the state equation (8.1). The output of the cell is:

$$y_c(n) = f(x_c(n)) = \frac{1}{2}\bigg(|x_c(n)+1| - |x_c(n)-1|\bigg) \tag{8.6}$$

Two cascode current mirrors shown in Fig. 8.4 easily obtain this.

The bias current I_{dd} of the NMOSTs working as cascode current sources fixes the saturation current. Therefore:

1. if $|i_i| < I_{\text{dd}}$ then $i_0 = i_i$,
2. if $i_i \geq I_{\text{dd}}$ the transistors of the input branch of the first current mirror turn off and so $i_0 = I_{\text{dd}}$,
3. if $i_i \leq -I_{\text{dd}}$ the transistors of the input branch of the second current mirror will now turn off making $i_0 = I_{\text{dd}}$.

The sizes of the transistors are analogous to those used for the cascoded current placed between the shift register cells. However it is worth noting that, while in the case of the analog delay line only one variable is passed through

Fig. 8.3. Layout of the delay line

the stages, here the value of the state can be much larger in magnitude because of the many contributions arising from (8.1). This means that the mirrors in the shift register will never clip the signal, provided that the input range satisfies the allowed limits. Conversely the output function will saturate when the bound of $I_{dd} = 5\mu A$ is exceeded, as desired.

8.3.1 Multiplier and Ancillary Circuitry

The multiplier used to implement the synaptic connections is the one discussed in Chap. 7. However, the actual implementation has a much smaller

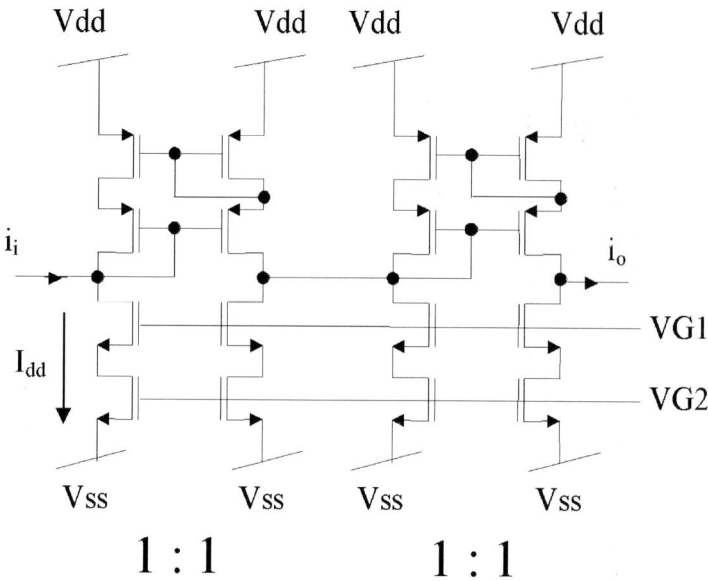

Fig. 8.4. Nonlinear output function circuit

area. Specifically, the size of the the transistors constituting the S^2I memory cells is four times smaller (each W and L is half as large as that reported in Chap. 7).

This means that the aspect ratio and hence the quiescent current are unchanged. However, a decrease in accuracy has to be expected since the smaller cell is more sensitive to process variations, size inaccuracies, and uncompensated charge injection[2].

On the other hand, the total available area of the chip (1.8mm×1.8mm) fixes a limit to the number of modules that can be placed on it. A trade-off is implied.

To avoid confusion in the remainder of this chapter, we will refer to the four main phases of the multiplier (called ϕ_1–ϕ_4 in Chap. 7) as θ_1–θ_4.

In order to bias the current squarer, the multiplier's input node must be kept at $V_{dd}/2 = 1.5V$. Therefore a common gate amplifier was used as level shifter. Moreover, to have the same order of magnitude for the two operands of the multiplier, a current amplifier is placed at the signal input. The layout of one complete multiplier, the corresponding level shifter and other circuitry is shown in Fig. 8.5.

On-chip linear $I-V$ converters allow the external observation of the data in the delay line. These are based on the square-law characteristic of the MOS transistor in saturation [180].

---

[2] This although effectively minimized through the S^2I technique, can never be fully canceled.

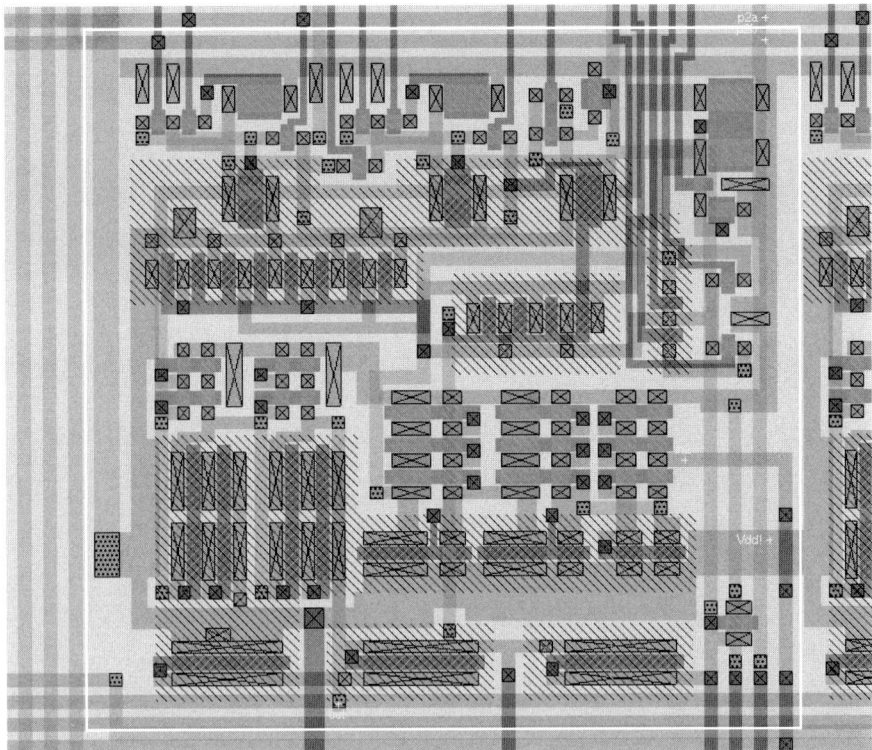

Fig. 8.5. Layout of the multiplier and its ancillary circuitry

The phases and control signal for the switches are obtained by on-chip digital circuitry. In particular an eight-stage dynamic ring counter generates eight non-overlapping phases. These are eventually fed to combinatorial and sequential logic (including logic gates, J-K master slave flip-flops and other cross-coupled latches) to obtain the required signals. The latter are driven throughout the chip's digital buses and re-generated at regular intervals by line-drivers. The layout of the whole digital part of the chip is shown in Fig. 8.6.

The final layout of the chip is shown in Fig. 8.7. The long horizontal strip at the top is the digital circuitry. The long vertical strip on the right-hand side is the tapped delay line. The large rectangular area in the middle contains the abutted multipliers of the six cells. Finally the nonlinearities and the additional S^2I memories completing the cells are shown on the left-hand side.

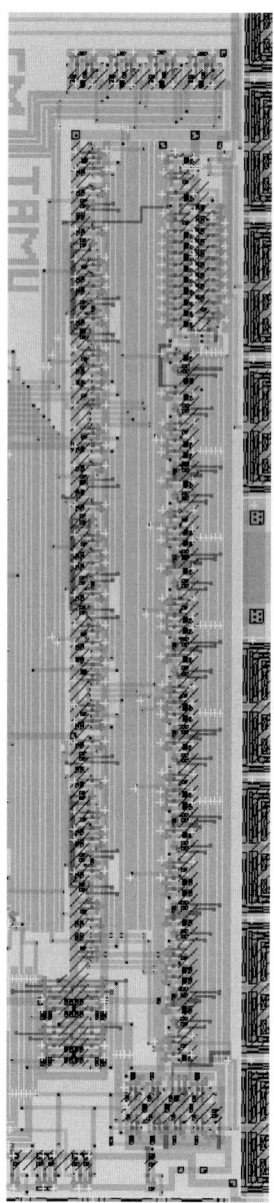

Fig. 8.6. Digital circuits for the generation of the control phases

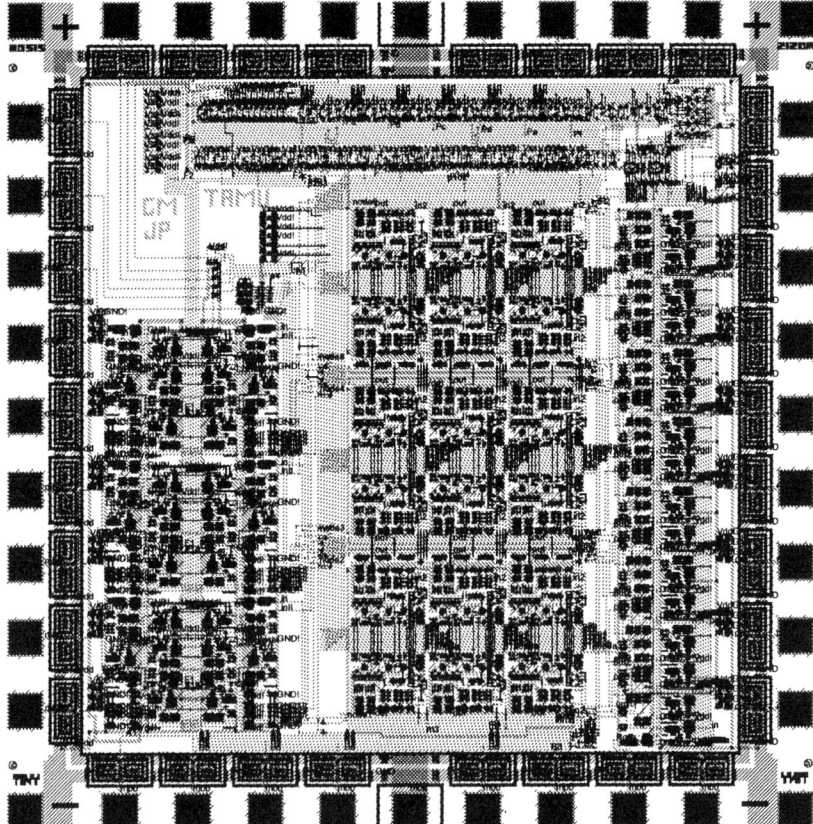

Fig. 8.7. Final layout of the cellular neural network chip

8.4 Cell Behavior and Hardware Multiplexing

From what was discussed in the above sections two remarks can be made. First of all, the shift register cells allow new data to enter and retrieve the stored data only during ϕ_1.

Data is internally exchanged between the two half-delay cells during ϕ_2. The full delay cell is completely isolated by the rest of the system in this phase. During this second phase the rest of the hardware would be essentially idle. These two phases determine the time synchronization represented by the time index of the state equation (8.1).

Secondly, the result of the multiplication process is required before the end of a clock cycle. Therefore, we now outline a strategy that permits one to exploit the available hardware during the idle phase and to make savings in terms of area.

Let us refer to the block diagram shown in Fig. 8.8 depicting a CNN cell.

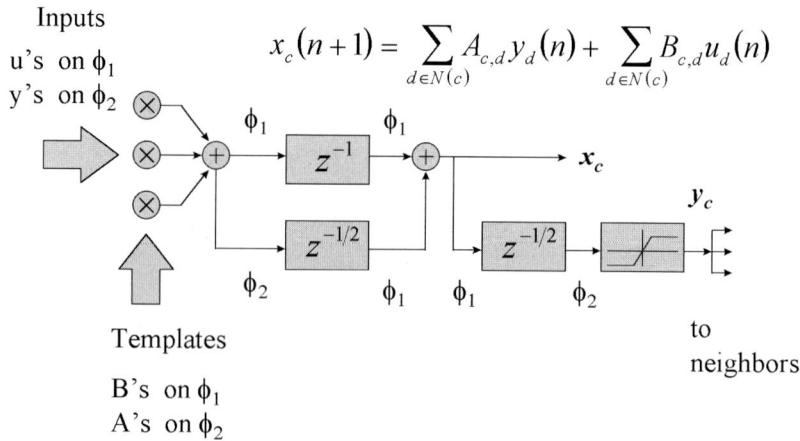

Fig. 8.8. Cell block diagram

The programmable synaptic connections using the multipliers are drawn on the left-hand side of the figure. These multipliers accept the outputs of the shift register u_d during ϕ_1 with the corresponding weights (namely the control templates $B_{c,d}$) provided by off chip currents.

During ϕ_2, however, the neighbor outputs y_d are fed into the inputs of the multipliers in place of u_d, while the feedback template $A_{c,d}$ is fed instead of the control template $B_{c,d}$ [3].

In other words, during ϕ_1 the weighted sum of the shift register outputs $\sum_{d \in N(c)} B_{c,d} u_d(n+1)$ is present on the summing node at the output of the multipliers.

This sum enters the full delay cell. At the same time, the previous value of this sum is present at the output of the delay cell (because it entered the delay cell during the previous period n). Moreover, let us assume that the output of the half-delay cell, depicted below the full delay cell, is providing the weighted sum $\sum_{d \in N(c)} A_{c,d} y_d(n)$ at its output. (this assumption will soon be proven).

Therefore, the state $x_c(n+1)$ is obtained at the summing node of the two previous outputs, in accordance with the state equation (8.1). This enters the second half-delay cell. Its output however is zero and thus so is the output of the nonlinearity.

On ϕ_2 the weighted sum $\sum_{d \in N(c)} B_{c,d} u_d(n+1)$ is stored in the full delay cell. It is passing from the first half-delay cell to the second half-delay cell which constitutes the full delay cell itself. It is therefore completely isolated from the rest of the system. The cells outputs y_d and the feedback templates $A_{c,d}$ are fed to the synapses. So the weighted sum $\sum_{d \in N(c)} A_{c,d} y_d(n+1)$ is

---

[3] The alternate inputs are obtained by on-chip current-steering switches controlled by ϕ_1 and ϕ_2.

fed to the input of the half-delay cell on the left-hand side of the figure. This value is available at its output during ϕ_1 of the next clock period (namely during period $n + 2$).

This proves the above assumption about the output of this half-delay cell.

During ϕ_2 there are no currents at the summing node on the right of the full delay cell. The half-delay cell on the left-hand side releases its stored value (namely $x_c(n+1)$) and hence, due to the nonlinear block, the output $y_c(n+1)$ of the cell is available. This is consistent with the fact that the outputs $y_d(n+1)$ are provided as inputs of the cells during this phase. The above approach allow us to

1. exploit the rest of the hardware during the idle phase ϕ_2 of the shift register and
2. use only three multipliers instead of six, saving on area and power.

To obtain all of this, only two half-delay cells have been added to the classical CNN cell architecture. From this scheme it can be seen that while the outputs of the delay cells are available during the whole corresponding phases, the sampling itself is accomplished only during $\phi_{1a} - \phi_{1b}$ (or $\phi_{2a} - \phi_{2b}$ for other half-delay cells) corresponding to θ_4. A whole multiplication cycle is performed during ϕ_1 and another one during ϕ_2.

8.5 Results and Example

A prototype chip has been designed and fabricated using the CMOS N-well MOSIS Orbit 2μ technology. HSPICE simulation results, at transistor and functional level, are presented here together with corresponding experimental results. Let us first consider some transistor level simulation results.

As a first simulation scenario a sinusoidal input is fed into the input of the tapped delay line and shifted along it. The outputs of the on-chip linear $I - V$ converters corresponding to the delay stages sr1, sr2, sr3 and sr7 respectively, are shown in Fig. 8.9.

The initial values present in the delays before the first sample reaches the stage can also be noticed.

Two of these voltages are observed at the oscilloscope image shown in Fig. 8.10.

Figure 8.11 shows the node voltage at the summing node where all the multiplier contributions are added up (refer also to the block diagram of Fig. 8.8) under the condition that $A_{c,d} = 0$, $B_{c,d} = \delta_{c,d} \cdot (u(t + \tau) + 1) \cdot k$, where $\delta_{c,d}$ is the *Dirac delta* function, $u(t)$ the *Heaviside function* and $k \in \mathbb{R}$ a scale factor.

It is clearly seen how the abrupt change in the control template value at a certain time τ implies an abrupt change in the amplitude of the signal.

The two input currents of one of the multipliers are shown in Fig. 8.12. One of the two inputs is fed alternately with B and A while the other one is fed

200   8. A Discrete-Time CNN Chip for Audio Signal Processing

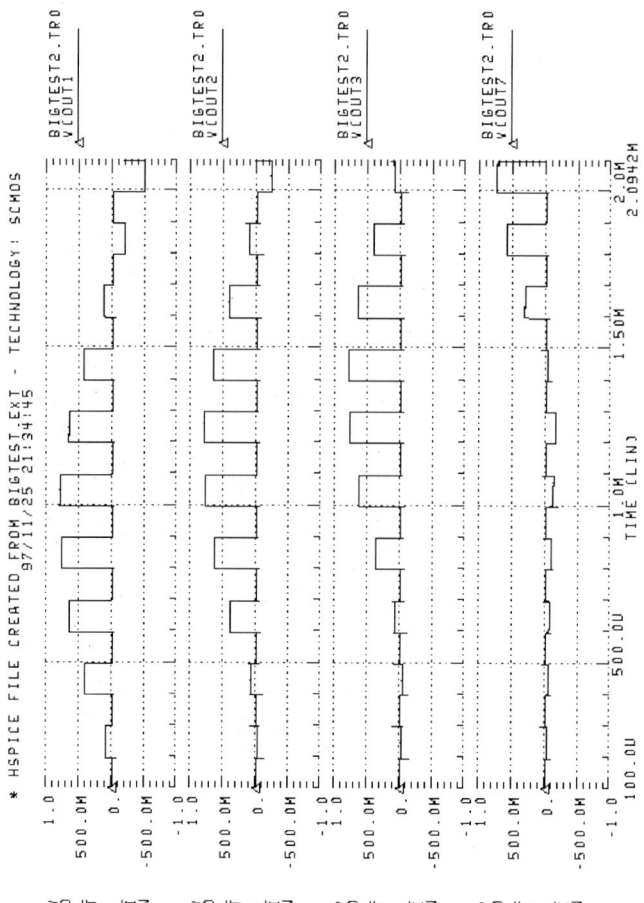

Fig. 8.9. Tapped delay line outputs for sr1, sr2, sr3, and sr7

with u and y. The different signals on the distinct phases can be distinguished. It is seen that during phase ϕ_1 the corresponding B template coefficient is

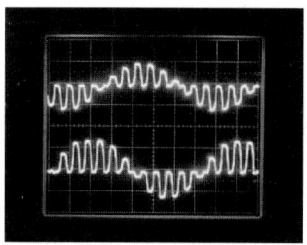

Fig. 8.10. Tapped delay line outputs for sr6 (*upper trace*) and sr2 (*lower trace*)

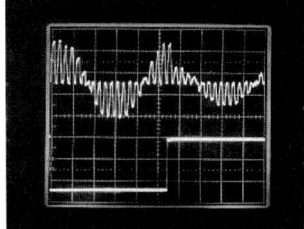

Fig. 8.11. Node voltage at the summing node of one of the cells (*upper trace*) and corresponding control template (*lower trace*)

provided at one of the inputs while the output current coming from the delay line (u) is fed to the other one.

In addition, the 4 phases θ_1–θ_4 of the multiplier are distinguished by the fact that one of the inputs (the template coefficient) enters during $\theta_1+\theta_2$ while the other input is accepted in θ_1 and θ_4. Conversely, the template coefficient A is provided during phase ϕ_2 both with the corresponding output of the cell.

We now give a practical example of the functionality of the proposed architecture. The voice of the author has been sampled at 8kHz. White noise has been added to yield a (4000 samples) noisy signal with signal-to-noise ratio SNR=1.1965 .

A SPICE macromodel [191] of the chip was simulated to process a wavelet decomposition of this signal (top of Fig. 8.13) according to the algorithm described in [185, 186].

The wavelet coefficients are $\phi = [d_0\ d_1\ d_2\ d_3]$ with $d_0 = (1+\sqrt{3})/4\sqrt{2}$, $d_1 = (3+\sqrt{3})/4\sqrt{2}$, $d_2 = (3-\sqrt{3})/4\sqrt{2}$, $d_3 = (1-\sqrt{3})/4\sqrt{2}$.

The corresponding control templates for the algorithm are $B_1 = [d_0\ d_1\ 0]$ and $B_2 = [d_3\ d_4\ 0]$ (see [186] for a detailed description of the algorithm).

The filtered signal (bottom of Fig. 8.13) has a SNR=6.4288, an improvement of 14.6043dB.

In Fig. 8.13 the time shift between the two signals cannot be seen at this level of magnification. The noise reduction, however, is visible, particularly where the low frequency components of the vocal signal are dominant (at the beginning, in the middle, and at the end).

Finally, a square wave is fed into the input of the chip and is low-pass filtered to retrieve the dominant harmonics. The filtered output is shown in the upper trace of the oscilloscope photo in Fig. 8.14. It is seen how the output waveform is smooth and essentially reduced to the fundamental sinusoidal component of the original square wave. The lower trace of the oscilloscope, on the other hand, shows the activity at the summing node of the corresponding cell.

It is seen that the corresponding node voltage is composed of an interlaced pattern of inputs (weighted by the control template B) and feedback signals

202   8. A Discrete-Time CNN Chip for Audio Signal Processing

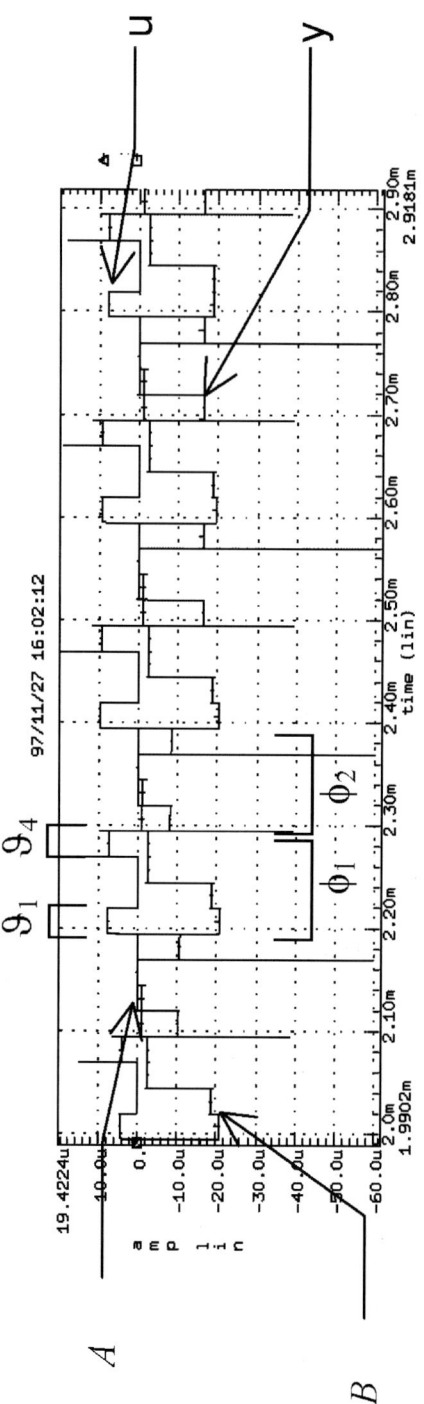

Fig. 8.12. Inputs of one of the multipliers

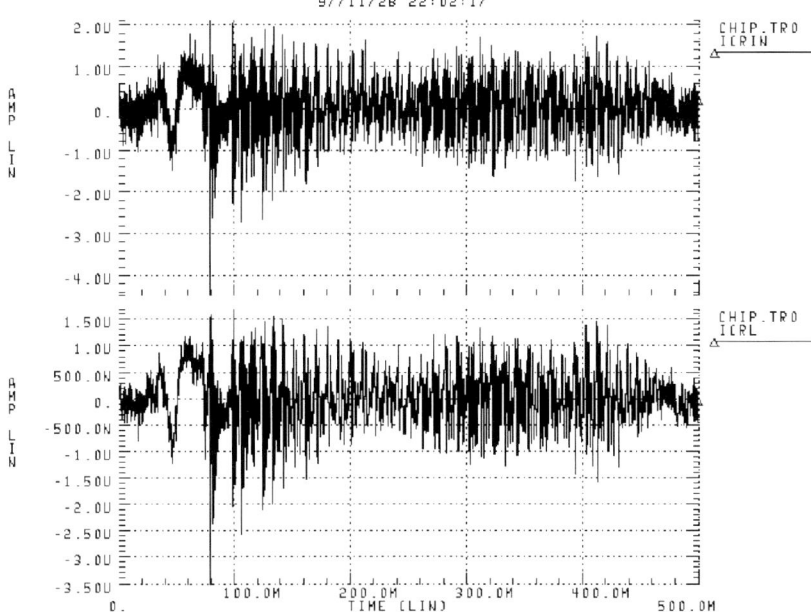

Fig. 8.13. Noisy vocal signal (*top*) and filtered output (*bottom*)

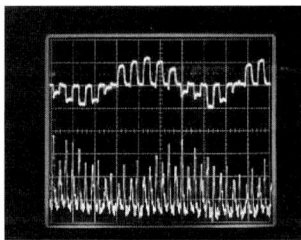

Fig. 8.14. Output of one of the cells (*upper trace*) and corresponding voltage at the summing node (*lower trace*)

(weighted by the feedback template A). In fact, the same set of multipliers are working, alternately, during phases ϕ_1 and ϕ_2 in the multiplexing scheme explained before.

8.6 Summary

The VLSI implementation of a discrete-time one-dimensional cellular neural network has been discussed. One of the peculiarities of the proposed architecture is a hardware-multiplexing strategy. This allows one to use the hardware efficiently, halving the number of multipliers and storing the intermediate results in temporary memories.

Simulation results at transistor and functional levels have been reported together with some experimental results.

A. Mathematical Background

This appendix summarizes the relevant mathematical background and provides convenient source for quick reference to known facts. It is not a tutorial and is far from being complete. It is assumed that the reader is fairly familiar with this material. Moreover linear system theory is assumed to be known and will not be presented.

The topics discussed are covered in detail in many textbooks on ordinary differential equations (ODEs) and also in the specialized literature [51,67,76, 77, 192–194].

In the following the notation "iff" stands for "if and only if". Moreover, the open ball with centre x and radius $\epsilon > 0$ is denoted by $B_\epsilon(x)$ or by $V(x)$ without explicit mention of the radius.

A.1 Topology

Definition A.1.1 (Topological space). *A topological space is a set X together with a collection of subsets of X, called open sets, satisfying the axioms:*

1. *The empty set $\emptyset$ and X are open sets.*
2. *The union of any family of open sets is an open set.*
3. *The intersection of finitely many open sets is an open set.*

Definition A.1.2 (Closed set). *A subset A of a topological space X is called closed if the complement $X - A$ is open.*

Definition A.1.3 (Interior). *The interior of a subset $A \subset X$, denoted $\overset{\circ}{A}$, is the largest (possibly empty) open subset of X which is contained in A. A is open iff $A = \overset{\circ}{A}$.*

Definition A.1.4 (Boundary). *The boundary of a subset $A \subset X$, denoted ∂A, is the set of points of A which are not in the interior of A. Namely $\partial A \doteq A - \overset{\circ}{A}$.*

Definition A.1.5 (Closure). *The closure of A in X, denoted by $\bar{A}$, is the smallest closed subset of X which contains A. A is closed iff $A = \bar{A}$.*

Definition A.1.6 (Dense subset). *A subset $A \subset X$ is called dense in X if $\bar{A} \equiv X$.*

Definition A.1.7 (Hausdorff space). *A space X is called Hausdorff, if*

$$\forall x_1, x_2 \in X,\ x_1 \neq x_2,\ \Rightarrow \exists V(x_1), V(x_2)\ :\ V(x_1) \cap V(x_2) = \emptyset. \qquad (A.1)$$

Definition A.1.8 (Covering). *Let $A \subset X$. A collection $\mathcal{C}$ of subsets of X is said to be a covering of A if A is contained in the union of the elements of $\mathcal{C}$.*

Definition A.1.9 (Compact). *The subset $A \subset X$ is called compact if every covering of A by open subsets of X contains a finite sub-collection covering A.*

Every closed subset of a compact space is compact. Every compact subset of a Hausdorff space is closed.

A.2 Operations and Functions

Definition A.2.1 (Vector field). *A vector field $\boldsymbol{f}\colon U \to \mathbb{R}^n$ is a vector function defined on some subset $U \subseteq \mathbb{R}^m$; $n,m \in \mathbb{N}$.*

Definition A.2.2 (Lipschitz functions). *A vector field $\boldsymbol{f}$ is Lipschitz if*

$$\exists k \in \mathbb{R}, 0 < k < \infty,\ :\ \|\boldsymbol{f}(\boldsymbol{x}) - \boldsymbol{f}(\boldsymbol{x}')\| \leq k\,\|\boldsymbol{x} - \boldsymbol{x}'\|,\quad \forall \boldsymbol{x}, \boldsymbol{x}' \in \mathbb{R}^n. \qquad (A.2)$$

Definition A.2.3 (Affine function). *A function $\boldsymbol{f}\colon U \subseteq \mathbb{R}^m \to \mathbb{R}^n$ that can be written as $\boldsymbol{f}(x) = A\boldsymbol{x} + \boldsymbol{b}$ with $A \in \mathbb{R}^{n \times m}, \boldsymbol{b} \in \mathbb{R}^n$ is called affine. It is a nonlinear function*[1].

Definition A.2.4 (Piecewise-Linear (PWL) fontinuous function). *A Piecewise-Linear (PWL) continuous function is a continuous function that is locally affine.*

Definition A.2.5 (Einstein summation rule). *Let $S \subseteq \Omega$ be a sphere of influence with center $c \in \Omega$ and radius $r > 0$. The Einstein summation rule is defined as*

$$a_d^c y^d \doteq \sum_{\forall d \in S} a(d;c) y^d \qquad (A.3)$$

where $a(d;c)$ are the weights corresponding to the couples $(d,c) \in S^2$.

---

[1] In fact, it is easy to verify that it does not satisfy the superposition principle.

Definition A.2.6 (C^k functions). *A function is C^k if it is k-times differentiable.*

Definition A.2.7 (Diffeomorphism). *A C^k diffeomorphism $\boldsymbol{f}\colon M \to N$ is a mapping $\boldsymbol{f}$ which is $1-1$, onto, and has the property that both $\boldsymbol{f}$ and $\boldsymbol{f}^{-1}$ are k-times differentiable.*

Definition A.2.8 (Homeomorphism). *A homeomorphism is a C^0 diffeomorphism, i.e. a continuous mapping $\boldsymbol{f}\colon M \to N$ with a continuous inverse.*

A.3 Matrices

Definition A.3.1 (Irreducible matrix). *$A = [A_{ij}] \in \mathbb{R}^{n \times n}$ is irreducible if*

$$\forall i \neq j, \quad \exists \text{ a chain of indices } i = k_0, \ldots, k_m = j : \\ A_{k_r, k_{r-1}} \neq 0, \quad r = 1, 2, \ldots, m\,. \tag{A.4}$$

A.4 Dimension

Only metric dimensions are hereby considered.

Definition A.4.1 (Capacity or fractal dimension). *Let us consider a covering $\mathcal{C}$ of $A \subset \mathbb{R}^n$. Let $N(\epsilon)$ be the minimum number of n-dimensional cubes of side length ϵ needed to cover A. The capacity or fractal dimension[2] of A is defined as*

$$D_{\text{cap}} \doteq \lim_{\epsilon \to 0} \frac{\ln N(\epsilon)}{\ln 1/\epsilon} \tag{A.5}$$

if the limit exists.

Definition A.4.2 (Hausdorff dimension). *Let $\mathcal{C}$ be a covering of $A \subset \mathbb{R}^n$ of n-dimensional cubes of variable edge lengths $\{\epsilon_i\}$. We define the quantity $l_d(\epsilon)$ by*

$$l_d(\epsilon) \doteq \inf_{\mathcal{C}} \sum_i \epsilon_i^d \tag{A.6}$$

where $d \in \mathbb{R}^+$ which is still to be specified. Now let

$$l_d = \lim_{\epsilon \to 0} l_d(\epsilon)\,. \tag{A.7}$$

The Hausdorff dimension of A is the value $d = D_H$ above which $l_d = 0$ and below which $l_d = \infty$.

---
[2] In the event that D_{cap} is not integer.

A.5 Dynamical Systems: Basic Definitions

Definition A.5.1 (State equation of a continuous-time system). *A continuous time dynamical system can be described by a system of ordinary differential equations:*

$$\frac{d\boldsymbol{x}}{dt} \doteq \dot{\boldsymbol{x}} = \boldsymbol{f}(\boldsymbol{x}), \tag{A.8}$$

where $\boldsymbol{x} = \boldsymbol{x}(t) \in \mathbb{R}^n$ *(called the state) is a vector-valued function of an independent variable (usually time) and* $\boldsymbol{f}$ *is a smooth vector field defined on some subset* $U \subseteq \mathbb{R}^n$. *We say that the vector field* $\boldsymbol{f}$ *generates a flow* $\phi_t \colon U \to \mathbb{R}^n$, *where* $\phi_t(\boldsymbol{x}) = \phi(\boldsymbol{x}, t)$ *is a smooth function defined* $\forall \boldsymbol{x} \in U$ *and* t *in some interval* $I = (a, b) \subseteq \mathbb{R}$, *and* ϕ *satisfies (A.8) in the sense that*

$$\frac{d}{dt}\Big(\phi(\boldsymbol{x}, t)\Big)\bigg|_{t=\tau} = \boldsymbol{f}\Big(\phi(\boldsymbol{x}, \tau)\Big), \quad \forall \boldsymbol{x} \in U, \ \tau \in I. \tag{A.9}$$

It should be noted that, in its domain of definition, ϕ_t satisfies the group properties (i) $\phi_0 = \mathrm{id}$, and (ii) $\phi_{t+s} = \phi_t \circ \phi_s$.

In the following, unless stated otherwise, the vector field $\boldsymbol{f}$ will be assumed to be *smooth*, namely C^∞.

Definition A.5.2 (Autonomous/nonautonomous systems). *Systems of the form (A.8), in which the vector field does not contain time explicitly, are called* autonomous.

Otherwise they are called nonautonomous.

Definition A.5.3 (Trajectory). *Often an initial condition* $\boldsymbol{x}(0) = \boldsymbol{x}_0 \in U$ *is given. In this case a solution* $\phi(\boldsymbol{x}_0, t)$ *such that* $\phi(\boldsymbol{x}_0, 0) = \boldsymbol{x}_0$ *is sought. This solution is often written as* $\boldsymbol{x}(\boldsymbol{x}_0, t)$ *or simply* $\boldsymbol{x}(t)$. *The function* $\phi(\boldsymbol{x}_0, \cdot) \colon I \to \mathbb{R}^n$ *defines a solution curve, trajectory, or orbit of the differential equation (A.8) based at* $\boldsymbol{x}_0$.

Since, for an autonomous system, the vector field $\boldsymbol{f}$ is invariant with respect to translations in time, solutions based at times $t_0 \neq 0$ can always be translated to $t_0 = 0$.

Conversely, a nonautonomous n-dimensional system can be transformed to an $(n+1)$-dimensional *autonomous* system by augmenting it with an additional dummy state variable $\theta(t) = t$:

$$\begin{cases} \dot{\boldsymbol{x}}(t) = \boldsymbol{f}\Big(\boldsymbol{x}(t), \theta(t)\Big) \\ \dot{\theta}(t) = 1 \end{cases} \tag{A.10}$$

Definition A.5.4 (State equation of a discrete-time system). *A discrete time dynamical system can be defined by a system of difference equations:*

$$\boldsymbol{x}(k+1) = \boldsymbol{G}(\boldsymbol{x}(k)) \quad \text{or} \quad \boldsymbol{x}_{k+1} = \boldsymbol{G}(\boldsymbol{x}_k) \quad \text{or} \quad \boldsymbol{x} \mapsto \boldsymbol{G}(\boldsymbol{x}) \quad k \in \mathbb{N}, \quad \text{(A.11)}$$

in which $\boldsymbol{G}(\cdot)$ is a nonlinear vector-valued function, called a map. $\boldsymbol{x}(k) = \boldsymbol{x}_k \in \mathbb{R}^n$ is the state, $\boldsymbol{x}(k_0) = \boldsymbol{x}_0$ is the initial condition.

Sometimes it is assumed that $k \in \mathbb{Z}$.

In analogy with the continuous-time counterparts, it is possible to define the flow ϕ_k, the orbit [3] $\phi_k(\boldsymbol{x}_0)$. Finally, it is possible to distinguish between autonomous and nonautonomous systems through the explicit independence/dependence of $\boldsymbol{G}$ from k respectively.

The kth iterate of a point $\boldsymbol{p}$ under $\boldsymbol{G}$ is indicated as $\boldsymbol{G}^k(\boldsymbol{p})$:

$$\boldsymbol{G}^k(\boldsymbol{p}) \doteq \boldsymbol{G}(\boldsymbol{G}^{k-1}(\boldsymbol{p})) = \underbrace{\boldsymbol{G}\left(\boldsymbol{G}\left(\boldsymbol{G}\left(\cdots\left(\boldsymbol{G}\left(\boldsymbol{x}\right)\right)\cdots\right)\right)\right)}_{k \text{ times}} \quad \text{(A.12)}$$

Definition A.5.5 (Contracting and expanding flow). ϕ_t *is contracting if*

$$\|\phi_t(\boldsymbol{x}_0) - \phi_t(\hat{\boldsymbol{x}}_0)\| < \|\boldsymbol{x}_0 - \hat{\boldsymbol{x}}_0\|, \quad \forall \boldsymbol{x}_0 \neq \hat{\boldsymbol{x}}_0, \forall t > 0. \quad \text{(A.13)}$$

It is called expanding otherwise.

A.6 Steady-State Behavior

A trajectory based at $\boldsymbol{x}_0$ settles, possibly after some transient, onto a set of points called a *limit set*. A number of general definitions are now given for the purpose of identifying the possible asymptotic behaviors of nonlinear systems.

Definition A.6.1 (Invariant set). *An invariant set S for a flow ϕ_t or map $\boldsymbol{G}$ on $\mathbb{R}^n$ is a subset $S \subset \mathbb{R}^n$ such that*

$$\phi_t(\boldsymbol{x}) \in S \ (\text{or } \boldsymbol{G}(\boldsymbol{x}) \in S), \quad \forall \boldsymbol{x} \in S, \forall t \in \mathbb{R}. \quad \text{(A.14)}$$

Definition A.6.2 (Non-wandering states). *A state $\boldsymbol{x} \in \mathbb{R}^n$ of the flow ϕ_t (or the map $\boldsymbol{G}$) is called non-wandering if*

$$\forall B_\epsilon(\boldsymbol{x}), \forall T > 0, \exists t > T \ : \ \phi_t(B_\epsilon(\boldsymbol{x})) \cap B_\epsilon(\boldsymbol{x}) \neq \emptyset, \quad \text{(A.15a)}$$

or

$$\forall B_\epsilon(\boldsymbol{x}), \forall K > 0, \exists k > K \ : \ \boldsymbol{G}^n(B_\epsilon(\boldsymbol{x})) \cap B_\epsilon(\boldsymbol{x}) \neq \emptyset. \quad \text{(A.15b)}$$

---

[3] The term *trajectory* is more commonly used in the context of continuous time systems, while the term *orbit* is appropriate to discrete time maps.

Conversely, a point x' that is not non-wandering is called *wandering*.

Definition A.6.3 (Non-wandering set). *The set Ω of all the non-wandering states is the non-wandering set.*

Definition A.6.4 (ω-limit point). *A point p is an ω-limit point of x if*

$$\exists \{\phi_{t_k}(x)\} \;:\; \phi_{t_i}(x) \to p \text{ and } t_i \to \infty. \tag{A.16}$$

Definition A.6.5 (α-limit point). *A point q is an α-limit point of x if*

$$\exists \{\phi_{t_k}(x)\} \;:\; \phi_{t_i}(x) \to q \text{ and } t_i \to -\infty. \tag{A.17}$$

For maps G the t_i are integers.

The α- and ω-limit sets $\alpha(x)$, $\omega(x)$ are the sets of α and ω limit points of x.

Definition A.6.6 (Attracting set). *A closed invariant set $A \subset \mathbb{R}^n$ is called an attracting set if*

$$\exists B_\epsilon(A) \;:\; \phi_t(x) \in B_\epsilon(A),\; \forall t \geq 0 \text{ and } \phi_t(x) \to A \\ \text{as } t \to \infty, \forall x \in B_\epsilon(A). \tag{A.18}$$

Definition A.6.7 (Domain/basin of attraction). *The set $\bigcup_{t \leq 0} \phi_t(B_\epsilon(A))$ is the domain of attraction of A.*

When all the trajectories, and not just those which start in a neighborhood, converge to the attracting set, then the attracting set is a *global attracting set*. A *repelling set* is defined analogously, replacing t by $-t$.

Definition A.6.8 (Attractor). *An attractor is an attracting set which contains a dense orbit.*

In an asymptotically stable linear system the limit set is independent of the initial condition and is unique. So it makes sense to talk about of *the* steady-state behavior. But, a nonlinear system can have different limit sets and, therefore, it can show different asymptotic behaviors. Which of these occurs depends on the initial conditions.

It is almost superfluous to mention that, since non-attracting limit sets cannot be experimentally observed in physical systems, the asymptotic behavior of a real circuit corresponds to motion on an attracting limit set.

A.6.1 Classification of Asymptotic Behavior

Definition A.6.9 (Equilibria). *An equilibrium point or stationary point (fixed point) of a vector field f (a map G) is a state x_q such that $f(x_q) = 0$ ($G(x_q) = x_q$)*

In the state space, the limit set of an equilibria consists of a single non-wandering point x_q. An equilibria is said to have *dimension zero*.

Definition A.6.10 (Periodic point). *A state p is periodic if $\exists 0 < T < \infty : \phi_T(p) = p$.*

An alternative, equally important, definition of a periodic solution is the following:

Definition A.6.11 (Periodic solution). *$\phi_t(p)$ is a periodic solution if*

$$\phi_t(p) = \phi_{t+T}(p), \quad \forall t \in \mathbb{R} \tag{A.19}$$

and some minimal period $T > 0$.

Definition A.6.12 (Cycle). *A periodic orbit which is not a stationary point is called a cycle.*

Definition A.6.13 (Limit cycle). *A limit cycle Γ is an isolated periodic orbit.*

The limit cycle Γ is thus a closed curve in the state space where each point (all are non-wandering ones) is periodically visited with period T. It has *dimension one*.

Definition A.6.14 (Period-K solution). *A subharmonic periodic solution or period-K orbit of a discrete time system is the set*

$$\{x_k, 1 \le k \le K \mid x_k = G^K(x_k)\}. \tag{A.20}$$

Definition A.6.15 (Quasi-periodic solution). *An N-frequency quasiperiodic solution $\phi_t(x)$ is one that can be written as a function of N independent variables and it is periodic in each of these variables with incommensurate frequencies*

$$\phi_t = h(t, t, \ldots, t),$$
$$h(t_1, t_2, \ldots, t_i + T_i, \ldots, t_N) = h(t_1, t_2, \ldots, t_i, \ldots, t_N), \tag{A.21a}$$
$$1 \le i \le N,$$

and

$$\Omega_i \doteq \frac{2\pi}{T_i}, \quad 1 \le i \le N,$$
$$m_1 \Omega_1 + m_2 \Omega_2 + \cdots + m_N \Omega_N = 0, \tag{A.21b}$$

does not hold for any set of integers, $m_1, m_2, \ldots, m_N$ (negative integers are allowed)[4].

---

[4] except for the trivial case $m_1 = m_2 = \cdots = m_N = 0$. In other words, the frequencies Ω_i are linearly independent.

The limit set of a quasi-periodic motion has an integer dimension greater than 1.

The final behavior admitted by in a nonlinear dynamical system is that known as *chaos*. A strict mathematical definition of chaotic behavior is still debated. Chaos is a bounded, low-dimensional, non-wandering motion which exhibits both "randomness" and "order" [68]. It can be defined by negation. Namely: it is an asymptotic behavior that is *not* an equilibria nor a periodic orbit nor a quasi-periodic motion.

Chaos is characterized by some peculiar features such as:

1. Continuous, broad-band, noise-like spectrum of every component of the state vector $x(t)$.
2. Sensitive dependence on initial conditions.
3. Fractional dimension[5] limit set.

The term "strange attractor" is often encountered in descriptions of chaotic (or supposedly chaotic) behavior. However, it has to be noted that it is very difficult to show that a dense orbit exists. In fact, many of the observed "strange attractors" may not be true attractors but merely attracting sets, since they may contain stable periodic orbits [67]. More on chaos and the above enumerated features will be discussed in Sect. A.7.3.

A.7 Stability

Let us first consider the stability of fixed points of vector fields. Maps will be considered later.

A.7.1 Stability of equilibrium points

Definition A.7.1 (Stable equilibrium point). *A fixed point x_q of f is said to be stable if*

$$\forall V(x_q) \subset U, \ \exists V_1 \subset V \ : \ x_0 \in V_1, \ \phi_t(x_0) \subset V, \ \forall t > 0. \tag{A.22}$$

Definition A.7.2 (Asymptotically stable equilibrium point). *A stable equilibria x_q is asymptotically stable if, in addition, V_1 can be chosen such that*

$$\lim_{t \to \infty} x(t) = x_q. \tag{A.23}$$

The two above definitions are *local* because they concern only the behavior near the fixed point x_q. In this case the linearization method can be applied. Let

---

[5] Commonly called *Fractal*.

$$\dot{\boldsymbol{\xi}} = D\boldsymbol{f}(\boldsymbol{x}_q)\boldsymbol{\xi}, \quad \boldsymbol{\xi} \in \mathbb{R}^n \tag{A.24}$$

be the linearized system, where $D\boldsymbol{f} = [\partial f_i/\partial x_i]$ is the Jacobian matrix of $\boldsymbol{f}$ and $\boldsymbol{x} = \boldsymbol{x}_q + \boldsymbol{\xi}$, $|\boldsymbol{\xi}| \ll 1$.

Therefore the linearized flow $D\phi_t(\boldsymbol{x}_q)\boldsymbol{\xi}$ arising from $\dot{\boldsymbol{x}} = \boldsymbol{f}(\boldsymbol{x})$ at the fixed point $\boldsymbol{x}_q$ is obtained by integration of (A.24):

$$D\phi_t(\boldsymbol{x}_q)\boldsymbol{\xi} = e^{tD\boldsymbol{f}(\boldsymbol{x}_q)}\boldsymbol{\xi} \tag{A.25}$$

Let us just briefly recall that, in a linear system, depending on the eigenvalues associated with the fixed point, six possible classes of equilibria can be distinguished:

1. *Attracting sink*: stable equilibria with (some) complex eigenvalues; all eigenvalues have negative real part; it corresponds to a stable under-damped response.
2. *Repelling sink*: unstable equilibria with (some) complex eigenvalues; at least one eigenvalue has positive real part; it corresponds to an unstable under-damped response.
3. *Attracting node*: stable equilibria with (all) real negative eigenvalues; it corresponds to a stable over-damped response.
4. *Repelling node*: unstable equilibria with (all) real positive eigenvalues; it corresponds to an unstable exponentially growing response.
5. *Saddle*: unstable equilibria with (all) real (some positive, some negative) eigenvalues; it corresponds to an unstable exponentially growing response for some, but not all, the components.
6. *Center*: marginally stable[6] equilibria with (all) zero and/or imaginary eigenvalues; it corresponds to undamped sustained oscillations that must not be confused with a limit cycle[7].

The same classes of qualitative behavior can be distinguished in nonlinear systems. Of course it does not make any sense to talk about eigenvalues. It makes sense, however, to consider the eigenvalues of the linearized system near the fixed point (A.24). In particular because of the following theorem.

Theorem A.7.1 (Hartman–Großman). *If $D\boldsymbol{f}(\boldsymbol{x}_q)$ has no zero or purely imaginary eigenvalues then there is a homeomorphism $\boldsymbol{h}$ defined on some neighborhood $V(\boldsymbol{x}_q) \subset \mathbb{R}^n$ locally taking orbits of the nonlinear flow ϕ_t of $\dot{\boldsymbol{x}} = \boldsymbol{f}(\boldsymbol{x})$, to those of the linear flow $e^{tD\boldsymbol{f}(\boldsymbol{x}_q)}$ of (A.24). The homeomorphism preserves the sense of orbits and can also be chosen to preserve parametrization by time.*

Definition A.7.3 (Hyperbolic equilibria). *The equilibria $\boldsymbol{x}_q$ is called hyperbolic or non-degenerate if $D\boldsymbol{f}(\boldsymbol{x}_q)$ has no eigenvalues with zero real part.*

---

[6] Generally acknowledged to be unstable.
[7] The amplitude of a limit cycle oscillation does not depend on the initial conditions. But that of a center does.

214   A. Mathematical Background

The stability of a degenerate equilibrium point cannot be determined by linearization. Hence, a saddle or a center in a nonlinear system, being degenerate fixed points, cannot be simply identified by linearization.

Definition A.7.4 (Local manifolds). *The local stable and unstable manifolds,*
$W_{\text{loc}}^{\text{s}}(\boldsymbol{x}_q)$ *and* $W_{\text{loc}}^{\text{u}}(\boldsymbol{x}_q)$ *of the equilibria* $\boldsymbol{x}_q$ *are the sets:*

$$W_{\text{loc}}^{\text{s}}(\boldsymbol{x}_q) \doteq \sec \boldsymbol{x} \in U \mid \lim_{t \to \infty} \boldsymbol{\phi}_t(\boldsymbol{x}) = \boldsymbol{x_q} \text{ and } \boldsymbol{\phi}_t(\boldsymbol{x}) \in U \; \forall t \geq 0,$$
$$W_{\text{loc}}^{\text{u}}(\boldsymbol{x}_q) \doteq \sec \boldsymbol{x} \in U \mid \lim_{t \to -\infty} \boldsymbol{\phi}_t(\boldsymbol{x}) = \boldsymbol{x_q} \text{ and } \boldsymbol{\phi}_t(\boldsymbol{x}) \in U \; \forall t \leq 0,$$
(A.26)

where $U \subset \mathbb{R}^n$ *is a neighborhood of the fixed point* $\boldsymbol{x}_q$.

The invariant manifolds $W_{\text{loc}}^{\text{s}}(\boldsymbol{x}_q)$ and $W_{\text{loc}}^{\text{u}}(\boldsymbol{x}_q)$ provide nonlinear analogues of the flat stable and unstable eigenspaces E^{s}, E^{u} of the linear systems.

Theorem A.7.2 (Stable manifold theorem for a fixed point). *Let $\boldsymbol{x}_q$ be a hyperbolic fixed point of $\dot{\boldsymbol{x}} = \boldsymbol{f}(\boldsymbol{x})$. Then there exist local stable and unstable manifolds $W_{\text{loc}}^{\text{s}}(\boldsymbol{x}_q)$, $W_{\text{loc}}^{\text{u}}(\boldsymbol{x}_q)$, of the same dimensions n_{s}, n_{u} as those of the eigenspaces E^{s}, E^{u} of the linearized system (A.24), and tangent to E^{s}, E^{u} at $\boldsymbol{x}_q$. $W_{\text{loc}}^{\text{s}}(\boldsymbol{x}_q)$, $W_{\text{loc}}^{\text{u}}(\boldsymbol{x}_q)$ are as smooth as the function $\boldsymbol{f}$.*

The local invariant manifolds $W_{\text{loc}}^{\text{s}}(\boldsymbol{x}_q)$, $W_{\text{loc}}^{\text{u}}(\boldsymbol{x}_q)$ have global analogs $W^{\text{s}}(\boldsymbol{x}_q)$, $W^{\text{u}}(\boldsymbol{x}_q)$ obtained by letting points in $W_{\text{loc}}^{\text{s}}(\boldsymbol{x}_q)$ flow backwards in time and those in $W_{\text{loc}}^{\text{u}}(\boldsymbol{x}_q)$ flow forwards.

Definition A.7.5 (Global manifolds). *The global stable and unstable manifolds, $W^{\text{s}}(\boldsymbol{x}_q)$ and $W^{\text{u}}(\boldsymbol{x}_q)$ of the equilibria $\boldsymbol{x}_q$ are the sets*

$$W^{\text{s}}(\boldsymbol{x}_q) \doteq \bigcup_{t \leq 0} \boldsymbol{\phi}_t(W_{\text{loc}}^{\text{s}}(\boldsymbol{x}_q)),$$
$$W^{\text{u}}(\boldsymbol{x}_q) \doteq \bigcup_{t \geq 0} \boldsymbol{\phi}_t(W_{\text{loc}}^{\text{u}}(\boldsymbol{x}_q)).$$
(A.27)

Existence and uniqueness of solutions of the initial value problem $\dot{\boldsymbol{x}} = \boldsymbol{f}(\boldsymbol{x})$, $\boldsymbol{x}(0) = \boldsymbol{x}_0$ ensure that two stable (or unstable) manifolds of distinct fixed points cannot intersect, nor can $W^{\text{s}}(\boldsymbol{x}_q)$ (or $W^{\text{u}}(\boldsymbol{x}_q)$) intersect itself. However, intersections of stable and unstable manifolds of distinct fixed points or the same fixed point can occur.

Definition A.7.6 (Homoclinic orbit). *A E-homoclinic orbit (or, simply, homoclinic orbit) is a trajectory $\boldsymbol{\Gamma}$, which is asymptotic to a fixed point $\boldsymbol{x}_q$ in both positive and negative time.*

Definition A.7.7 (Heteroclinic connection). *If a nonconstant solution is asymptotic to $\overline{x}_i$ in negative time, and to $\overline{x}_{i+1}$ in positive time, then we have an heteroclinic connection $\Gamma_i \doteq W^{\mathrm{u}}(\overline{x}_i) \cap W^{\mathrm{s}}(\overline{x}_{i+1})$.*

Definition A.7.8 (Heteroclinic orbit). *If there is a loop, which connects m fixed points $\overline{x}_1, \ldots, \overline{x}_m$ in one direction, the common set*

$$\Lambda \doteq \bigcup_{i=1}^{m} (\Gamma_i \cup \overline{x}_i) \tag{A.28}$$

is called a heteroclinic orbit.

Theorem A.7.3 (Lyapunov stability). *Let x_q be a fixed point for $\dot{x} = f(x)$ and $V: W \to \mathbb{R}$ be a differentiable function defined on some neighborhood $W \subseteq U$ of x_q such that*

1. *$V(x_q) = 0$ and $V(x) > 0$ if $x \neq x_q$; and*
2. *$\dot{V}(x) \leq 0$ in $W - \{x_q\}$;*
   then x_q is stable. Moreover if
3. *$\dot{V}(x) < 0$ in $W - \{x_q\}$;*

then x_q is asymptotically stable.

Here

$$\dot{V}(x) = \sum_{j=1}^{n} \frac{\partial V}{\partial x_j} \dot{x}_j = \sum_{j=1}^{n} \frac{\partial V}{\partial x_j} f_j(x) \tag{A.29}$$

is the derivative of V along solution curves of $\dot{x} = f(x)$.

Definition A.7.9 (Completely stable systems). *Let X be the state space of the dynamical system $\dot{x} = f(x)$ and $\hat{x} \in X$ a constant vector. The system $\dot{x} = f(x)$ is completely stable or convergent iff*

$$\lim_{t \to \infty} x(t) = \hat{x}, \quad \lim_{t \to \infty} \dot{x}(t) = 0, \quad \forall x(0) \in X. \tag{A.30}$$

Definition A.7.10 (Globally stable system). *The system $\dot{x} = f(x)$ is said to be globally asymptotically stable or globally convergent iff*

$$\forall x(0) = x_0 \in \mathbb{R}^n, \quad \lim_{t \to \infty} x(t) = \hat{x} \in \mathbb{R}^n \text{ (and } \hat{x} \text{ is unique)} \lim_{t \to \infty} \dot{x}(t) = 0 \tag{A.31}$$

Observe that, in a globally stable system, the fixed point $\hat{x}$ is unique and independent of the initial condition. In other words, its basin of attraction is the whole state space.

Definition A.7.11 (Stability almost everywhere). *The system $\dot{x} = f(x)$ is said stable almost everywhere or almost convergent if the set of initial values in which the system does not converge to a fixed point has zero Lebesgue measure.*

Definition A.7.12 (Complete instability almost everywhere). *The system $\dot{x} = f(x)$ is said completely unstable almost everywhere if $\forall x(0) = x_0$ (except possibly a set of zero Lebesgue measure) does not converge to an equilibrium.*

Definition A.7.13 (Cooperative systems). *The system $\dot{x} = f(x)$, $f \in C^1$ is said to be cooperative iff the off-diagonal elements of the Jacobian matrix $J = Df(x)$ are non-negative, i.e. $J_{ij} = \partial f_i/\partial x_j \geq 0$, $i \neq j$.*

Definition A.7.14 (Irreducible systems). *The system $\dot{x} = f(x)$, $f \in C^1$ is said to be irreducible iff the Jacobian matrix $J = Df(x)$ is irreducible $\forall x$.*

Let us now consider the discrete time maps. All the concepts discussed for the vector fields can easily be generalized to maps with little formal modification. It is almost redundant to recall that, as regards the eigenvalues corresponding to fixed points of linear systems, the stable region of the complex plane, for maps, is the unit circle centered on the origin. Therefore it is not the sign of the real part of the eigenvalues that is going to determine the stability, but the modulus of the eigenvalues. Hence, for example, the definition of *hyperbolic* fixed point (A.7.3) is modified as follows:

Definition A.7.15 (Hyperbolic fixed point). *A fixed point x_q for G (i.e. $G(x_q) = x_q$) is said to be hyperbolic if $DG(x_q)$ has no unit modulus eigenvalues.*

Analogously there exists a theory for diffeomorphisms parallel to that for flows. In particular there are analogous Hartman-Großman and invariant manifolds theorems.

Theorem A.7.4 (Hartman–Großman). *Let $G\colon \mathbb{R}^n \to \mathbb{R}^n$ be a (C^1) diffeomorphism with a hyperbolic fixed point x_q. Then there exists a homeomorphism h defined on some neighborhood $U(x_q)$ such that $h(G(\xi)) = DG(x_q)h(\xi), \forall \xi \in U$.*

Definition A.7.16 (Local manifolds). *The local stable and unstable manifolds, $W^s_{\text{loc}}(x_q)$ and $W^u_{\text{loc}}(x_q)$ of the fixed point x_q are the sets:*

$$W^s_{\text{loc}}(x_q) \doteq \left\{ x \in U : \lim_{n \to +\infty} G^n(x) = x_q, \text{ and } G^n(x) \in U, \forall n \geq 0 \right\}$$

$$W^u_{\text{loc}}(x_q) \doteq \left\{ x \in U : \lim_{n \to +\infty} G^{-n}(x) = x_q, \text{ and } G^{-n}(x) \in U, \forall n \geq 0 \right\}, \tag{A.32}$$

where $U \subset \mathbb{R}^n$ is a neighborhood of the fixed point x_q.

Theorem A.7.5 (Stable manifold theorem). *Let $G\colon \mathbb{R}^n \to \mathbb{R}^n$ be a (C^1) diffeomorphism with a hyperbolic fixed point $\boldsymbol{x}_q$. Then there are local stable and unstable manifolds $W^{\mathrm{s}}_{\mathrm{loc}}(\boldsymbol{x}_q)$, $W^{\mathrm{u}}_{\mathrm{loc}}(\boldsymbol{x}_q)$, tangent to the eigenspaces $E^{\mathrm{s}}_{\boldsymbol{x}_q}$, $E^{\mathrm{u}}_{\boldsymbol{x}_q}$ of $D\boldsymbol{G}(\boldsymbol{x}_q)$ at $\boldsymbol{x}_q$ and of corresponding dimensions. $W^{\mathrm{s}}_{\mathrm{loc}}(\boldsymbol{x}_q)$, $W^{\mathrm{u}}_{\mathrm{loc}}(\boldsymbol{x}_q)$ are as smooth as the map $\boldsymbol{G}$.*

Definition A.7.17 (Global manifolds). *The global stable and unstable manifolds, $W^{\mathrm{s}}(\boldsymbol{x}_q)$ and $W^{\mathrm{u}}(\boldsymbol{x}_q)$ of the equilibria $\boldsymbol{x}_q$ are the sets*

$$W^{\mathrm{s}}(\boldsymbol{x}_q) \doteq \bigcup_{n \geq 0} \boldsymbol{G}^{-n}(W^{\mathrm{s}}_{\mathrm{loc}}(\boldsymbol{x}_q))$$
$$W^{\mathrm{u}}(\boldsymbol{x}_q) \doteq \bigcup_{n \geq 0} \boldsymbol{G}^{n}(W^{\mathrm{u}}_{\mathrm{loc}}(\boldsymbol{x}_q)) \,. \tag{A.33}$$

It is clear that flows and maps differ from the fact that, while the *trajectory* $\phi_t(\boldsymbol{p})$ of a flow is a *curve* in $\mathbb{R}^n$, the *orbit* $\{\boldsymbol{G}^n(\boldsymbol{p})\}$ of a map is a *sequence of points*. Thus, while the invariant manifolds of flows are composed of unions of solution curves, those of maps are unions of discrete orbit points.

Similarly, it is possible to talk about homoclinic and heteroclinic orbits.

A.7.2 Stability of Limit Cycles

The study of the stability of limit cycles can be converted into the study of the stability of fixed points by means of the *Poincaré sections and maps*.

Let $\boldsymbol{\gamma}$ be a periodic orbit of the flow $\phi_t \in \mathbb{R}^n$ arising from a nonlinear vector field $\boldsymbol{f}(\boldsymbol{x})$.

Definition A.7.18 (Poincaré section). *Let us consider a local $(n-1)$-dimensional hypersurface $\Sigma \subset \mathbb{R}^n$. Let $\boldsymbol{n}(\boldsymbol{x})$ be the unit normal to Σ at $\boldsymbol{x}$. Let Σ be a transverse cross section of $\boldsymbol{f}$, i.e. $\boldsymbol{f}(\boldsymbol{x}) \cdot n(\boldsymbol{x}) \neq 0$, $\forall \boldsymbol{x} \in \Gamma$. Γ is called a Poincaré section.*

Definition A.7.19 (Poincaré map). *Let $\boldsymbol{p}$ be the (unique) intersection point of $\boldsymbol{\gamma}$ and Γ ($\boldsymbol{p} = \boldsymbol{\gamma} \cap \Gamma$). Let $U \subset \Gamma$ be some neighborhood of $\boldsymbol{p}$*[8]. *The first return or Poincaré map $P\colon U \to \Gamma$ is defined for a point $\boldsymbol{q} \in U$ by*

$$P(\boldsymbol{q}) \doteq \phi_\tau(\boldsymbol{q}) \,, \tag{A.34}$$

where $\tau = \tau(\boldsymbol{q})$ is the time taken for $\phi_t(\boldsymbol{q})$ to first return to Γ.

Note that τ generally depends upon $\boldsymbol{q}$ and need not to be equal to $T = T(\boldsymbol{p})$, the period of $\boldsymbol{\gamma}$, although $\lim_{\boldsymbol{q} \to \boldsymbol{p}} \tau = T$.

Clearly, $\boldsymbol{p}$ is a fixed point for the map P and the stability of $\boldsymbol{p}$ reflects the stability of $\boldsymbol{\gamma}$ for the flow ϕ_t. If $\boldsymbol{p}$ is hyperbolic, and $DP(\boldsymbol{p})$, the linearized

---
[8] If $\boldsymbol{\gamma}$ has multiple intersections with Γ, then shrink Γ until there is only one intersection.

map, has n_s eigenvalues with modulus less than one and n_u with modulus greater than one ($n_s + n_u = n - 1$), then $\dim W^s(\boldsymbol{p}) = n_s$, and $\dim W^u(\boldsymbol{p}) = n_u$ for the map. Since the orbits of P lying in W^s and W^u are formed by intersections of orbits (solution curves) of ϕ_t with Γ, the dimensions of $W^s(\boldsymbol{\gamma})$ and $W^u(\boldsymbol{\gamma})$ are each one greater than those for the map.

Analogously, closed orbits $\{\boldsymbol{x}_k^*\}_{k=1}^K$ of P with period K correspond to Kth-order sub-harmonics of the underlying dynamical system.

Let $\hat{\boldsymbol{x}}(t) = \hat{\boldsymbol{x}}(t+T)$ be a solution lying on the closed orbit $\boldsymbol{\gamma}$, based at $\boldsymbol{x}(0) = \boldsymbol{p} \in \Sigma$. Consider the linearization around $\boldsymbol{\gamma}$:

$$\dot{\boldsymbol{\xi}} = D\boldsymbol{f}(\hat{\boldsymbol{x}}(t))\boldsymbol{\xi}\,, \tag{A.35}$$

where $D\boldsymbol{f}(\hat{\boldsymbol{x}}(t))$ is an $n \times n$, T-periodic matrix. The corresponding Poincaré map $P(\boldsymbol{x})$ has the fixed point $\boldsymbol{p}$. The linearized map corresponding to (A.35) and Σ has the form

$$\dot{\boldsymbol{\Phi}} = D\boldsymbol{f}(\phi_t(\boldsymbol{x}_0))\boldsymbol{\Phi}\,, \tag{A.36}$$

with solution $\boldsymbol{\Phi}_t(\boldsymbol{x}_0)$ and $\boldsymbol{\Phi}_T(\boldsymbol{p}) = DP(\boldsymbol{p}) = D_{\boldsymbol{x}_0}\phi_T(\boldsymbol{p})$. Observe, in the last expression, the partial derivative with respect to the initial condition. Equation (A.36) is called *the variational equation*. The solution matrix of (A.35) is

$$X(t) = Z(t)e^{tR}, \quad Z(t) = Z(t+T)\,, \tag{A.37}$$

where $X, Z, R \in \mathbb{R}^{n,n}$. In particular we can choose $X(0) = Z(0) = I$, hence

$$X(T) = Z(T)e^{TR} = Z(0)e^{TR} = e^{TR}\,. \tag{A.38}$$

It follows that the behavior of the solutions in the neighborhood of $\boldsymbol{\gamma}$ is determined by the eigenvalues of the constant matrix e^{TR}. These ones are also the eigenvalues of the solution $\boldsymbol{\Phi}_t(\boldsymbol{x}_0)$ of the variational equation (A.36). These eigenvalues, $m_1, \ldots, m_n$, are called the *characteristic (Floquet) multipliers* of $\boldsymbol{\gamma}$. The eigenvalues $\mu_1, \ldots, \mu_n$, of R are the *characteristic exponents* of $\boldsymbol{\gamma}$. One of the Floquet multipliers is always unity[9]. The moduli of the remaining $n - 1$, if none are unity[10], determine the stability of $\boldsymbol{\gamma}$. These ones are independent of the chosen Poincaré map.

In a nonautonomous system, according to (A.10) and completely consistent with definition A.7.19, the Poincaré map can be obtained by a periodic sampling along t of the system trajectory. Furthermore, in this case, the definition is *global* and not local (as in the case of Def. A.7.19). Let

---

[9] Not true in the non-autonomous case.

[10] If there is at least one that lies on the unit circle then an exception similar to the case of the Hartman-Großman theorem with non-hyperbolic points applies. Hence the stability cannot be simply determined by the Floquet multipliers.

$S^1 = \mathbb{R} \pmod T$ be the circular component reflecting the periodicity[11] of the vector field $\boldsymbol{f}$ in θ. The global cross-section is then

$$\Sigma = \{(\boldsymbol{x}, \theta) \in \mathbb{R}^n \times S^1 \;:\; \theta = \theta_0\}, \tag{A.39}$$

while the global Poincaré map $P\colon \Sigma \to \Sigma$ is

$$P(\boldsymbol{x}_0) = \boldsymbol{\pi} \cdot \boldsymbol{\phi}_T(\boldsymbol{x}_0, \theta_0), \tag{A.40}$$

where $\boldsymbol{\phi}_t \colon \mathbb{R}^n \times S^1 \to \mathbb{R}^n \times S^1$ is the flow of (A.10) and $\boldsymbol{\pi}$ denotes projection onto the first factor.

In the case of Kth-order sub-harmonics, the discussion can be repeated by considering the eigenvalues of the solution $DP^K(\boldsymbol{p}_1)$ ($\boldsymbol{p}_1$ is any one of the K periodic points) of the variational equation corresponding to the (Kth iterate) Poincaré map P^K.

Definition A.7.20 (P-homoclinic orbit). *If $\boldsymbol{x}_q$ happens to be a homoclinic fixed point of a discrete map (e.g. Poincaré) corresponding to a periodic orbit γ in the flow $\boldsymbol{\phi}_t$, then the homoclinic orbit $\boldsymbol{\Gamma}$ is a P-homoclinic orbit.*

A.7.3 Lyapunov Exponents

Lyapunov exponents are a generalization of the eigenvalues and the Floquet multipliers. They allow one to characterize the stability of any type of steady-state behavior including quasi-periodic and chaotic motion. Let $\boldsymbol{\Phi}_t(\boldsymbol{x}_0)$ be the solution of the variational equation.

Definition A.7.21 (Lyapunov exponents). *Let $\{m_i(t)\}_{i=1}^n$ be the eigenvalues of $\boldsymbol{\Phi}_t(\boldsymbol{x}_0)$. The Lyapunov exponents are*

$$\lambda_i \doteq \lim_{t \to \infty} \frac{1}{t} \ln |m_i(t)|, \quad i = 1, \dots, n \tag{A.41}$$

if the limit exists[12].

Given that the definition is for a limit $t \to \infty$, any point belonging to the basin of attraction of an attractor has the same Lyapunov exponents[13]. Moreover, it can be seen that the Lyapunov exponents (LEs) reduce to the real part of the eigenvalues in the case of a fixed point. Meanwhile, the following relationship holds with the Floquet multipliers $m_1, \dots, m_n$ of a limit cycle:

---

[11] Usually, T is the period of the forcing input or an integer multiple of it.
[12] lim can be replaced by lim sup to guarantee the existence. However the interpretation of the Lyapunov exponents is correct only when the limit exists.
[13] In the case of some strange attractors, this is true for *almost every point*.

$$\lambda_i = \frac{1}{T} \ln |m_i|, \quad i = 1, \ldots, n. \qquad (A.42)$$

Of course, one of the LEs is zero[14]. In general, for any bounded attractor of an autonomous system, except an equilibrium point, one LE is always zero.

The Lyapunov exponents give the average rate of contraction ($\lambda_i < 0$) or expansion ($\lambda_i > 0$) in a particular direction near a particular trajectory. For the trajectory to remain bounded, the contraction must outweigh expansion, and this implies the property

$$\sum_{i=0}^{n} \lambda_i < 0. \qquad (A.43)$$

Ignoring the case of non-hyperbolic attractors, the following classification based on the LEs holds ($\lambda_1 \geq \lambda_2 \geq \cdots \geq \lambda_n$):

1. *Stable fixed point:* $\lambda_i < 0, \forall i$.
2. *Stable limit cycle:* $\lambda_1 = 0, \lambda_i < 0$ for $i = 2, \ldots, n$.
3. *Stable torus*[15]: $\lambda_1 = \lambda_2 = 0, \lambda_i < 0$ for $i = 3, \ldots, n$.
4. *Stable K-torus:*[16]: $\lambda_1 = \cdots = \lambda_K = 0, \lambda_i < 0$ for $i = K+1, \ldots, n$.
5. *Chaos:* $\lambda_1 > 0$, $\sum_i \lambda_i < 0$.
6. *Hyperchaos:* $\lambda_1, \ldots, \lambda_K > 0$, $\sum_i \lambda_i < 0$.

A.8 Topological Equivalence and Conjugacy, Structural Stability and Bifurcations

Definition A.8.1 (Equivalent maps). *Two C^r maps F and G are C^k equivalent or C^k conjugate ($k \leq r$) if there exists a C^k diffeomorphism h such that $h \circ F = G \circ h$. C^0 equivalence is called topological equivalence.*

Definition A.8.2 (Equivalent vector fields). *Two C^r vector fields f and g are said to be C^k equivalent ($k \leq r$) if there exists a C^k diffeomorphism h which takes orbits $\phi_t^f(x)$ of f to orbits $\phi_t^g(x)$ of g, preserving sense but not necessarily parametrization by time. If h does preserve parametrization by time, then it is called a conjugacy.*

Definition A.8.3 (Structural stability). *A map $F \in C^r(\mathbb{R})$ (or a C^r vector field f) is structurally stable if $\exists \epsilon > 0$ such that all C^1, ϵ perturbations of F (or f) are topologically equivalent to F (or f).*

---

[14] The one corresponding to the unit multiplier.
[15] Quasi-periodic motion.
[16] Quasi-periodic motion as well.

It is clear that a vector field (or map) that has a non-hyperbolic fixed point cannot be structurally stable because any perturbation can remove it or turn it into a hyperbolic one. It is straightforward that the same observation applies to periodic orbits. It comes out that having all hyperbolic fixed points and closed orbits is a necessary but not sufficient condition for the structural stability.

Homoclinic and heteroclinic orbits are not structurally stable.

When, by changing one of the system parameters, the dynamic system undergoes an abrupt qualitative change of behavior (e.g. a stable fixed point becomes unstable and a stable limit cycle appear), we say that the system undergoes a *bifurcation*. More precisely, the value of the parameter $\epsilon = \epsilon_0$ for which the system is not structurally stable is called the *bifurcation value* of ϵ, where ϵ is the so-called *bifurcation parameter*.

There are a certain number of known bifurcations: *Hopf*, *saddle-node*, *period-doubling*. These are called *local* bifurcations because they may be understood by linearization [67].

A.9 Šilnikov Method

In this section, the Šilnikov method is considered in its restriction to three-dimensional dissipative continuous systems with homoclinic trajectories.

Definition A.9.1 (Šilnikov map). *A Poincaré map defined in a neighbor U of a homoclinic trajectory $\mathcal{H}$ based at the fixed point $\boldsymbol{x}_q$ and such that $\boldsymbol{x}_q \notin U$ is called the Šilnikov map.*

Theorem A.9.1 (Šilnikov theorem). *Consider the third order autonomous system*

$$\dot{\boldsymbol{x}} = \boldsymbol{f}(\boldsymbol{x}), \tag{A.44}$$

where $\boldsymbol{f}$ is a C^2 vector field on $\mathbb{R}^3$. Let $\boldsymbol{x}_q$ be a fixed point for (A.44) and suppose that:

1. *$\boldsymbol{x}_q$ is a saddle focus, whose eigenvalues of the corresponding linearized system are of the form*

   $$\gamma, \sigma \pm j\omega, \qquad \gamma, \sigma, \omega \in \mathbb{R} \tag{A.45a}$$

   with $\omega \neq 0$, and satisfy the Šilnikov inequality

   $$|\gamma| > |\sigma| > 0. \tag{A.45b}$$

2. *There exists a homoclinic trajectory $\mathcal{H}$ based at $\boldsymbol{x}_q$.*

Then:

1. *The Šilnikov map defined in a neighbor of $\mathcal{H}$ possesses a countable number of Smale horseshoes in its discrete dynamics.*
2. *For any sufficiently small C^1-perturbation $\boldsymbol{g}$ of $\boldsymbol{f}$, the perturbed system*

$$\dot{\boldsymbol{x}} = \boldsymbol{g}(\boldsymbol{x}) \tag{A.46}$$

   has at least a finite number of Smale horseshoes in the discrete dynamics of the Šilnikov map defined near $\mathcal{H}$.
3. *Both the unperturbed system (A.44) and the perturbed system (A.46) exhibit horseshoe chaos.*

Some important points deserve mention. First of all, results (1) and (2) imply the structural stability of the horseshoe chaos. In other words, chaos persists in spite of perturbations, although homoclinic trajectories are not structurally stable.

Secondly, if $|\gamma| \leq |\sigma|$ then chaos is extinguished. Therefore $|\sigma| = \gamma$ is the bifurcation point between regular and chaotic behavior. In fact, perhaps the most difficult aspect of this method is the proof of existence of $\mathcal{H}$.

Finally, the Šilnikov method has been extended to PWL C^2 vector fields for which:

1. $\boldsymbol{x}_q$ is in the interior of one of the domains into which the state space can be partitioned according to the PWL nonlinearity;
2. $\mathcal{H}$ is bounded away from all other fixed points and it is not tangent to any of the boundary surfaces.

A.10 Particular Results for Two-Dimensional Flows

The nature of admissible solutions for two-dimensional (planar) flows is rather limited. In fact, the choice of possible limit sets is restricted to fixed points, cycles, homoclinic, and heteroclinic orbits only. This inherent relative simplicity allows some further general results to hold. The topic, however, is far from trivial. For instance, Andronov and co-workers [194] and Hirsch and Smale [77] have written well over a thousand of pages on this subject.

Theorem A.10.1 (Poincaré–Bendixon theorem). *A nonempty compact ω-limit or α-limit set of a planar flow, which contains no fixed points, is a closed orbit.*

Very useful is also the following well-known criterion.

Theorem A.10.2 (Negative Bendixon's criterion). *If on a simply connected region $D \subseteq \mathbb{R}^n$ the divergence of the vector field*

$$\frac{\partial f_1}{\partial x_1} + \frac{\partial f_2}{\partial x_2} \quad (A.47)$$

is not identically zero and does not change sign, then the system $\dot{\boldsymbol{x}} = \boldsymbol{f}(\boldsymbol{x})$ has no closed orbits lying entirely in D.

A generalization of this is

Theorem A.10.3 (Negative Dulac's criterion). *If a continuous function $B(\boldsymbol{x})$ with continuous derivatives exists, such that in simply connected region $D \subseteq \mathbb{R}^n$ the expression*

$$\frac{\partial(Bf_1)}{\partial x_1} + \frac{\partial(Bf_2)}{\partial x_2} \quad (A.48)$$

is not identically zero and does not change sign, then the system $\dot{\boldsymbol{x}} = \boldsymbol{f}(\boldsymbol{x})$ has no closed orbits lying entirely in D.

A limit cycle can appear as a consequence of a *Hopf Bifurcation*:

Theorem A.10.4 (Hopf bifurcation theorem). *Let us consider the dynamic system*

$$\dot{\boldsymbol{x}} = \boldsymbol{f}(\boldsymbol{x}, \mu), \quad (A.49)$$

where $\boldsymbol{x} \in \mathbb{R}^2$, $\boldsymbol{f} \in C^k$ with $k \geq 4$, while $\mu \in \mathbb{R}$ is a system parameter. Suppose that (A.49) has an equilibrium point at the origin $\forall \mu$. Furthermore, suppose the eigenvalues $\lambda_1(\mu)$ and $\lambda_2(\mu)$ are purely imaginary for $\mu = \mu_0$. If the real part of the eigenvalues, $\mathrm{Re}\{\lambda_1(\mu)\}$, satisfies

$$\left. \frac{d}{d\mu}\left(\mathrm{Re}\{\lambda_1(\mu)\}\right) \right|_{\mu=\mu_0} > 0 \quad (A.50)$$

and the origin is an asymptotically stable equilibrium point for $\mu = \mu_0$ then:

1. *$\mu = \mu_0$ is a bifurcation point of the system;*
2. *for $\mu \in (\mu_1, \mu_0)$ and some $\mu_1 < \mu_0$, the origin is a stable focus;*
3. *for $\mu \in (\mu_0, \mu_2)$ and some $\mu_2 > \mu_0$, the origin is an unstable focus surrounded by a stable limit cycle, whose size increases with μ.*

B. Library of Templates

In this appendix the most frequently used CNN templates are presented. For each template, its name, the problem it solves (indicated as "function"), a flow chart (where needed) and the template numerical values are reported. The sources used are: [195], [12], [17], [18], [196], [197], [198].

Name: AVERAGE
Function: Spatial averaging of pixel intensities over the $r = 1$ convolutional window with binary output.

$$A = \begin{pmatrix} 0 & 1 & 0 \\ 1 & 2 & 1 \\ 0 & 1 & 0 \end{pmatrix}, \quad B = \begin{pmatrix} 0 & 0 & 0 \\ 0 & 0 & 0 \\ 0 & 0 & 0 \end{pmatrix}, \quad I = 0. \tag{B.1}$$

Name: AND
Function: Logical "AND" function of the input and the initial state pictures.

$$A = \begin{pmatrix} 0 & 0 & 0 \\ 0 & 1.5 & 0 \\ 0 & 0 & 0 \end{pmatrix}, \quad B = \begin{pmatrix} 0 & 0 & 0 \\ 0 & 1.5 & 0 \\ 0 & 0 & 0 \end{pmatrix}, \quad I = -1. \tag{B.2}$$

Name: BANK-NOTE RECOGNITION
Function: This algorithm identifies Canadian banknotes on color images. The bank-notes can be placed into the frame with arbitrary shift and rotation. The algorithm finds the green and the black circles on the bank-notes. It analyses its color, shape, and size. The algorithm can be separated into three parts. These parts are indicated in the flow-chart. The templates can be found in this template library.
This chart, shown in Fig. B.1, contains only half of the algorithm, and finds only one circle. The other one can be found with a similar method, but with different parameters in the color filtering.

Name: BLACK
Function: Drives the whole network into black.

$$A = \begin{pmatrix} 0 & 0 & 0 \\ 0 & 2 & 0 \\ 0 & 0 & 0 \end{pmatrix}, \quad B = \begin{pmatrix} 0 & 0 & 0 \\ 0 & 0 & 0 \\ 0 & 0 & 0 \end{pmatrix}, \quad I = 4. \tag{B.3}$$

226 B. Library of Templates

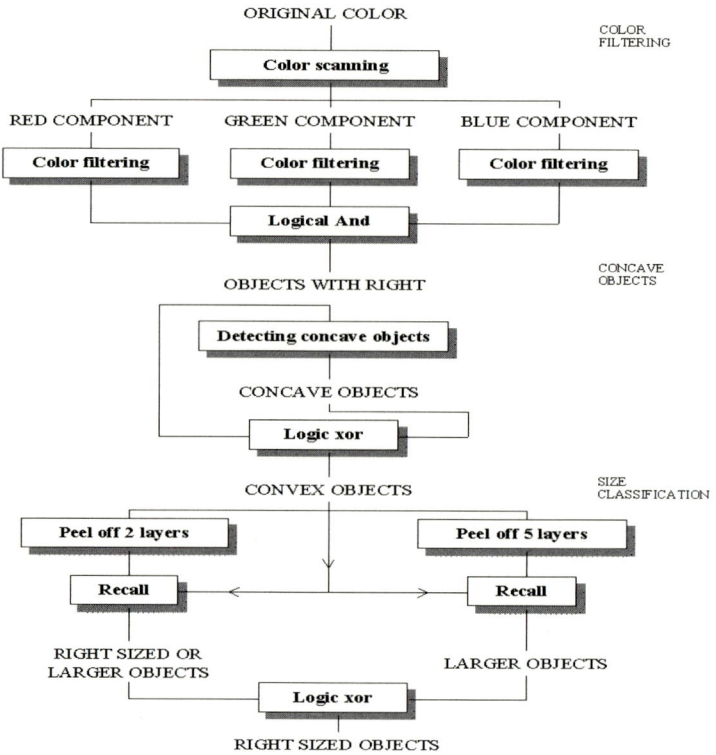

Fig. B.1. Flow-chart of the algorithm for the template for bank-note recognition

Name: BLACK AND WHITE SKELETONIZATION
Function: The algorithm finds the skeleton of a black-and-white object. The 8 templates should be applied circularly, always feeding the output back to the input before using the next template.

$$A1 = [3] \ , \ B1 = \begin{pmatrix} 0.25 & 0.25 & 0 \\ 0.25 & -0.25 & -0.25 \\ 0 & -0.25 & 0 \end{pmatrix} , \ I1 = -0.75 \tag{B.4}$$

$$A2 = [3] \ , \ B2 = \begin{pmatrix} 0.25 & 0.25 & 0.25 \\ 0 & -0.25 & 0 \\ -0.125 & -0.25 & -0.125 \end{pmatrix} , \ I2 = -0.85 \tag{B.5}$$

$$A3 = [3] \ , \ B3 = \begin{pmatrix} 0 & 0.25 & 0.25 \\ -0.25 & -0.25 & 0.25 \\ 0 & -0.25 & 0 \end{pmatrix} , \ I3 = -0.75 \tag{B.6}$$

$$A4 = [3] \quad , \quad B4 = \begin{pmatrix} -0.125 & 0 & 0.25 \\ -0.25 & -0.25 & 0.25 \\ -0.125 & 0 & 0.25 \end{pmatrix} , \quad I4 = -0.75 \quad \text{(B.7)}$$

$$A5 = [3] \quad , \quad B5 = \begin{pmatrix} 0 & & \\ -0.25 & 0 & \\ -0.25 & -0.25 & 0.25 \\ 0 & 0.25 & 0.25 \end{pmatrix} , \quad I5 = -0.75 \quad \text{(B.8)}$$

$$A6 = [3] \quad , \quad B6 = \begin{pmatrix} -0.125 & -0.25 & -0.125 \\ 0 & -0.25 & 0 \\ 0.25 & 0.25 & 0.25 \end{pmatrix} , \quad I6 = -0.85 \quad \text{(B.9)}$$

$$A7 = [3] \quad , \quad B7 = \begin{pmatrix} 0 & -0.25 & 0 \\ 0.25 & -0.25 & -0.25 \\ 0.25 & 0.25 & 0 \end{pmatrix} , \quad I7 = -0.75 \quad \text{(B.10)}$$

$$A8 = [3] \quad , \quad B8 = \begin{pmatrix} 0.25 & 0 & -0.125 \\ 0.25 & -0.25 & -0.25 \\ 0.25 & 0 & -0.25 \end{pmatrix} , \quad I8 = -0.85. \quad \text{(B.11)}$$

Name: CCD DIAG
Function: Diagonal connected component detector.

$$A = \begin{pmatrix} 1 & 0 & 0 \\ 0 & 2 & 0 \\ 0 & 0 & -1 \end{pmatrix} , \quad B = \begin{pmatrix} 0 & 0 & 0 \\ 0 & 0 & 0 \\ 0 & 0 & 0 \end{pmatrix} , \quad I = 0. \quad \text{(B.12)}$$

Name: CCD HOR
Function: Horizontal connected component detector.

$$A = \begin{pmatrix} 0 & 0 & 0 \\ 1 & 2 & -1 \\ 0 & 0 & 0 \end{pmatrix} , \quad B = \begin{pmatrix} 0 & 0 & 0 \\ 0 & 0 & 0 \\ 0 & 0 & 0 \end{pmatrix} , \quad I = 0. \quad \text{(B.13)}$$

Name: CCD VERT
Function: Vertical connected component detector.

$$A = \begin{pmatrix} 0 & 1 & 0 \\ 0 & 2 & 0 \\ 0 & -1 & 0 \end{pmatrix} , \quad B = \begin{pmatrix} 0 & 0 & 0 \\ 0 & 0 & 0 \\ 0 & 0 & 0 \end{pmatrix} , \quad I = 0. \quad \text{(B.14)}$$

Name: CENTER POINT DETECTION
Function: The algorithm identifies the center point of the black-and-white input object. This is always a point of the object halfway between the farthestmost points of it. Here a time-varying DTCNN template is given, each element of which should be used for a single step. It is easily transformed to a continuous-time network.

$$A1 = \begin{pmatrix} 0.25 & 0 & 0 \\ 0.25 & 1.75 & -0.25 \\ 0.25 & 0 & 0 \end{pmatrix} , \quad I1 = -1 \tag{B.15}$$

$$A2 = \begin{pmatrix} 0.25 & 0.25 & 0.25 \\ 0.25 & 1.75 & 0 \\ 0.25 & 0 & -0.25 \end{pmatrix} , \quad I2 = -0.5 \tag{B.16}$$

$$A3 = \begin{pmatrix} 0.25 & 0.25 & 0.25 \\ 0 & 1.75 & 0 \\ 0 & -0.25 & 0 \end{pmatrix} , \quad I3 = -1.0 \tag{B.17}$$

$$A4 = \begin{pmatrix} 0.25 & 0.25 & 0.25 \\ 0 & 1.75 & 0.25 \\ -0.25 & 0 & 0.25 \end{pmatrix} , \quad I4 = -0.5 \tag{B.18}$$

$$A5 = \begin{pmatrix} 0 & 0 & 0.25 \\ -0.25 & 1.75 & 0.25 \\ 0 & 0 & 0.25 \end{pmatrix} , \quad I5 = -1.0 \tag{B.19}$$

$$A6 = \begin{pmatrix} -0.25 & 0 & 0.25 \\ 0 & 1.75 & 0.25 \\ 0.25 & 0.25 & 0.25 \end{pmatrix} , \quad I6 = -0.5 \tag{B.20}$$

$$A7 = \begin{pmatrix} 0 & -0.25 & 0 \\ 0 & 1.75 & 0 \\ 0.25 & 0.25 & 0.25 \end{pmatrix} , \quad I7 = -1.0 \tag{B.21}$$

$$A8 = \begin{pmatrix} 0.25 & 0 & -025 \\ 0.25 & 1.75 & 0 \\ 0.25 & 0.25 & 0.25 \end{pmatrix} , \quad I8 = -0.5 \tag{B.22}$$

Name: SOME COLOR VISION PHENOMENA: SINGLE AND DOUBLE OPPONENCIES

In the retina and in the visual cortex there are single and double color opponent cells. Their receptive fields are shown in Fig. B.2, where (a) belongs to single and (b) belongs to the double opponent cell. The template simulating the single opponent cell has two layers. The input of the first layer is the monochromatic red map, while the second layer gets the green map. The result appears on the second layer. The template is the following:

$$B12 = \begin{pmatrix} 0 & 0 & 0 \\ 0 & 2 & 0 \\ 0 & 0 & 0 \end{pmatrix} , \quad B22 = \begin{pmatrix} -0.25 & -0.25 & -0.25 \\ -0.25 & 0 & -0.25 \\ -0.25 & -0.25 & -0.25 \end{pmatrix} , \tag{B.23}$$

By swapping the layers we get the template that generates the G+R− single opponents. The output of the R+G− and G+R− layers provide the input for the first and the second layer of the double opponent structure respectively. The output appears on the second layer. The template is as follows:

B. Library of Templates   229

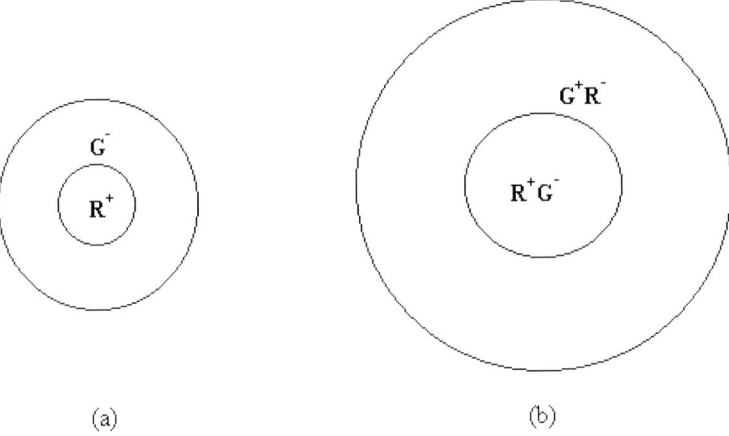

Fig. B.2a,b. Receptive field scheme

$$A = \begin{pmatrix} 0&0&0&0&0&0&0 \\ 0&0&0&0&0&0&0 \\ 0&0&0&0&0&0&0 \\ 0&0&0&0&0&0&0 \\ 0&0&0&2&0&0&0 \\ 0&0&0&0&0&0&0 \\ 0&0&0&0&0&0&0 \\ 0&0&0&0&0&0&0 \\ 0&0&0&0&0&0&0 \end{pmatrix}, \tag{B.24}$$

$$B = \begin{pmatrix} 0.02&0.02&0.02&0.02&0.02&0.02&0.02 \\ 0.02&0&0&0&0&0&0.02 \\ 0.02&0&0&0&0&0&0.02 \\ 0.02&0&0&0&0&0&0.02 \\ 0.02&0&0&2&0&0&0.02 \\ 0.02&0&0&0&0&0&0.02 \\ 0.02&0&0&0&0&0&0.02 \\ 0.02&0&0&0&0&0&0.02 \\ 0.02&0.02&0.02&0.02&0.02&0.02&0.02 \end{pmatrix}. \tag{B.25}$$

Name: CONCCONT
Function: Concentric contour detection for DTCNN: the templates give rise to a set of alternating black and white rings running from the boundary to the interior.

$$A = \begin{pmatrix} 0 & -1 & 0 \\ -1 & 3.5 & -1 \\ 0 & -1 & 0 \end{pmatrix}, \quad B = \begin{pmatrix} 0&0&0 \\ 0&4&0 \\ 0&0&0 \end{pmatrix}, \quad I = -4. \tag{B.26}$$

230   B. Library of Templates

Name: CONTOUR
Function: Grey-scale contou detector.

$$A = \begin{pmatrix} 0 & 0 & 0 \\ 0 & 2 & 0 \\ 0 & 0 & 0 \end{pmatrix}, \quad B = \begin{pmatrix} a & a & a \\ a & a & a \\ a & a & a \end{pmatrix}, \quad I = 0.7, \tag{B.27}$$

where

$$a = \begin{cases} 0.5 & \text{if } (V_{xij} - V_{xkl}) < -0.18 \\ -1 & \text{if } -0.18 < (V_{xij} - V_{xkl}) < 0.18 \\ 0.5 & \text{if } (V_{xij} - V_{xkl}) > 0.18 \end{cases}.$$

Name: CONTOUR EXTRACTION II
Function: Grey-scale contour detector.

$$A = \begin{pmatrix} 0 & 0 & 0 \\ 0 & 0 & 0 \\ 0 & 0 & 0 \end{pmatrix}, \quad B = \begin{pmatrix} 0 & -1 & 0 \\ -1 & 4.5 & -1 \\ 0 & -1 & 0 \end{pmatrix}, \quad I = 0.25. \tag{B.28}$$

Name: CORNER
Function: Convex corner detector.

$$A = \begin{pmatrix} 0 & 0 & 0 \\ 0 & 2 & 0 \\ 0 & 0 & 0 \end{pmatrix}, \quad B = \begin{pmatrix} -0.25 & -0.25 & -0.25 \\ -0.25 & 2 & -0.25 \\ -0.25 & -0.25 & -0.25 \end{pmatrix}, \quad I = -3. \tag{B.29}$$

Name: CUT7V
Function: Cuts two pixels from the top and the bottom of everything, and if the height of the remaining segment is less than 4 pixels, it deletes it. In other words, all vertical lines not longer than 7 pixels are deleted.

$$A = \begin{pmatrix} 0 & 0 & 1 & 0 & 0 \\ 0 & 0 & 0.5 & 0 & 0 \\ 0 & 0 & 2 & 0 & 0 \\ 0 & 0 & 0.5 & 0 & 0 \\ 0 & 0 & 1 & 0 & 0 \end{pmatrix}, \quad B = \begin{pmatrix} 0 & 0 & 1 & 0 & 0 \\ 0 & 0 & 1 & 0 & 0 \\ 0 & 0 & 1 & 0 & 0 \\ 0 & 0 & 1 & 0 & 0 \\ 0 & 0 & 1 & 0 & 0 \end{pmatrix}, \quad I = -5.5. \tag{B.30}$$

Name: DELDIAG1
Function: Deletes one pixel wide diagonal lines.

$$A = \begin{pmatrix} 0 & 0 & 0 \\ 0 & 2 & 0 \\ 0 & 0 & 0 \end{pmatrix}, \quad B = \begin{pmatrix} -0.25 & 0 & -0.25 \\ 0 & 0 & 0 \\ -0.25 & 0 & -0.25 \end{pmatrix}, \quad I = -2. \tag{B.31}$$

Name: DELVERT1
Function: Deletes one pixel vertical lines.

$$A = \begin{pmatrix} 0 & 0 & 0 \\ 0 & 1 & 0 \\ 0 & 0 & 0 \end{pmatrix}, \ B = \begin{pmatrix} 0 & -0.25 & 0 \\ 0 & 0 & 0 \\ 0 & -0.25 & 0 \end{pmatrix}, \ I = -1.5. \tag{B.32}$$

Name: DEPTH CLASSIFICATION

Function: The algorithm determines the depth of black-and-white objects based on a pair of stereo images. It determines whether an object is closer than a given distance or not. The first step of the algorithm is to reduce the objects in both input images to a single pixel; then the distance between corresponding points is calculated; this distance can be a threshold to determine whether the object is too close or not. As a first preprocessing step, grey-scale images can be converted to black-and-white using the AVERAGE template. The flow-chart of the algorithm is reported in Fig. B.3.

List of the templates used:

Elongate objects: add pixels to the top and bottom of each object (use the left image as input)

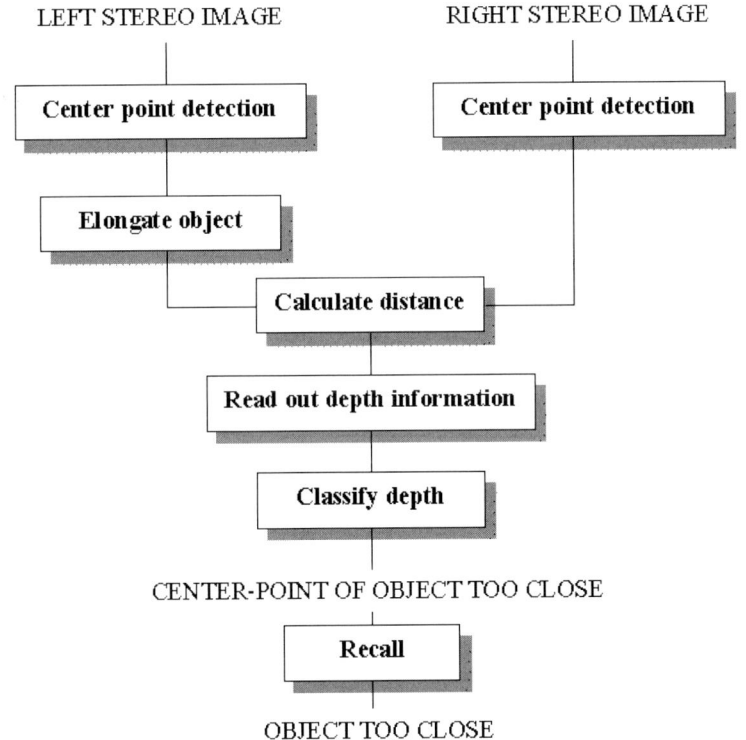

Fig. B.3. Flow-chart of the algorithm for depth classification

232   B. Library of Templates

$$A = \begin{pmatrix} 1 \end{pmatrix}, \quad B = \begin{pmatrix} 3 \\ 3 \\ 3 \end{pmatrix}, \quad I = 4.5. \tag{B.33}$$

Calculate depth: (use the elongated left image as initial state)

$$A = \begin{pmatrix} a & 1 & a \end{pmatrix}, \tag{B.34}$$

where

$$a = \begin{cases} \frac{-(Y_{ij} - Y_{kl} + 0.05)}{2} & \text{if } (Y_{ij} - Y_{kl}) < -0.05 \\ 0 & \text{if } (Y_{ij} - Y_{kl}) > -0.05 \end{cases}.$$

Read out depth: (use the right center points as a fixed state map)

$$A = \begin{pmatrix} 1 \end{pmatrix}, \quad I = -2. \tag{B.35}$$

Classify depth:

$$A = \begin{pmatrix} 2 \end{pmatrix}, \quad I = -\text{threshold}. \tag{B.36}$$

Name: DIAG
Function: Detects approximately diagonal lines in the SW–NE direction.

$$A = \begin{pmatrix} 0 & 0 & 0 & 0 & 0 \\ 0 & 0 & 0 & 0 & 0 \\ 0 & 0 & 2 & 0 & 0 \\ 0 & 0 & 0 & 0 & 0 \\ 0 & 0 & 0 & 0 & 0 \end{pmatrix}, \quad B = \begin{pmatrix} -1 & -1 & -0.5 & 0.5 & 1 \\ -1 & -0.5 & 1 & 1 & 0.5 \\ -0.5 & 1 & 5 & 1 & -0.5 \\ 0.5 & 1 & 1 & -0.5 & -1 \\ 1 & 0.5 & -0.5 & -1 & -1 \end{pmatrix}. \tag{B.37}$$

$$I = -9. \tag{B.38}$$

Name: DIAG1LIU
Function: Detects every diagonal, one pixel wide line in the SW–NE direction (like /). Modifying the position of the ±2 values of the B template, the template can be sensitized to other directions as well (vertical, horizontal or NW–SE diagonal).

$$A = \begin{pmatrix} 0 & 0 & 0 \\ 0 & 2 & 0 \\ 0 & 0 & 0 \end{pmatrix}, \quad B = \begin{pmatrix} -2 & 0 & 2 \\ 0 & 2 & 0 \\ 2 & 0 & -2 \end{pmatrix}. \tag{B.39}$$

$$I = -9. \tag{B.40}$$

Name: DIAGGREY
Function: Detects lines (contours) in a grey-scale image that are approximately in the NW–SE direction.

$$A = \begin{pmatrix} 0 & 0 & 0 & 0 & 0 \\ 0 & 0 & 0 & 0 & 0 \\ 0 & 0 & 1 & 0 & 0 \\ 0 & 0 & 0 & 0 & 0 \\ 0 & 0 & 0 & 0 & 0 \end{pmatrix}, \quad B = \begin{pmatrix} b & b & a & a & a \\ b & b & b & a & a \\ a & b & 0 & b & a \\ a & a & b & b & b \\ a & a & a & b & b \end{pmatrix}, \quad I = -2.3, \tag{B.41}$$

where

$$a = \begin{cases} -0.5 & \text{if } (V_{uij} - V_{ukl}) < 0.18 \\ 0 & \text{if } (V_{uij} - V_{ukl}) > 0.18 \end{cases}.$$

$$b = \begin{cases} 0.5 & \text{if } (V_{uij} - V_{ukl}) < 0.18 \\ 0 & \text{if } (V_{uij} - V_{ukl}) > 0.18 \end{cases}.$$

Name: DIFFUSION
Function: Simulates the well known heat-diffusion.

$$A = \begin{pmatrix} 0.1 & 0.15 & 0.1 \\ 0.15 & 0 & 0.15 \\ 0.1 & 0.15 & 0.1 \end{pmatrix}, \quad B = \begin{pmatrix} 0 & 0 & 0 \\ 0 & 0 & 0 \\ 0 & 0 & 0 \end{pmatrix}, \quad I = 0. \tag{B.42}$$

Name: DILATATION
Function: Increases the size of an object by one pixel in the horizontal and vertical directions.

$$A = \begin{pmatrix} 0 & 0 & 0 \\ 0 & 1 & 0 \\ 0 & 0 & 0 \end{pmatrix}, \quad B = \begin{pmatrix} 0 & 1 & 0 \\ 1 & 1 & 1 \\ 0 & 1 & 0 \end{pmatrix}, \quad I = -4. \tag{B.43}$$

Name: DIRECTIONAL SELECTIVITY
Function: Realizes the directional selectivity function that is a phenomenon of the visual pathway.

$$A = (0), \quad B = \begin{pmatrix} 0 & 0 & 0 \\ 0 & a & b \\ 0 & 0 & 0 \end{pmatrix} \tag{B.44}$$

$$A^t = (0), \quad B^t = \begin{pmatrix} 0 & 0 & 0 \\ -c & 0 & 0 \\ 0 & 0 & 0 \end{pmatrix}, \quad I = 0 \; t = 1 \tag{B.45}$$

$a, b, c > 0$.

Name: EDGE
Function: Black and white edge detector.

$$A = \begin{pmatrix} 0 & 0 & 0 \\ 0 & 2 & 0 \\ 0 & 0 & 0 \end{pmatrix}, \quad B = \begin{pmatrix} -0.25 & -0.25 & -0.25 \\ -0.25 & 2 & -0.25 \\ -0.25 & -0.25 & -0.25 \end{pmatrix}, \quad I = -1.5. \tag{B.46}$$

Name: ELEMENTARY TEXTON DETECTION
Function: Detects some primitive features of textons.

$$A = \begin{pmatrix} 1 \end{pmatrix}, \quad B = \begin{pmatrix} -0.25 & -0.25 & -0.25 \\ -0.25 & 1 & -0.25 \\ -0.25 & -0.25 & -0.25 \end{pmatrix}, \quad I = -0.5. \tag{B.47}$$

Name: EROSION
Function: Decreases the size of an object by one pixel in the horizontal and vertical directions.

$$A = \begin{pmatrix} 0 & 0 & 0 \\ 0 & 1 & 0 \\ 0 & 0 & 0 \end{pmatrix}, \quad B = \begin{pmatrix} 0 & 1 & 0 \\ 1 & 1 & 1 \\ 0 & 1 & 0 \end{pmatrix}, \quad I = 4. \tag{B.48}$$

Name: EXTRACTING HORIZONTAL LINES
Function: Extracts all horizontal lines of an image.

$$A = \begin{pmatrix} -1 & -1 & -1 \\ 1 & 1 & 1 \\ -1 & -1 & -1 \end{pmatrix}, \quad B = \begin{pmatrix} 0 & 0 & 0 \\ 0 & 0 & 0 \\ 0 & 0 & 0 \end{pmatrix}, \quad I = -8. \tag{B.49}$$

Name: EXTREME
Function: Find the locations where the gradient of the field is smaller than a given threshold value, i.e. the extremities of the picture.

$$A = \begin{pmatrix} 0 & 0 & 0 \\ 0 & 1 & 0 \\ 0 & 0 & 0 \end{pmatrix}, \quad B = \begin{pmatrix} a & a & a \\ a & a & a \\ a & a & a \end{pmatrix}, \quad I = threshold. \tag{B.50}$$

where:

$$a = \begin{cases} -2 & \text{if } (V_{xij} - V_{xkl}) < -0.2 \\ 1 + 40 * (V_{xij} - V_{xkl}) & \text{if } -0.2 < (V_{xij} - V_{xkl}) < 0 \\ 1 - 40 * (V_{xij} - V_{xkl}) & \text{if } 0 < (V_{xij} - V_{xkl}) < 0.2 \\ -2 & \text{if } (V_{xij} - V_{xkl}) > 0.2. \end{cases}$$

Name: FIGDEL
Function: Converts into white every black pixel with at least one black neighbor. Leaves only isolated pixels on the screen.

$$A = \begin{pmatrix} 0 & 0 & 0 \\ 0 & 2 & 0 \\ 0 & 0 & 0 \end{pmatrix}, \quad B = \begin{pmatrix} -0.25 & -0.25 & -0.25 \\ -0.25 & 1 & -0.25 \\ -0.25 & -0.25 & -0.25 \end{pmatrix}, \quad I = -4. \tag{B.51}$$

Name: FIGEXTR
Function: Converts into white every black pixel having no black neighbor. Leaves only connected components on the screen. Opposite to FIGDEL.

$$A = \begin{pmatrix} 0 & 0 & 0 \\ 0 & 2 & 0 \\ 0 & 0 & 0 \end{pmatrix}, \quad B = \begin{pmatrix} -0.25 & -0.25 & -0.25 \\ -0.25 & 0.5 & -0.25 \\ -0.25 & -0.25 & -0.25 \end{pmatrix}, \quad I = 0. \tag{B.52}$$

B. Library of Templates

Name: FIGREC
Function: Reconstructs blobs from a part of them. Feed the original image on the input and a picture containing just a part of it on the initial state. Those blobs which have pixels on the initial state will appear on the output.

$$A = \begin{pmatrix} 0.25 & 0.25 & 0.25 \\ 0.25 & 1 & 0.25 \\ 0.25 & 0.25 & 0.25 \end{pmatrix}, \quad B = \begin{pmatrix} 0 & 0 & 0 \\ 0 & 1.75 & 0 \\ 0 & 0 & 0 \end{pmatrix}, \quad I = 0. \tag{B.53}$$

Name: FILTER
Function: Convolution low-pass filter.

$$A = \begin{pmatrix} 0 & 0 & 0 \\ 0 & 2 & 0 \\ 0 & 0 & 0 \end{pmatrix}, \quad B = \begin{pmatrix} -0.25 & -0.25 & -0.25 \\ -0.25 & 2 & -0.25 \\ -0.25 & -0.25 & -0.25 \end{pmatrix}, \quad I = -2. \tag{B.54}$$

Name: FILLED AREA
Function: To find areas in the initial state totally fitting or fitting into a closed curve in the input picture.

$$A = \begin{pmatrix} 0 & 1 & 0 \\ 1 & 5 & 1 \\ 0 & 1 & 0 \end{pmatrix}, \quad B = \begin{pmatrix} 0 & 0 & 0 \\ 0 & 2 & 0 \\ 0 & 0 & 0 \end{pmatrix}, \quad I = -5.25. \tag{B.55}$$

Name: GRADIENT
Function: Finds the locations where the gradient of the field is higher than a given threshold value.

$$A = \begin{pmatrix} 0 & 0 & 0 \\ 0 & 1 & 0 \\ 0 & 0 & 0 \end{pmatrix}, \quad B = \begin{pmatrix} a & a & a \\ a & a & a \\ a & a & a \end{pmatrix}, \quad I = threshold. \tag{B.56}$$

where:

$$a = \begin{cases} -2 & \text{if } (V_{xij} - V_{xkl}) < -0.2 \\ -1 + (V_{xij} - V_{xkl}) & \text{if } -2 < (V_{xij} - V_{xkl}) < 0 \\ (V_{xij} - V_{xkl}) & \text{if } 0 < (V_{xij} - V_{xkl}) < 2 \\ 2 & \text{if } (V_{xij} - V_{xkl}) > 2. \end{cases}$$

Name: GRADIENT CONTROLLED DIFFUSION I
Function: This algorithm simulates a simpler form of anisotropic diffusion, given by the equation:

$$\frac{\partial I}{\partial t} = \Delta[k* \mid \text{grad}(G(s_1) * I(x,y,t)) \mid *(I(x,y,t) \tag{B.57}$$
$$- G(s) * I(x,y,t) + G(s) * I(x,y,t)].$$

Here $I(x,y,t)$ is the image varying in time, $G(s)$ and $G(s1)$ are Gaussian filters with apertures s and $s1$, k is a constant value between 1 and 3. Both

Gaussian filtering and the Laplace operation (delta) is carried out by DIFFUSION templates with different diffusion coefficient. The GRADIENT template can also be found in this library. This equation can be used for noise filtering without decreasing the sharpness of edges.

A flow-chart of the algorithm is shown in Fig. B.4:

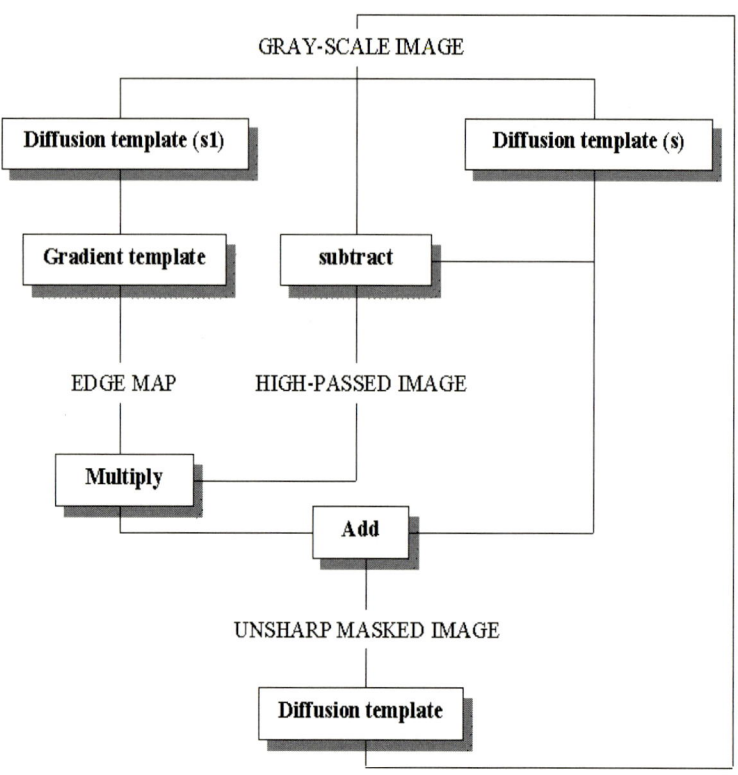

Fig. B.4. Flow-chart of the algorithm for the Gradient Controlled Diffusion I

Name: GRADIENT CONTROLLED DIFFUSION II
Function: A simple structure for edge-enhancement during noise elimination. The method is similar to the above algorithm. Only the equation used for filtering is different. Here

$$\frac{\partial I}{\partial t} = \Delta[I(x,y,z) * (1 - k* \mid grad(G(s) * I(x,y,z)) \mid)))]. \tag{B.58}$$

A flow-chart of this algorithm is shown in Fig. B.5.

Name: GREY-SCALE SKELETONIZATION
Function: The algorithm finds the skeleton of grey-scale objects. The algorithm uses eight sets of templates, skeletonizing the objects circularly. Each

B. Library of Templates

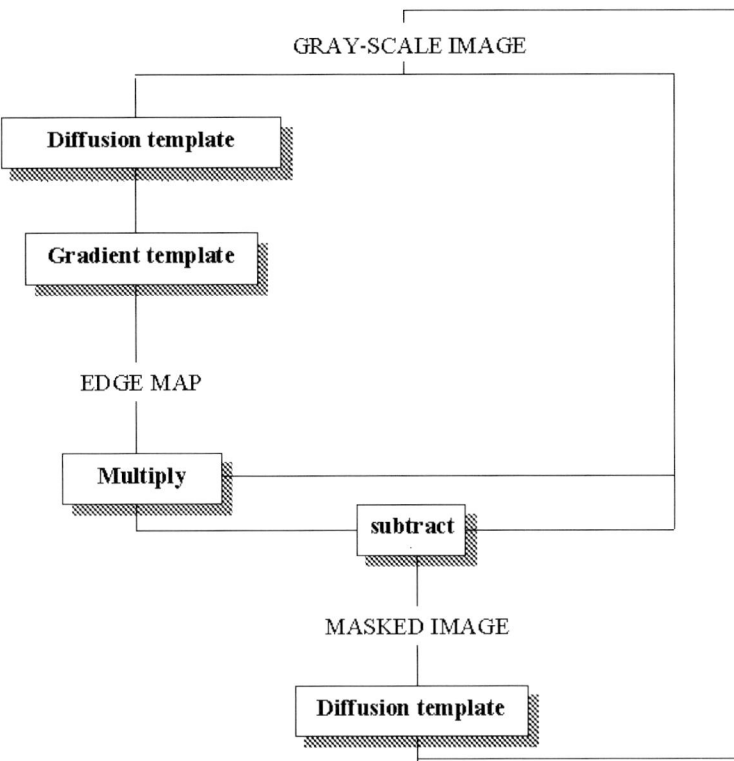

Fig. B.5. Flow-chart of the algorithm for the Gradient Controlled Diffusion II

step contains two templates. First appropriate pixels are selected by the selection templates, afterwards these are used as fixed state masks for the replacement. The result of the replacement is fed back to the input.

Templates of the algorithm.
Selection templates:

$$A_1 = \begin{pmatrix} 1 \end{pmatrix}, \quad B_1 = \begin{pmatrix} a & a & 0 \\ a & 0 & b \\ 0 & b & 0 \end{pmatrix}, \quad I_1 = -4.5 \qquad (B.59)$$

$$A_2 = \begin{pmatrix} 1 \end{pmatrix}, \quad B_2 = \begin{pmatrix} a & a & a \\ 0 & 0 & 0 \\ b & b & 0 \end{pmatrix}, \quad I_2 = -4.5 \qquad (B.60)$$

$$A_3 = \begin{pmatrix} 1 \end{pmatrix}, \quad B_3 = \begin{pmatrix} 0 & a & a \\ b & 0 & a \\ 0 & b & 0 \end{pmatrix}, \quad I_3 = -4.5 \qquad (B.61)$$

$$A_4 = \begin{pmatrix} 1 \end{pmatrix}, \quad B_4 = \begin{pmatrix} b\,0\,a \\ b\,0\,a \\ 0\,0\,a \end{pmatrix}, \quad I_4 = -4.5 \tag{B.62}$$

$$A_5 = \begin{pmatrix} 1 \end{pmatrix}, \quad B_5 = \begin{pmatrix} 0\,b\,0 \\ b\,0\,a \\ 0\,a\,a \end{pmatrix}, \quad I_5 = -4.5 \tag{B.63}$$

$$A_6 = \begin{pmatrix} 1 \end{pmatrix}, \quad B_6 = \begin{pmatrix} 0\,b\,b \\ 0\,0\,0 \\ a\,a\,a \end{pmatrix}, \quad I_6 = -4.5 \tag{B.64}$$

$$A_7 = \begin{pmatrix} 1 \end{pmatrix}, \quad B_7 = \begin{pmatrix} 0\,b\,0 \\ a\,0\,b \\ a\,a\,0 \end{pmatrix}, \quad I_7 = -4.5 \tag{B.65}$$

$$A_8 = \begin{pmatrix} 1 \end{pmatrix}, \quad B_8 = \begin{pmatrix} a\,0\,0 \\ a\,0\,b \\ a\,0\,b \end{pmatrix}, \quad I_8 = -4.5. \tag{B.66}$$

$$\tag{B.67}$$

Replacement templates:

$$A_1 = \begin{pmatrix} c\,c\,0 \\ c\,1\,0 \\ 0\,0\,0 \end{pmatrix}, \quad A_2 = \begin{pmatrix} c\,c\,c \\ 0\,1\,0 \\ 0\,0\,0 \end{pmatrix}, \quad A_3 = \begin{pmatrix} 0\,c\,0 \\ 0\,1\,c \\ 0\,0\,0 \end{pmatrix}, \tag{B.68}$$

$$A_4 = \begin{pmatrix} 0\,0\,c \\ 0\,1\,c \\ 0\,0\,c \end{pmatrix}, \quad A_5 = \begin{pmatrix} 0\,0\,0 \\ 0\,1\,c \\ 0\,c\,c \end{pmatrix}, \quad A_6 = \begin{pmatrix} 0\,0\,0 \\ c\,1\,0 \\ c\,c\,c \end{pmatrix}, \tag{B.69}$$

$$A_7 = \begin{pmatrix} 0\,0\,0 \\ c\,1\,0 \\ c\,c\,0 \end{pmatrix}, \quad A_8 = \begin{pmatrix} c\,0\,0 \\ c\,1\,0 \\ c\,0\,0 \end{pmatrix}, \tag{B.70}$$

where

$$a = \begin{cases} 0 & \text{if } (U_{ij} - U_{kl}) < 0 \\ 1 & \text{if } (U_{ij} - U_{kl}) > 0 \end{cases}$$

$$b = \begin{cases} 1 & \text{if } (U_{ij} - U_{kl}) < 0 \\ 0 & \text{if } (U_{ij} - U_{kl}) > 0 \end{cases}$$

$$c = \begin{cases} -0.33 * (Y_{ij} - Y_{kl}) & \text{if } (Y_{ij} - Y_{kl}) < 0 \\ 0 & \text{if } (Y_{ij} - Y_{kl}) > 0. \end{cases}$$

Name: HERRING
Function: Simulates the herring-grid illusion: grey patches appear at the intersections of a grid of black squares on a white background.

$$A^t = \begin{pmatrix} -0.1 & -0.1 & -0.1 & -0.1 & -0.1 \\ -0.1 & -0.3 & 0 & -0.3 & -0.1 \\ -0.1 & -0.3 & 0 & -0.3 & -0.1 \\ -0.1 & -0.3 & 0 & -0.3 & -0.1 \\ -0.1 & -0.1 & -0.1 & -0.1 & -0.1 \end{pmatrix}. \quad (B.71)$$

$$B = \begin{pmatrix} -0.16 & -0.16 & -0.16 & -0.16 & -0.16 \\ -0.16 & -0.40 & -0.40 & -0.40 & -0.16 \\ -0.16 & -0.40 & 4.00 & -0.40 & -0.16 \\ -0.16 & -0.40 & -0.40 & -0.40 & -0.16 \\ -0.16 & -0.16 & -0.16 & -0.16 & -0.16 \end{pmatrix}, \ I=0, \ t=3. \quad (B.72)$$

Name: HLF33
Function: 3*3 image half–toning: converts the grey-scale input image to black and white, still preserving the main features of the input. The speed of convergence is controlled by $e = [0.1 \cdots 1]$. The greater the e, the faster the process and the rougher the result. The inverse of the template is INVHLF33. The result is good in the square error measure, but the pattern is not necessarily pleasant to see.

$$A = \begin{pmatrix} -0.07 & -0.1 & -0.07 \\ -0.1 & 1+e & -0.1 \\ -0.07 & -0.1 & -0.07 \end{pmatrix} \ B = \begin{pmatrix} 0.07 & 0.1 & 0.07 \\ 0.1 & 0.32 & 0.1 \\ 0.07 & 0.1 & 0.07 \end{pmatrix}, \ I=0. \quad (B.73)$$

Name: HLF55
Function: 5*5 image half–toning: converts the grey-scale input image to black and white, still preserving the main features of the input. It is optimal in square error measure.

$$A = \begin{pmatrix} -0.02 & -0.07 & -0.10 & -0.07 & -0.02 \\ -0.07 & -0.32 & -0.46 & -0.32 & -0.07 \\ -0.10 & -0.46 & 1.05 & -0.46 & -0.10 \\ -0.07 & -0.32 & -0.46 & -0.32 & -0.07 \\ -0.02 & -0.07 & -0.10 & -0.07 & -0.02 \end{pmatrix}. \quad (B.74)$$

$$B = \begin{pmatrix} 0.02 & 0.07 & 0.10 & 0.07 & 0.02 \\ 0.07 & 0.32 & 0.46 & 0.32 & 0.07 \\ 0.10 & 0.46 & 0.81 & 0.46 & 0.10 \\ 0.07 & 0.32 & 0.46 & 0.32 & 0.07 \\ 0.02 & 0.07 & 0.10 & 0.07 & 0.02 \end{pmatrix}, \ I=0. \quad (B.75)$$

Name: HLF55-KC
Function: 5∗5 image halftoning: converts the grey-scale input image to black and white, still preserving the main features of the input. The output should be looked at from a distance. The output image quality is optimized for human vision. If simulating the behavior of the CNN using the forward Euler integration form, the time step should be less than 0.4.

$$A = \begin{pmatrix} -0.03 & -0.09 & -0.13 & -0.09 & -0.03 \\ -0.09 & -0.36 & -0.60 & -0.36 & -0.09 \\ -0.13 & -0.60 & 1.05 & -0.60 & -0.13 \\ -0.09 & -0.36 & -0.60 & -0.36 & -0.09 \\ -0.03 & -0.09 & -0.13 & -0.09 & -0.03 \end{pmatrix} \quad (B.76)$$

$$B = \begin{pmatrix} 0.00 & 0.00 & 0.07 & 0.00 & 0.00 \\ 0.00 & 0.36 & 0.76 & 0.36 & 0.00 \\ 0.07 & 0.76 & 2.12 & 0.76 & 0.07 \\ 0.00 & 0.36 & 0.76 & 0.36 & 0.00 \\ 0.00 & 0.00 & 0.07 & 0.00 & 0.00 \end{pmatrix}, \quad I = 0. \quad (B.77)$$

Name: HOLE

Function: Performs hole filling. The picture should be fed into the input, while the initial state should be black.

$$A = \begin{pmatrix} 0 & 1 & 0 \\ 1 & 2 & 1 \\ 0 & 1 & 0 \end{pmatrix}, \quad B = \begin{pmatrix} 0 & 0 & 0 \\ 0 & 4 & 0 \\ 0 & 0 & 0 \end{pmatrix}, \quad I = -1. \quad (B.78)$$

Name: HOLE 2

Function: Perform hole filling. The picture should be fed into the input, while the initial state should be black.

$$A = \begin{pmatrix} 0 & 1 & 0 \\ 1 & 2.5 & 1 \\ 0 & 1 & 0 \end{pmatrix}, \quad B = \begin{pmatrix} 0 & 0 & 0 \\ 0 & 5 & 0 \\ 0 & 0 & 0 \end{pmatrix}, \quad I = 0. \quad (B.79)$$

Name: HOLLOW

Function: This diffusion type template fills the concave location of object.

$$A = \begin{pmatrix} 0.5 & 0.5 & 0.5 \\ 0.5 & 2 & 0.5 \\ 0.5 & 0.5 & 0.5 \end{pmatrix} \quad B = \begin{pmatrix} 0 & 0 & 0 \\ 0 & 2 & 0 \\ 0 & 0 & 0 \end{pmatrix} \quad I = 3.5. \quad (B.80)$$

Name: HAMMING DISTANCE COMPUTATION

Function: In the theory of information processes it is a common problem that, given a code received on a noisy channel and the set of legal code words, we have to determine the code word nearest in some metric to the one received. In the case of binary codes, the Hamming distance is the most common choice to measure the distance. Here a four–step method is given to select the legal code closest to the input. The first step compares the input to all legal code words. For this the m legal code words should be put in a single image, each code constituting a separate row, while the input should be written in another image m times. This step can be performed by the logical XOR template. In the second step the number of differences is calculated, after feeding the output of the previous step back to the input and setting the initial state to 0. In the third step the minimum distance is determined, and finally the

best matching code word is selected. For this, the output of the previous step should be used as initial state, and that of step 2 as input.
Templates "Difference":
$$A_2 = \begin{pmatrix} 0 & 0 & 1 \end{pmatrix}, \ B_2 = \begin{pmatrix} a \end{pmatrix}, \ I = 0.02. \tag{B.81}$$
Templates "Minimum Distance":
$$A_3 = \begin{pmatrix} b \\ 1 \\ b \end{pmatrix}, \ I = 0.02. \tag{B.82}$$
Templates "Batch Matching":
$$A_4 = \begin{pmatrix} 2 \end{pmatrix}, \ A_3 = \begin{pmatrix} -1 \end{pmatrix}, \ I = 0.02. \tag{B.83}$$
where
$$a = \begin{cases} 0.06 & \text{if } U_{ij} < 0 \\ 0 & \text{if } U_{ij} > 0 \end{cases}$$

$$b = \begin{cases} 0.06 & \text{if } U_{ij} < 0 \\ -0.5 * U_{ij} & \text{if } U_{ij} > 0 \end{cases}.$$

Name: HISTOGRAM
Function: The template computes the one-dimensional histogram of a black-and-white image, i.e. shifts all black pixels to the left, leaving all white ones on the right.
$$A = \begin{pmatrix} a & 1 & b \end{pmatrix}, \tag{B.84}$$
where
$$a = \begin{cases} 0 & \text{if } V_{yij} - V_{ykl} < 1.5 \\ -3 & \text{if } V_{xij} - V_{xkl} > 1.5 \end{cases}$$

$$b = \begin{cases} 3 & \text{if } V_{yij} - V_{ykl} < 1.5 \\ 0 & \text{if } V_{xij} - V_{xkl} > 1.5 \end{cases}.$$

Name: INCREASE
Function: Increases the size of an object by one pixel in all directions in an iteration step of a DTCNN.
$$A = \begin{pmatrix} 0.5 & 0.5 & 0.5 \\ 0.5 & 0.5 & 0.5 \\ 0.5 & 0.5 & 0.5 \end{pmatrix}, \ B = \begin{pmatrix} 0 & 0 & 0 \\ 0 & 0 & 0 \\ 0 & 0 & 0 \end{pmatrix}, \ I = 4. \tag{B.85}$$

Name: INTERPOL
Function: This algorithm fits a surface that is as smooth as possible on the given points. If the altitude of the surface is known at the point (i,j), then this altitude is filled into the state at the position (i,j) and the state is kept fix during the iteration. If the altitude is not known, then zero is filled into the state. For exact solution of the problem, the A template is space variant. If the following template is used, then the edges of the result will be a bit deformed but the network is space invariant.

$$A = \begin{pmatrix} 0 & 0 & -2 & 0 & 0 \\ 0 & -4 & 16 & -4 & 0 \\ -2 & 16 & -39 & 16 & -2 \\ 0 & -4 & 16 & -4 & 0 \\ 0 & 0 & -2 & 0 & 0 \end{pmatrix}, \quad B = \begin{pmatrix} 0 \end{pmatrix}, \quad I = 0. \tag{B.86}$$

Name: INV
Function: Logical "NOT". Inverts the input image.

$$A = \begin{pmatrix} 0 & 0 & 0 \\ 0 & 1 & 0 \\ 0 & 0 & 0 \end{pmatrix}, \quad B = \begin{pmatrix} 0 & 0 & 0 \\ 0 & -2 & 0 \\ 0 & 0 & 0 \end{pmatrix}, \quad I = 0. \tag{B.87}$$

Name: INV-OR
Function: Logical "OR" function of the initial state and the logical "NOT" of the input.

$$A = \begin{pmatrix} 0 & 0 & 0 \\ 0 & 1 & 0 \\ 0 & 0 & 0 \end{pmatrix}, \quad B = \begin{pmatrix} 0 & 0 & 0 \\ 0 & -1 & 0 \\ 0 & 0 & 0 \end{pmatrix}, \quad I = 1. \tag{B.88}$$

Name: INVHLF
Function: Inverse of the 3*3 halftoned image. The result is lack of fine edges, as the control of HLF33 smoothes the input. The time step should be equal to 1, the number of steps is 1.

$$A = \begin{pmatrix} 0 \end{pmatrix}, \quad B = \begin{pmatrix} 0.07 & 0.1 & 0.07 \\ 0.1 & 0.32 & 0.1 \\ 00.07 & 0.1 & 0.07 \end{pmatrix}, \quad I = 0. \tag{B.89}$$

Name: INVHLF55
Function: Inverse of the 5*5 halftoned image. The result is lack of fine edges, as the control of HLF55 smoothes the input.

$$A = \begin{pmatrix} 0 \end{pmatrix}, \quad B = \begin{pmatrix} 0 & 0.01 & 0.02 & 0.01 & 0 \\ 0.01 & 0.06 & 0.09 & 0.06 & 0.01 \\ 0.02 & 0.09 & 0.16 & 0.09 & 0.02 \\ 0.01 & 0.06 & 0.09 & 0.06 & 0.01 \\ 0 & 0.01 & 0.02 & 0.01 & 0 \end{pmatrix}, \quad I = 0. \tag{B.90}$$

Name: JUNCTION
Function: This template extract the junctions of a skeleton.
$$A = \begin{pmatrix} 0 & 0 & 0 \\ 0 & 2 & 0 \\ 0 & 0 & 0 \end{pmatrix}, \quad B = \begin{pmatrix} 0.5 & 0.5 & 0.5 \\ 0.5 & 2 & 0.5 \\ 0.5 & 0.5 & 0.5 \end{pmatrix}, \quad I = 3.5. \tag{B.91}$$

Name: LAPLACE
Function: Solves the Laplace PDE: $\nabla^2 x = 0$.
$$A = \begin{pmatrix} 0 & \frac{v}{h^2} & 0 \\ \frac{v}{h^2} & -4 + \frac{v}{h^2} + \frac{1}{R} & \frac{v}{h^2} \\ 0 & \frac{v}{h^2} & 0 \end{pmatrix}, \quad B = \begin{pmatrix} 0 \end{pmatrix}, \quad I = 0, \tag{B.92}$$

where v is the diffusion constant, h the spatial step, and R the value of the resistor of the CNN cell. If $v = h = R = 1$, our template is as follows:
$$A = \begin{pmatrix} 0 & 1 & 0 \\ 1 & -3 & 1 \\ 0 & 1 & 0 \end{pmatrix}, \quad B = \begin{pmatrix} 0 \end{pmatrix}, \quad I = 0. \tag{B.93}$$

Name: LCP
Function: Extracts local concave place, i.e. pixels having no southern neighbor, but both eastern, western and either south-western, or south-eastern neighbor.
$$A = \begin{pmatrix} 0 & 0 & 0 \\ 0 & 2 & 0 \\ 0 & 1 & 0 \end{pmatrix}, \quad B = \begin{pmatrix} 0 & 0 & 0 \\ 1 & 2 & 1 \\ 0.5 & -1 & 0.5 \end{pmatrix}, \quad I = -5.5. \tag{B.94}$$

Name: LGN RELAY FUNCTION 2L
Function: Realizes one of the most important properties of the lateral geniculate nucleus (LGN): the relay function. This model is realized on a two layer CNN.
$$A11^{\tau=2} = \begin{pmatrix} a \end{pmatrix}, \quad B11 = \begin{pmatrix} 1 \end{pmatrix}, \quad I_1 = 0.8, \tag{B.95}$$

$$A12 = \begin{pmatrix} b \end{pmatrix} \quad B22 = \begin{pmatrix} -0.1 & -0.2 & 0.3 & 0.2 & 0.1 \\ -0.2 & -0.2 & -0.3 & 0.3 & 0.2 \\ 0.3 & -0.3 & -0.6 & -0.3 & 0.3 \\ 0.2 & 0.2 & -0.3 & -0.2 & -0.2 \\ 0.1 & 0.2 & 0.3 & -0.2 & -0.1 \end{pmatrix}. \tag{B.96}$$

where
$$a = \begin{cases} 0 & \text{if } V_x < 0.9 \\ -8 & \text{if } V_x > 0.9 \end{cases}$$

$$b = \begin{cases} 0 & \text{if } V_x < -0.4 \\ -0.8 & \text{if } V_x > -0.4 \end{cases}.$$

Name: LGN CENTER SURROUND EFFECT
Function: Realizes the response of the off-center and on-center type LGN.
Off-center:

$$A = \begin{pmatrix} 1 \end{pmatrix}, \quad B = \begin{pmatrix} 0 & -0.35 & -1 \\ -0.35 & 2.5 & -0.35 \\ -1 & -0.35 & 0 \end{pmatrix}, \quad I = -2. \tag{B.97}$$

On-center:

$$A = \begin{pmatrix} 1 \end{pmatrix}, \quad B = \begin{pmatrix} 0 & -0.35 & 1 \\ 0.35 & -2.5 & 0.35 \\ 1 & 0.35 & 0 \end{pmatrix}, \quad I = 2. \tag{B.98}$$

Name: LGN WALSH PATTERNS
Function: Realizes the response of the Walsh patterns type LGN.

$$A = \begin{pmatrix} 0 \end{pmatrix}, \quad B = \begin{pmatrix} 0 & 0.1 & 0.1 & 0.1 & 0.1 \\ 0 & 0.1 & -0.45 & -0.15 & 0.1 \\ 0 & 0.1 & -0.65 & -0.15 & 0.1 \\ 0 & 0.1 & 0.1 & 0.1 & 0.1 \\ 0 & 0 & 0 & 0 & -0 \end{pmatrix}. \tag{B.99}$$

Name: LGTHTUNE
Function: Length-tuning: only lines no longer than 3 pixels remain on the screen, while all others disappear, regardless of their orientation.

$$A = \begin{pmatrix} -3 & 0 & -3 & 0 & -3 \\ 0 & 1 & 1 & 1 & 0 \\ -3 & 1 & a & 1 & -3 \\ 0 & 1 & 1 & 1 & 0 \\ -3 & 0 & -3 & 0 & -3 \end{pmatrix}, \quad I = -1, \tag{B.100}$$

where

$$a = \begin{cases} -3 & \text{if } V_{xij} < 0.2 \\ 1 & \text{if } V_{xij} > 0.2 \end{cases}.$$

Name: LIFE-DT
Function: Simulates the game of life on a 3-layer Discrete-Time CNN (DTCNN). The original life pattern should be fed to the initial state of all three layers, while the third layer provides the output. A new output appears in every second iteration.

$$A_{31} = \begin{pmatrix} 0\,0\,0 \\ 0\,1\,0 \\ 0\,0\,0 \end{pmatrix} , \quad A_{13} = \begin{pmatrix} 0.3\,0.3\,0.3 \\ 0.3\,0.3\,0.3 \\ 0.3\,0.3\,0.3 \end{pmatrix} , \quad I_1 = 1. \quad (\text{B.101})$$

$$A_{32} = \begin{pmatrix} 0\,0\,0 \\ 0\,1\,0 \\ 0\,0\,0 \end{pmatrix} , \quad A_{23} = \begin{pmatrix} -0.6\,-0.6\,-0.6 \\ -0.6\,0\,-0.6 \\ -0.6\,-0.6\,-0.6 \end{pmatrix} . \quad (\text{B.102})$$

$$I_2 = -0.8 , \quad I_3 = -1.5. \quad (\text{B.103})$$

Name: LIFE-1
Function: Simulates one step of the game of life on a 2-layer CNN. The original life pattern should be fed to the input of the first layer, while the output will appear on the second layer.

$$A_{11} = (1) , \quad A_{12} = (a) , \quad A_{22} = (1) , \quad B_{11} = \begin{pmatrix} 0.3\,0.3\,0.3 \\ 0.3\,0.3\,0.3 \\ 0.3\,0.3\,0.3 \end{pmatrix} , \quad (\text{B.104})$$

$$I_1 = 1 , \quad I_2 = -0.8 , \quad B_{21} = \begin{pmatrix} -0.6\,-0.6\,-0.6 \\ -0.6\,0\,-0.6 \\ -0.6\,-0.6\,-0.6 \end{pmatrix} , \quad (\text{B.105})$$

where

$$a = \begin{cases} -4 & \text{if } Y_{2ii} < -0.9 \\ 0 & \text{if } Y_{2ii} > -0.9 \end{cases}.$$

Name: LIFE-1L
Function: Simulates the game of life on a single-layer discrete-time CNN with piecewise-linear thresholding, or in other words, a CNN approximated by the forward Euler integration form using time step 1. The original life pattern should be fed to the initial state, and a new state appears after every second iteration.

$$A = \begin{pmatrix} a\,a\,a \\ a\,b\,a \\ a\,a\,a \end{pmatrix} , \quad B = (0) , \quad I = 0, \quad (\text{B.106})$$

where

$$a = \begin{cases} 0 & \text{if } x < 0.9 \\ 0.1 & \text{if } x > 0.9 \end{cases},$$

$$b = \begin{cases} -0.5 & \text{if } x < -0.9 \\ -1 & \text{if } -0.9 < x < -0.27 \\ 1 & \text{if } -0.27 < x < -0.13 \\ -1 & \text{if } -0.13 < x < 0.9 \\ -0.45 & \text{if } x > 0.9 \end{cases}.$$

Name: LINE3060
Function: Detects grey-scale lines within a slope range of approximately 30° (30°–60°).

$$A = \begin{pmatrix} 0 & 0 & 0 \\ 0 & 2 & 0 \\ 0 & 0 & 0 \end{pmatrix}, \quad B = \begin{pmatrix} b & b & b \\ b & 0 & b \\ a & b & b \end{pmatrix}, \quad I = 0, \tag{B.107}$$

where

$$a = \begin{cases} -0.5 & \text{if } U_{ij} - U_{kl} < -0.85 \\ 2 & \text{if } -0.85 < U_{ij} - U_{kl} < 0.85 \\ -0.5 & \text{if } U_{ij} - U_{kl} > 0.85 \end{cases},$$

$$b = \begin{cases} 0 & \text{if } U_{ij} - U_{kl} < -0.4 \\ -1 & \text{if } U_{ij} - U_{kl} > -0.4 \end{cases}.$$

Name: LOCAL MAXIMA
Function: Detects local maxima in a certain area in grey-scale image. The process of the minima requires the inverted and reflected nonlinear function. $Amax, C$ and $Amax, D$ use two different fixed-state patterns to solve the problem. In case of the $Amax, D/Amax, C$ template, the resulting mask has a line/checkerboard pattern.

$$Amax, C = \begin{pmatrix} 0 & a_{max} & 0 \\ a_{max} & 1 & a_{max} \\ 0 & a_{max} & 0 \end{pmatrix}, \quad B = \begin{pmatrix} 0 & 0 & 0 \\ 0 & 0 & 0 \\ 0 & 0 & 0 \end{pmatrix}, \quad I = 0, \tag{B.108}$$

$$Amax, D = \begin{pmatrix} a_{max} & 0 & a_{max} \\ 0 & 1 & 0 \\ a_{max} & 0 & a_{max} \end{pmatrix}, \quad B = \begin{pmatrix} 0 & 0 & 0 \\ 0 & 0 & 0 \\ 0 & 0 & 0 \end{pmatrix}, \quad I = 0, \tag{B.109}$$

where

$$a_{max} = \begin{cases} 0 & \text{if } Y_{ij} - Y_{kl} < 0 \\ \frac{Y_{ij} - Y_{kl}}{8} & \text{if } 0 < Y_{ij} - Y_{kl} < 2 \\ 0.25 & \text{if } Y_{ij} - Y_{kl} > 2 \end{cases}.$$

Name: LSE
Function: Extracts local southern elements, i.e. pixels having neither south-western, nor southern, nor south-eastern neighbors.

$$A = \begin{pmatrix} 0 & 0 & 0 \\ 0 & 2 & 0 \\ 0 & 0 & 0 \end{pmatrix}, \quad B = \begin{pmatrix} 0 & 0 & 0 \\ 0 & 2 & 0 \\ -1 & -1 & -1 \end{pmatrix}, \quad I = -5.5, \tag{B.110}$$

Name: MAJORITY VOTE-TAKER
Function: The goal of the one–dimensional majority vote-taker template is to decide whether a row of the input image contains more black or more white pixels, or whether their number is equal. The effect is realized in two phases. The first template (setting the initial state to 0) gives rise to an image in which the sign of the rightmost pixel corresponds to the dominant color, namely positive if there are more black pixels than white ones, negative in the opposite case, and 0 in the case of equality. Using the second template this information can be extracted, driving the whole network to black or white, depending on the dominant color, or leaving the rightmost pixel unchanged otherwise. The method can easily be extended to two or even three dimensions.

$$A_1 = \begin{pmatrix} 1 & 0 & 0 \end{pmatrix}, \quad B_1 = \begin{pmatrix} 0.05 \end{pmatrix}, \quad A_2 = \begin{pmatrix} 0 & a & 2 \end{pmatrix}, \tag{B.111}$$

where

$$a = \begin{cases} -1 & \text{if } V_y < -0.025 \\ 0 & \text{if } -0.025 < V_y < 0.025 \\ 1 & \text{if } V_y > 0.025. \end{cases}$$

Name: MASKED CONNECTED COMPONENT DETECTOR
Function: Horizontal and vertical CCD templates in which the pattern shifting is halted by a mask in the input picture. The direction of shift is from the negative to the positive non-zero off-center feedback template entries.
Horizontal shift:

$$A = \begin{pmatrix} 0 & 0 & 0 \\ \pm 1 & 2 & \mp 1 \\ 0 & 0 & 0 \end{pmatrix}, \quad B = \begin{pmatrix} 0 & 0 & 0 \\ 0 & -3 & 0 \\ 0 & 0 & 0 \end{pmatrix}, \quad I = -3. \tag{B.112}$$

Vertical shift:

$$A = \begin{pmatrix} 0 & \pm 1 & 0 \\ 0 & 2 & 0 \\ 0 & \mp 1 & 0 \end{pmatrix}, \quad B = \begin{pmatrix} 0 & 0 & 0 \\ 0 & -3 & 0 \\ 0 & 0 & 0 \end{pmatrix}, \quad I = -3. \tag{B.113}$$

Name: MASKED ERASE
Function: Directional erasure of initial state segment not masked by input image
Left-to-right:

$$A = \begin{pmatrix} 0 & 0 & 0 \\ 1.5 & 3 & 0 \\ 0 & 0 & 0 \end{pmatrix}, \quad B = \begin{pmatrix} 0 & 0 & 0 \\ 0 & 1.5 & 0 \\ 0 & 0 & 0 \end{pmatrix}, \quad I = -1.5. \tag{B.114}$$

Right-to-left:

$$A = \begin{pmatrix} 0 & 0 & 0 \\ 0 & 3 & 1.5 \\ 0 & 0 & 0 \end{pmatrix}, \quad B = \begin{pmatrix} 0 & 0 & 0 \\ 0 & 1.5 & 0 \\ 0 & 0 & 0 \end{pmatrix}, \quad I = 1.5. \tag{B.115}$$

Downwards:
$$A = \begin{pmatrix} 0 & 1.5 & 0 \\ 0 & 3 & 0 \\ 0 & 0 & 0 \end{pmatrix}, \quad B = \begin{pmatrix} 0 & 0 & 0 \\ 0 & 1.5 & 0 \\ 0 & 0 & 0 \end{pmatrix}, \quad I = -1.5. \tag{B.116}$$

Upwnwards:
$$A = \begin{pmatrix} 0 & 0 & 0 \\ 0 & 3 & 0 \\ 0 & 1.5 & 0 \end{pmatrix}, \quad B = \begin{pmatrix} 0 & 0 & 0 \\ 0 & 1.5 & 0 \\ 0 & 0 & 0 \end{pmatrix}, \quad I = -1.5. \tag{B.117}$$

Name: MASKED SHADOW
Function: Horizontal and vertical shadow templates in which the pattern propagation is halted by a mask in the input picture. The propagation goes from the direction of the non-zero off-center feedback template entry A_{ij}.
$$A = \begin{pmatrix} 0 & 1.8 & 1.5 \end{pmatrix}, \quad B = \begin{pmatrix} 0 & 1.2 & 0 \end{pmatrix}, \quad I = 0. \tag{B.118}$$

Name: MATCH
Function: Detects 3*3 patterns matching exactly the one prescribed by the template, namely having a black/white pixel where the template value is $+1/-1$, respectively.
$$A = \begin{pmatrix} 0 & 0 & 0 \\ 0 & 1 & 0 \\ 0 & 0 & 0 \end{pmatrix}, \quad B = \begin{pmatrix} v & v & v \\ v & v & v \\ v & v & v \end{pmatrix}, \quad I = -N + 0.5, \tag{B.119}$$

where

$$v = \begin{cases} 1 & \text{if the corresponding pixel is required to be black} \\ 0 & \text{if the corresponding pixel is "don't care"} \\ -1 & \text{if the corresponding pixel is required to be white} \end{cases}.$$

N = number of pixels required to be either black or white, i.e. the number of non-zero values in the B template.

Name: MODIFIED GRADIENT
Function: Marks regions of high level differences between adjacent pixels in grey-scale input image.
$$A = \begin{pmatrix} 0 & 0 & 0 \\ 0 & 1 & 0 \\ 0 & 0 & 0 \end{pmatrix}, \quad B = \begin{pmatrix} v & v & v \\ v & 0 & v \\ v & v & v \end{pmatrix}, \quad I = -0.5, \tag{B.120}$$

where

$$v = \begin{cases} 2 & \text{if } V_{uij} - V_{ukl} < -2 \\ -(2 * \frac{V_{uij} - V_{ukl}) + 1}{1.5} & \text{if } -2 < V_{uij} - V_{ukl} < -0.5 \\ 0 & \text{if } -0.5 < V_{uij} - V_{ukl} < 0.5 \\ 2 * \frac{(V_{uij} - V_{ukl}) + 1}{1.5} & \text{if } 0.5 < V_{uij} - V_{ukl} < 2 \\ 2 & \text{if } V_{uij} - V_{ukl} > 2. \end{cases}$$

B. Library of Templates   249

Name: MD-CONT
Function: Direction independent motion detection in continuous mode. Objects moving slower than one pixel/delay-time will stay on the screen, while others disappear.

$$A = \begin{pmatrix} 0 & 0 & 0 \\ 0 & 1 & 0 \\ 0 & 0 & 0 \end{pmatrix}, \quad B = \begin{pmatrix} 0 & 0 & 0 \\ 0 & 6 & 0 \\ 0 & 0 & 0 \end{pmatrix}, \quad I = -2 \tag{B.121}$$

$$A^t = \begin{pmatrix} 0.68 & 0.68 & 0.68 \\ 0.68 & 0.68 & 0.68 \\ 0.68 & 0.68 & 0.68 \end{pmatrix}, \quad B^t = \begin{pmatrix} 0 & 0 & 0 \\ 0 & 0 & 0 \\ 0 & 0 & 0 \end{pmatrix}, \quad t = 10\,.. \tag{B.122}$$

Name: MOTION DETECTION
Function: Detects the moving objects that have a certain speed and direction. Erase from the image the objects that move with the wrong speed or in the wrong direction. For the template shown, the speed is one pixel per sampling time, and direction is left-to-right.

$$A = \begin{pmatrix} 0 & 0 & 0 & 0 & 0 \\ 0 & 0.3 & 0.3 & 0.3 & 0 \\ 0 & 0.3 & 4.2 & 0.3 & 0 \\ 0 & 0.3 & 0.3 & 0.3 & 0 \\ 0 & 0 & 0 & 0 & 0 \end{pmatrix}, \quad B = \begin{pmatrix} 0 & 0 & 0 & 0 & 0 \\ -0.3 & -0.3 & -0.3 & 0 & 0 \\ -0.3 & 3.1 & -0.3 & 0 & 0 \\ -0.3 & -0.3 & -0.3 & 0 & 0 \\ 0 & 0 & 0 & 0 & 0 \end{pmatrix}, \tag{B.123}$$

$$I = -6\,. \tag{B.124}$$

Name: MOTION DIRECTION SENSITIVE
Function: Detects the motion direction of points moving in different directions.

Downwards:

$$A = \begin{pmatrix} 0 & 0 & 0 \\ 0 & 2 & 0 \\ 0 & 0 & 0 \end{pmatrix}, \quad B = \begin{pmatrix} 0 & 1 & 0 \\ 0 & 3 & 0 \\ 0 & 1 & 0 \end{pmatrix}, \tag{B.125}$$

$$A^t = \begin{pmatrix} 0 & 3 & 0 \\ 0 & 0 & 0 \\ 0 & 0 & 0 \end{pmatrix}, \quad B^t = [0], \quad I = -2.5, \quad t = 1. \tag{B.126}$$

Upwards:

$$A = \begin{pmatrix} 0 & 0 & 0 \\ 0 & 2 & 0 \\ 0 & 0 & 0 \end{pmatrix}, \quad B = \begin{pmatrix} 0 & 1 & 0 \\ 0 & 3 & 0 \\ 0 & 1 & 0 \end{pmatrix}. \tag{B.127}$$

$$A^t = \begin{pmatrix} 0 & 0 & 0 \\ 0 & 0 & 0 \\ 0 & 3 & 0 \end{pmatrix}, \quad B^t = [0], \quad I = -2.5, \quad t = 1. \tag{B.128}$$

Left-to-right:
$$A = \begin{pmatrix} 0 & 0 & 0 \\ 0 & 2 & 0 \\ 0 & 0 & 0 \end{pmatrix}, \quad B = \begin{pmatrix} 0 & 0 & 0 \\ 1 & 3 & 1 \\ 0 & 0 & 0 \end{pmatrix}. \quad (B.129)$$

$$A^t = \begin{pmatrix} 0 & 0 & 0 \\ 3 & 0 & 0 \\ 0 & 0 & 0 \end{pmatrix}, \quad B^t = [0], \quad I = -2.5, \quad t = 1. \quad (B.130)$$

Right-to-left:
$$A = \begin{pmatrix} 0 & 0 & 0 \\ 0 & 2 & 0 \\ 0 & 0 & 0 \end{pmatrix}, \quad B = \begin{pmatrix} 0 & 0 & 0 \\ 1 & 3 & 1 \\ 0 & 0 & 0 \end{pmatrix}. \quad (B.131)$$

$$A^t = \begin{pmatrix} 0 & 0 & 0 \\ 0 & 0 & 3 \\ 0 & 0 & 0 \end{pmatrix}, \quad B^t = [0], \quad I = -2.5, \quad t = 1. \quad (B.132)$$

Name: MOVEHOR
Function: Direction and speed dependent motion direction. Only objects moving horizontally to the right with a speed of 1 pixel/delay time remain on the screen.

$$A = \begin{pmatrix} -0.1 & -0.1 & -0.1 \\ -0.1 & 0 & -0.1 \\ -0.1 & -0.1 & -0.1 \end{pmatrix}, \quad B = \begin{pmatrix} 0 & 0 & 0 \\ 0 & 1.5 & 0 \\ 0 & 0 & 0 \end{pmatrix}, \quad I = -2. \quad (B.133)$$

$$A^t = \begin{pmatrix} 0 & 0 & 0 \\ 0 & 0 & 0 \\ 0 & 0 & 0 \end{pmatrix}, \quad B^t = \begin{pmatrix} 0 & 0 & 0 \\ 1.5 & 0 & 0 \\ 0 & 0 & 0 \end{pmatrix}, \quad t = 10. \quad (B.134)$$

Name: MULLER
Function: Simulates the Muller–Lyer illusion: although the horizontal lines between the arrows are of the same length, the one on the top seems to be longer.

$$A = \begin{pmatrix} 0 & 0 & 0 & 0 & 0 \\ 0 & 0 & 0 & 0 & 0 \\ 0 & 0 & 1.3 & 0 & 0 \\ 0 & 0 & 0 & 0 & 0 \\ 0 & 0 & 0 & 0 & 0 \end{pmatrix}, \quad (B.135)$$

$$B = \begin{pmatrix} -0.1 & -0.1 & -0.1 & -0.1 & -0.1 \\ -0.1 & -0.1 & -0.1 & -0.1 & -0.1 \\ -0.1 & -0.1 & 1.3 & -0.1 & -0.1 \\ -0.1 & -0.1 & -0.1 & -0.1 & -0.1 \\ -0.1 & -0.1 & -0.1 & -0.1 & -0.1 \end{pmatrix}, \quad I = 0. \quad (B.136)$$

Name: NARROW LINE DETECTION
Function: Detects narrow lines in an image.

$$A = \begin{pmatrix} -0.1 & 0.4 & -0.1 \\ 0.4 & 2.0 & 0.4 \\ -0.1 & 0.4 & -0.1 \end{pmatrix}, \quad B = \begin{pmatrix} 0.2 & 0.0 & 0.2 \\ -3.0 & 2.5 & -3.0 \\ 0.2 & 0.0 & 0.2 \end{pmatrix}, \quad I = -5.0. \quad (B.137)$$

Name: 3 NEIGHBORS
Function: Driving to black those white pixels in a black-and-white image that have which has at least three black neighbors.

$$A = [4], \quad B = \begin{pmatrix} 1.6 & 1.6 & 1.6 \\ 1.6 & 3 & 1.6 \\ 1.6 & 1.6 & 1.6 \end{pmatrix}, \quad I = 12.3. \quad (B.138)$$

Name: NOISE REMOVAL
Function: Filters "white noise" with small dispersion and impulsive noise with small dynamic range. Three variants of this filter may be implemented by the following templates.

Variant I

$$A = [0], \quad B = \begin{pmatrix} 0.11 & 0.11 & 0.11 \\ 0.11 & 0.11 & 0.11 \\ 0.11 & 0.11 & 0.11 \end{pmatrix}, \quad I = 0. \quad (B.139)$$

Variant II

$$A = [0], \quad B = \begin{pmatrix} 0.1 & 0.1 & 0.1 \\ 0.1 & 0.5 & 0.1 \\ 0.1 & 0.1 & 0.1 \end{pmatrix}, \quad I = 0. \quad (B.140)$$

Variant III

$$A = [0], \quad B = \begin{pmatrix} 0.0625 & 0.125 & 0.0625 \\ 0.125 & 0.25 & 0.125 \\ 0.0625 & 0.125 & 0.0625 \end{pmatrix}, \quad I = 0. \quad (B.141)$$

Name: OR
Function: Logical "OR" function of the input and the initial state.

$$A = \begin{pmatrix} 0 & 0 & 0 \\ 0 & 3 & 0 \\ 0 & 0 & 0 \end{pmatrix}, \quad B = \begin{pmatrix} 0 & 0 & 0 \\ 0 & 3 & 0 \\ 0 & 0 & 0 \end{pmatrix}, \quad I = 2. \quad (B.142)$$

Name: OBJECT COUNTING
Function: This algorithm counts the connected objects on a grey-scale image. The cited templates can be found in this template library.
   The flow-chart of the algorithm is shown in Fig. B.6:

252   B. Library of Templates

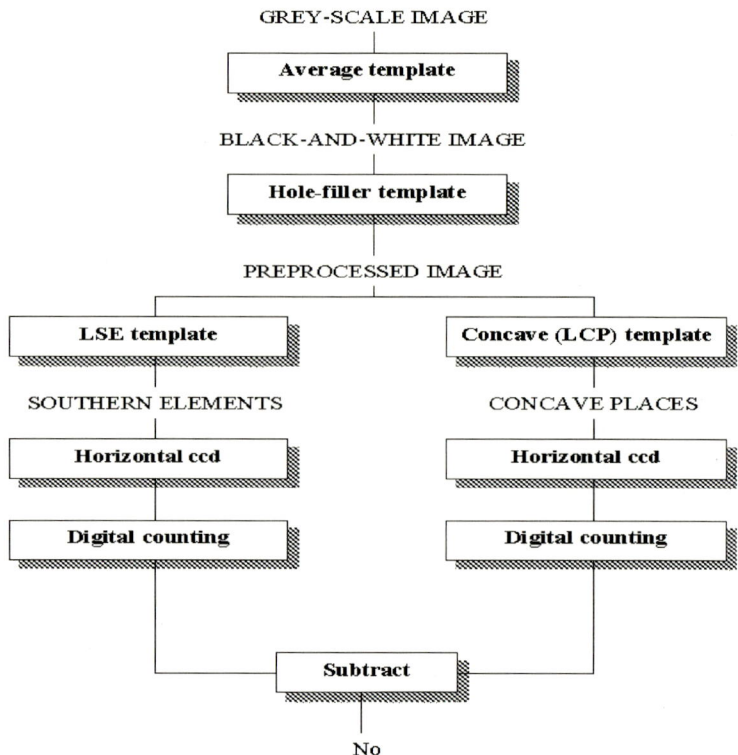

Fig. B.6. Flow-chart of the algorithm for "Object Counting": No=Number of objects

Name: OBJECT WITH SMALL HOLES
Function: Extract objects with small holes from a black-and-white image. The cited templates can be found in this template library.
A flow-chart of the algorithm is given in Fig. B.7.

Name: PARITY
Function: The template determines the parity of a row of the input image, i.e. it determines whether the number of black pixels in a row is even or odd. As a result, in the output image the leftmost pixel corresponds to the parity of the row, namely black representing odd, and white meaning even parity. It is also true that each pixel codes the parity of the pixels to the right of it, together with the pixel itself. Naturally, the parity of a column or a diagonal can be counted in the same manner. The parity of an array can also be determined if columnwise parity is computed on the result of the rowwise parity operation. The initial state should be set to -0.5.

$$A = \begin{pmatrix} 0\ a\ b \end{pmatrix},\ B = \begin{pmatrix} c \end{pmatrix}, \tag{B.143}$$

B. Library of Templates   253

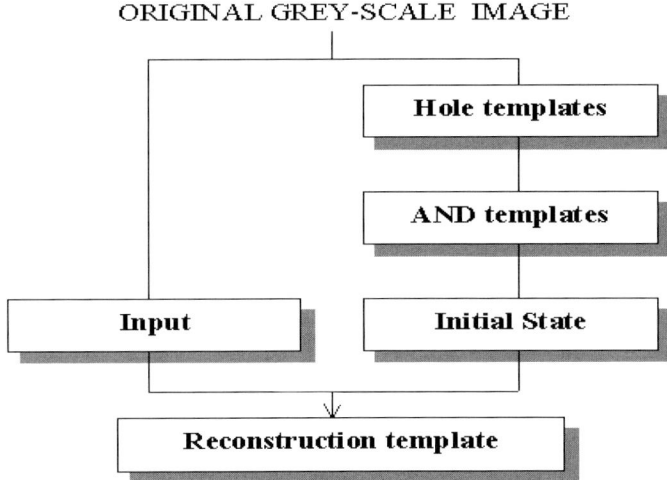

Fig. B.7. Flow-chart of the algorithm for "Object with small holes"

where

$$a = \begin{cases} -2 & \text{if } V_y < -0.7 \\ 0.2 & \text{if } -0.7 < V_y < -0.1 \\ -2 & \text{if } -0.1 < V_y < 0.1 \\ 2 & \text{if } 0.1 < V_y < 0.3 \\ -2 & \text{if } 0.3 < V_y < 0.7 \\ 2 & \text{if } V_y > 0.7 \end{cases}$$

$$b = \begin{cases} 0 & \text{if } V_y < -0.7 \\ -0.7 & \text{if } -0.7 < V_y < -0.1 \\ -0.1 & \text{if } -0.1 < V_y < 0.1 \\ 0.1 & \text{if } 0.1 < V_y < 0.3 \\ -0.1 & \text{if } 0.3 < V_y < 0.7 \\ 0 & \text{if } V_y > 0.7 \end{cases}$$

$$c = \begin{cases} -0.1 & \text{if } V_u < 0.1 \\ 0.1 & \text{if } V_u > 0.1 \end{cases}.$$

Name: PA-PB
Function: Takes the logical difference of the initial state and the input pictures.

$$A = \begin{pmatrix} 0\,0\,0 \\ 0\,1\,0 \\ 0\,0\,0 \end{pmatrix}, \quad B = \begin{pmatrix} 0\;\,0\;\,0 \\ 0\,{-1}\,0 \\ 0\;\,0\;\,0 \end{pmatrix}, \quad I = -1. \tag{B.144}$$

Name: PA-PB-F
Function: Takes the logical difference of the delayed and actual input in continuous mode together with noise filtering. Moving parts of an image can be extracted.

$$A = \begin{pmatrix} 0\,0\,0 \\ 0\,1\,0 \\ 0\,0\,0 \end{pmatrix}, \quad B = \begin{pmatrix} 0.25\;0.25\;0.25 \\ 0.25\;2.00\;0.25 \\ 0.25\;0.25\;0.25 \end{pmatrix}, \quad I = -4.75, \tag{B.145}$$

$$A^t = \begin{pmatrix} 0\,0\,0 \\ 0\,0\,0 \\ 0\,0\,0 \end{pmatrix}, \quad B^t = \begin{pmatrix} -0.25\;-0.25\;-0.25 \\ -0.25\;-2.00\;-0.25 \\ -0.25\;-0.25\;-0.25 \end{pmatrix}, \quad t = 10\;. \tag{B.146}$$

Name: PATCHMAKER
Function: This diffusion type template enlarges each object permanently until the transient is stopped. So the transient must be stopped after a certain time, otherwise it will drive all pixels to black.

$$A = \begin{pmatrix} 0\,1\,0 \\ 1\,2\,1 \\ 0\,1\,0 \end{pmatrix}, \quad B = \begin{pmatrix} 0\,0\,0 \\ 0\,2\,0 \\ 0\,0\,0 \end{pmatrix}, \quad I = 4.5. \tag{B.147}$$

Name: PEEL1PIX
Function: Peels off one pixel from all directions.

$$A = \begin{pmatrix} 0\;\;0.4\;\;0 \\ 0.4\;1.4\;0.4 \\ 0\;\;0.4\;\;0 \end{pmatrix}, \quad B = \begin{pmatrix} 4.6\;\,-2.8\;\,4.6 \\ -2.8\;\;\;1\;\;\;-2.8 \\ 4.6\;\,-2.8\;\,4.6 \end{pmatrix}, \quad I = -7.2\;. \tag{B.148}$$

Name: PEELHOR
Function: Peels off one pixel from the left.

$$A = \begin{pmatrix} 0\,0\,0 \\ 0\,2\,0 \\ 0\,0\,0 \end{pmatrix}, \quad B = \begin{pmatrix} 0\,0\,0 \\ 3\,3\,0 \\ 0\,0\,0 \end{pmatrix}, \quad I = -5. \tag{B.149}$$

Name: PHI
Function: Realizes the PHI phenomenon. The PHI phenomenon is a motion illusion: if a light dot in a picture is set off and at the same time or a bit later the light dot appears in another position, then the observer perceives fast motion. A two-layer CNN shows this phenomenon.

$$A_{11} = \begin{pmatrix} 0 & 0.3 & 0 \\ 0.1 & 0.2 & 0.1 \\ 0 & 0.3 & 0 \end{pmatrix}, \quad B_{11} = \begin{pmatrix} 0.1 & 0.1 & 0.1 \\ 0.1 & 0.3 & 0.1 \\ 0.1 & 0.1 & 0.1 \end{pmatrix}, \quad I = 0. \tag{B.150}$$

Name: RECONSTRUCTION
Function: Reconstructs an object in an image starting from some of its pixels. The input is the original image and the initial state is the set of relevant pixels.

$$A = \begin{pmatrix} 0 & 1 & 0 \\ 1 & 2 & 1 \\ 0 & 1 & 0 \end{pmatrix}, \quad B = \begin{pmatrix} 0 & 0 & 0 \\ 0 & 2 & 0 \\ 0 & 0 & 0 \end{pmatrix}, \quad I = -2.5. \tag{B.151}$$

Name: RIGHTCON
Function: Detects the right contour of the object.

$$A = \begin{pmatrix} 0 & 0 & 1 \\ 1 & 2 & -1 \\ 0 & 0 & 1 \end{pmatrix}, \quad B = \begin{pmatrix} 0 & 0 & 0 \\ -1 & 1 & -3 \\ 0 & 0 & 0 \end{pmatrix}, \quad I = -2. \tag{B.152}$$

Name: SCRATCH REMOVAL
Function: On photocopier machines the glass panel often gets scratched, and some scratches are then copied together with the material, resulting in a visually annoying copy. This algorithm is capable of removing such scratches, assuming that the location of the scratch is known in advance. This is a valid assumption, since the scratches can automatically be detected e.g. by copying a blank sheet of paper. The algorithm removes the scratches gradually, peeling off pixels circularly.

256   B. Library of Templates

The flow-chart of the algorithm is shown in Fig. B.8:

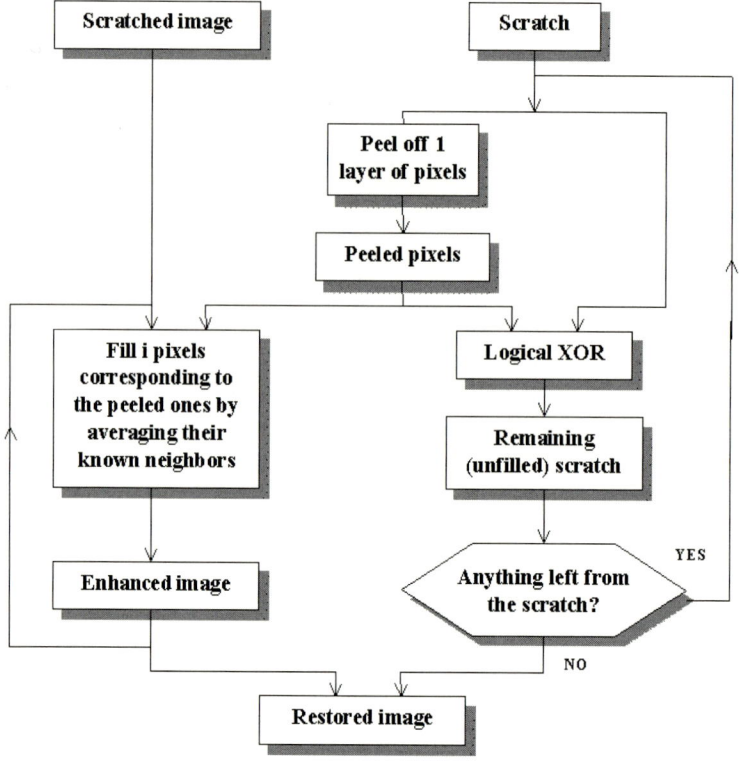

Fig. B.8. Flow-chart of the algorithm for "Scratch Removal"

Selection templates:

$$A_1 = \begin{pmatrix} 1 \end{pmatrix}, \quad B_1 = \begin{pmatrix} -0.5 & 0 & 0 \\ -0.5 & 0.5 & 0 \\ -0.5 & 0 & 0 \end{pmatrix}, \quad I_1 = -1.5 \qquad (B.153)$$

$$A_2 = \begin{pmatrix} 1 \end{pmatrix}, \quad B_2 = \begin{pmatrix} -0.5 & -0.5 & 0 \\ -0.5 & 0.5 & 0 \\ 0 & 0 & 0 \end{pmatrix}, \quad I_2 = -1.5 \qquad (B.154)$$

$$A_3 = \begin{pmatrix} 1 \end{pmatrix}, \quad B_3 = \begin{pmatrix} -0.5 & -0.5 & -0.5 \\ 0 & 0.5 & 0 \\ 0 & 0 & 0 \end{pmatrix}, \quad I_3 = -1.5 \qquad (B.155)$$

B. Library of Templates   257

$$A_4 = \begin{pmatrix} 1 \end{pmatrix}, \ B_4 = \begin{pmatrix} 0 & -0.5 & -0.5 \\ 0 & 0.5 & -0.5 \\ 0 & 0 & 0 \end{pmatrix}, \ I_4 = -1.5 \quad \text{(B.156)}$$

$$A_5 = \begin{pmatrix} 1 \end{pmatrix}, \ B_5 = \begin{pmatrix} 0 & 0 & -0.5 \\ 0 & 0.5 & -0.5 \\ 0 & 0 & -0.5 \end{pmatrix}, \ I_5 = -1.5 \quad \text{(B.157)}$$

$$A_6 = \begin{pmatrix} 1 \end{pmatrix}, \ B_6 = \begin{pmatrix} 0 & 0 & 0 \\ 0 & 0.5 & -0.5 \\ 0 & -0.5 & -0.5 \end{pmatrix}, \ I_6 = -1.5 \quad \text{(B.158)}$$

$$A_7 = \begin{pmatrix} 1 \end{pmatrix}, \ B_7 = \begin{pmatrix} 0 & 0 & 0 \\ 0 & 0.5 & 0 \\ -0.5 & -0.5 & -0.5 \end{pmatrix}, \ I_7 = -1.5 \quad \text{(B.159)}$$

$$A_8 = \begin{pmatrix} 1 \end{pmatrix}, \ B_8 = \begin{pmatrix} 0 & 0 & 0 \\ -0.5 & 0.5 & 0 \\ -0.5 & -0.5 & 0 \end{pmatrix}, \ I_8 = -1.5 \quad \text{(B.160)}$$

Fill templates:

$$B_1 = \begin{pmatrix} 0.33 & 0 & 0 \\ 0.34 & 0 & 0 \\ 0.33 & 0 & 0 \end{pmatrix}, \ B_2 = \begin{pmatrix} 0.34 & 0.33 & 0 \\ 0.33 & 0 & 0 \\ 0 & 0 & 0 \end{pmatrix}, \quad \text{(B.161)}$$

$$B_3 = \begin{pmatrix} 0.33 & 0.34 & 0.33 \\ 0 & 0 & 0 \\ 0 & 0 & 0 \end{pmatrix}, \ B_4 = \begin{pmatrix} 0 & 0.33 & 0.34 \\ 0 & 0 & 0.33 \\ 0 & 0 & 0 \end{pmatrix}, \quad \text{(B.162)}$$

$$B_5 = \begin{pmatrix} 0 & 0 & 0.33 \\ 0 & 0 & 0.34 \\ 0 & 0 & 0.33 \end{pmatrix}, \ B_6 = \begin{pmatrix} 0 & 0 & 0 \\ 0 & 0 & 0.33 \\ 0 & 0.33 & 0.34 \end{pmatrix}, \quad \text{(B.163)}$$

$$B_7 = \begin{pmatrix} 0 & 0 & 0 \\ 0 & 0 & 0 \\ 0.33 & 0.34 & 0.33 \end{pmatrix}, \ B_8 = \begin{pmatrix} 0 & 0 & 0 \\ 0.33 & 0 & 0 \\ 0.34 & 0.33 & 0 \end{pmatrix}. \quad \text{(B.164)}$$

Name: SMALL DIFF
Function: Extracting those pixels which have at least 4 neighboring pixels with maximum 0.1 difference in absolute value.

$$A = \begin{pmatrix} a & a & a \\ a & 0 & a \\ a & a & a \end{pmatrix}, \ B = 0, \ I = 6, \quad \text{(B.165)}$$

where

$$a = \begin{cases} -2 & \text{if } (V_{xij} - V_{xkl}) < -0.1 \\ 2 & \text{if } -0.1 < (V_{xij} - V_{xkl}) < 0.1 \\ -2 & \text{if } (V_{xij} - V_{xkl}) > 0.1 \end{cases}.$$

Name: UNPEEL
Function: Drive to black all that white pixels that have a direct black neighbor.

$$A = [2]\,,\quad B = \begin{pmatrix} 0 & 0.3 & 0 \\ 0.3 & 0 & 0.3 \\ 0 & 0.3 & 0 \end{pmatrix},\quad I = 2. \tag{B.166}$$

Name: UNCLOSED LINES
Function: Drives to white all those black pixels that have fewer than two direct black neighbors. The result is the elimination of the entire unclosed lines from a black-and-white image.

$$A = \begin{pmatrix} 0.5 & 0.5 & 0.5 \\ 0.5 & 2 & 0.5 \\ 0.5 & 0.5 & 0.5 \end{pmatrix},\quad B = \begin{pmatrix} 0 & 0 & 0 \\ 0 & 0 & 0 \\ 0 & 0 & 0 \end{pmatrix},\quad I = 1.5. \tag{B.167}$$

References

1. Leon O. Chua, Lin Yang, "Cellular neural networks," in *IEEE International Symposium on Circuits and Systems*, 1988, vol. 2, pp. 985–988.
2. Leon O. Chua, Lin Yang, "Cellular neural networks: Theory," *IEEE Transactions on Circuits and Systems*, vol. 35, no. 10, pp. 1257–1272, 1988.
3. Leon O. Chua, Lin Yang, "Cellular neural networks: Applications," *IEEE Transactions on Circuits and Systems*, vol. 35, no. 10, pp. 1273–1290, 1988.
4. Leon O. Chua, Lin Yang, "Cellular neural networks," *US Patent*, , no. 5140670, 1992.
5. S. Wolfram, "Computation theory of cellular automata," *Communications in Mathematical Physics*, vol. 96, pp. 15–57, 1984.
6. T. Toffoli, N. Margolus, *Cellular Automata Machines: A New Environment for Modeling*, MIT Press, Cambridge, MA, 1987.
7. J.J. Hopfield, "Neural networks and physical systems with emergent computational abilities," *Proc. Natl. Acad. Sci. USA.*, vol. 79, pp. 2554–2558, 1982.
8. Carver Mead, *Analog VLSI and Neural Systems*, Addison-Wesley, Reading, MA, 1989.
9. P. Arena, S. Baglio, L. Fortuna, G. Manganaro, "Cellular neural networks: A survey," in *Proc. of 7th IMACS-IFAC Symposium on Large Scale Systems Theory and Applications (LSS'95)*, 1995, vol. 1, pp. 53–58.
10. T. Roska, J. Vandewalle, Ed.s, "Special issue on CNNs," *International Journal on Circuit Theory and Applications*, vol. 20, no. sept./oct., 1992.
11. T. Roska, J. Vandewalle, Ed.s, "Special issue on CNNs," *International Journal on Circuit Theory and Applications*, vol. 24, no. 3, 1996.
12. T. Roska, J. A. Nossek, Ed.s, "Special issue on CNNs," *IEEE Transactions on Circuits and Systems–Part I: Fundamental Theory and Applications*, vol. 40, no. 3, 1993.
13. T. Roska, J.A. Nossek, Ed.s, "Special issue on CNNs," *IEEE Transactions on Circuits and Systems–Part II: Analog and Digital Signal Processing*, vol. 40, no. 3, 1993.
14. L.O. Chua, Ed., "Special issue on nonlinear waves, patterns and spatio-temporal chaos in dynamic arrays," *IEEE Transactions on Circuits and Systems–Part I: Fundamental Theory and Applications*, vol. 42, no. 10, 1995.
15. T. Roska, Ed., *Proc. Int. Workshop Cellular Neural Networks and Their Applications.* IEEE, 1990.
16. J.A. Nossek, Ed., *Proc. Int. Workshop Cellular Neural Networks and Their Applications.* IEEE, 1992.
17. V. Cimagalli, Ed., *Proc. Int. Workshop Cellular Neural Networks and Their Applications.* IEEE, 1994.
18. A. Rodríguez-Vázquez, Ed., *Proc. Int. Workshop Cellular Neural Networks and Their Applications.* IEEE, 1996.

19. V. Tavsanoglu, Ed., *Proc. Int. Workshop Cellular Neural Networks and Their Applications*. IEEE, 1998.
20. Leon O. Chua, Tamas Roska, "The cnn paradigm," *IEEE Transactions on Circuits and Systems–Part I: Fundamental Theory and Applications*, vol. 40, no. 3, pp. 147–156, 1993.
21. Leon O. Chua, Martin Hasler, George S. Moschytz, Jacques Neirynck, "Autonomous cellular neural networks: A unified paradigm for pattern formation and active wave propagation," *IEEE Transactions on Circuits and Systems–Part I: Fundamental Theory and Applications*, vol. 42, no. 10, pp. 559–577, 1995.
22. Patrick Thiran, "Influence of boundary conditions on the behavior of cellular neural networks," *IEEE Transactions on Circuits and Systems–Part I: Fundamental Theory and Applications*, vol. 40, no. 3, pp. 207–212, 1993.
23. B. Mirzai, G.S. Moschytz, "The influence of boundary conditions on the robustness of cellular neural networks," *IEEE Transactions on Circuits and Systems–Part I: Fundamental Theory and Applications*, vol. 45, no. 4, pp. 511–515, 1998.
24. Leon O. Chua, Chai Wah Wu, *On the universe of stable Cellular Neural Networks*, pp. 59–79, In J. Vandewalle [36], 1993.
25. Leon O. Chua, Tamás Roska, "Stability of a class of nonreciprocal cellular neural networks," *IEEE Transactions on Circuits and Systems*, vol. 37, no. 12, pp. 1520–1527, 1990.
26. Rafael C. Gonzalez, *Digital Image Processing*, Addison-Wesley, Reading, MA, 1992.
27. William K. Pratt, *Digital Image Processing*, John Wiley & Sons, New York, NJ, 1978.
28. D.P. Bertsekas, J.N. Tsitsiklis, *Parallel and distributed computations, Numerical methods*, Prentice Hall, New York, NJ, 1989.
29. Tamás Roska, Leon O.Chua, "Reprogrammable CNN and supercomputer," *US Patent*, , no. 5355528, 1994.
30. Tamás Roska, Leon O.Chua, "The CNN universal machine: An analogic array computer," *IEEE Transactions on Circuits and Systems–Part II: Analog and Digital Signal Processing*, vol. 40, no. 3, pp. 163–173, 1993.
31. Bertram E. Shi, "Gabor-type filtering in space and time with cellular neural networks," *IEEE Transactions on Circuits and Systems–Part I: Fundamental Theory and Applications*, vol. 45, no. 2, pp. 121–132, 1998.
32. Giuseppe Grassi, "A new approach to design cellular neural networks for associative memories," *IEEE Transactions on Circuits and Systems–Part I: Fundamental Theory and Applications*, vol. 44, no. 9, pp. 835–838, 1997.
33. Derong Liu, Anthony N.Michel, "Sparsely interconnected neural networks for associative memories with applications to cellular neural networks," *IEEE Transactions on Circuits and Systems–Part I: Fundamental Theory and Applications*, vol. 41, no. 4, pp. 295–307, 1994.
34. Derong Liu, "Cloning template design of cellular neural networks for associative memories," *IEEE Transactions on Circuits and Systems–Part I: Fundamental Theory and Applications*, vol. 44, no. 7, pp. 646–650, 1997.
35. K. Laszlo P. Szolgay, I. Szatmari, "A fast fixed point learning method to implement associative memory on cnn's," *IEEE Transactions on Circuits and Systems–Part I: Fundamental Theory and Applications*, vol. 44, no. 4, pp. 362–363, 1997.
36. T. Roska, J. Vandewalle, Ed., *Cellular Neural Networks*, John Wiley & Sons, New York, NJ, 1993.

37. F.A. Savaci, J. Vandewalle, "On the stability analysis of cellular neural networks," *IEEE Transactions on Circuits and Systems–Part I: Fundamental Theory and Applications*, vol. 40, no. 3, pp. 213–214, 1993.
38. N. Takahashi, L.O. Chua, "A new sufficient condition for nonsymmetric cnn's to have a stable equilibrium point," *IEEE Transactions on Circuits and Systems–Part I: Fundamental Theory and Applications*, vol. 44, no. 11, pp. 1092–1095, 1997.
39. Tamás Roska, Leon O. Chua, *Cellular Neural Networks with non-linear and delay-type template elements and non-uniform grids*, pp. 31–43, In J.Vandewalle [36], 1993.
40. M. Biey, F.Bonanni, M. Gilli, I. Maio, "Qualitative analysis of the dynamics of the time-delayed chua's circuit," *IEEE Transactions on Circuits and Systems–Part I: Fundamental Theory and Applications*, vol. 44, no. 6, pp. 486–500, 1997.
41. W. Heiligenberg, T. Roska, "On biological sensory information processing principles relevant to dually computing CNN's," *Rep. DNS-4-1992 , Dual and Neural Computing Systems Laboratory, Hungarian Academy of Science*, 1992.
42. T. Roska, J. Hámori, E. Lábos, K. Lotz, L. Orzó, J. Takács, P.L. Venetianer, Z. Vidnyánszky, A.Zarándy, "The use of CNN models in the subcortical visual pathway," *IEEE Transactions on Circuits and Systems–Part I: Fundamental Theory and Applications*, vol. 40, no. 3, pp. 182–195, 1993.
43. F. Werblin, A. Jacobs, "Using CNN to unravel space-time processing in the vertebrate retina," In V. Cimagalli [17], pp. 33–40.
44. Valerio Cimagalli, "A neural network architecture for detecting moving objects II," In T. Roska [15], pp. 124–126.
45. T. Roska, T. Boros, P. Thiran, L.O. Chua, "Detecting simple motion using cellular neural networks," In T. Roska [15], pp. 127–138.
46. T. Roska, T. Boros, A. Radványi, P. Thiran, L.O. Chua, *Detecting moving and standing objects using cellular neural networks*, pp. 175–190, In J. Vandewalle [36], 1993.
47. R. Madan, Ed., *Chua's circuit: a paradigm for chaos*, World Scientific, Singapore, 1993.
48. Hubert Harrer, Josef A. Nossek, *Discrete-time Cellular Neural Networks*, pp. 15–29, In J. Vandewalle [36], 1993.
49. C. Ho, S. Mori, "A systematic design method for discrete-time cellular neural networks," in *European Conference on Circuit Theory and Design*, 1993, pp. 693–698.
50. H. Magnussen, J.A. Nossek, "Global learning algorithms for discrete-time cellular neural networks," In V. Cimagalli [17], pp. 165–170.
51. John J. D'Azzo, Constantine H. Houpis, *Linear Control System Analysis and Design Convential and Modern*, Electrical and Electronic Engineering Series. McGraw-Hill, New York, NJ, 3rd edition, 1988.
52. Leon O. Chua, Tamás Roska, Peter L. Venetianer, "The CNN is universal as the Turing machine," *IEEE Transactions on Circuits and Systems–Part I: Fundamental Theory and Applications*, vol. 40, no. 4, pp. 289–291, 1993.
53. Kenneth R. Crounse, Leon O. Chua, "The CNN universal machine is as universal as a Turing machine," *IEEE Transactions on Circuits and Systems–Part I: Fundamental Theory and Applications*, vol. 43, no. 4, pp. 353–355, 1996.
54. Josef A. Nossek, "Design and learning with cellular neural networks," In V. Cimagalli [17], pp. 137–146.

55. P. Arena, S. Baglio, L. Fortuna, G. Manganaro, "Chua's circuit can be generated by CNN cells," *IEEE Transactions on Circuits and Systems–Part I: Fundamental Theory and Applications*, vol. 42, no. 2, pp. 123–125, 1995.
56. P. Arena, L. Fortuna, G. Manganaro, S. Spina, "CNN image processing for the automatic classification of oranges," In V. Cimagalli [17], pp. 463–467.
57. K. Wuthrich, *NMR of proteins and nucleic acids*, John Wiley & Sons, New York, NJ, 1986.
58. H. Kessler, M. Gehrke, C. Griesinger, ," *Adv. Chem. Int. Ed. Engl.*, vol. 27, pp. 490–536, 1988.
59. P. Catasti, E. Carrara, C. Nicolini, "PEPTO: an expert system for automatic peak assignment of two-dimensional nuclear magnetic resonance spectra of proteins," *Journal of Computational Chemistry*, 1992.
60. A. Friedman, *Mathematics in Industrial Problems*, vol. 31 of *IMA Vol. Math. Appl.*, Springer-Verlag, Berlin, Heidelberg, 1990.
61. P. Arena, S. Baglio, L. Fortuna, G. Manganaro, "Air quality modeling with CNN's," in *European Conference on Circuit Theory and Design*, 1995, vol. 2, pp. 885–888.
62. L. Fortuna, S. Graziani, G. Manganaro, G. Muscato, "The CNN's as innovative computing paradigm for modeling," in *Proc. IDA World Congress on Desalination and Water Sciences*, 1995, vol. 4, pp. 399–409.
63. T. Kozek, T. Roska, "A double time-scale CNN for solving 2-D Navier-Stokes equations," In V. Cimagalli [17], pp. 267–272.
64. Tamás Roska, Leon O. Chua, Dietrich Wolf, Tibor Kozek, Ronald Teztlaff, Frank Puffer, "Simulating nonlinear waves and partial differential equations via CNN-part I: Basic techniques," *IEEE Transactions on Circuits and Systems–Part I: Fundamental Theory and Applications*, vol. 42, no. 10, pp. 807–815, 1995.
65. Tibor Kozek, Leon O. Chua, Tamás Roska, Dietrich Wolf, Ronald Teztlaff, Frank Puffer, Karoly Lotz, "Simulating nonlinear waves and partial differential equations via CNN-part II: Typical examples," *IEEE Transactions on Circuits and Systems–Part I: Fundamental Theory and Applications*, vol. 42, no. 10, pp. 816–820, 1995.
66. Paolo Arena, *Reti neurali per modellistica e predizione in circuiti e sistemi*, Ph.D. thesis, University of Catania - Italy, 1994, ing. elettrotecnica VI ciclo, in italian.
67. John Guckenheimer, Philip Holmes, *Nonlinear Oscillations, dynamical systems, and bifurcations of vector fields*, vol. 42 of *Applied Mathematical Sciences*, Springer-Verlag, Berlin, Heidelberg, 1983.
68. Thomas S. Parker, Leon O. Chua, *Practical numerical algorithms for chaotic systems*, Springer-Verlag, Berlin, Heidelberg, 1989.
69. M.A. van Wyk, W.-H. Steeb, *Chaos in Electronics*, vol. 2 of *Mathematical Modelling, Theory and Applications*, Kluwer, Dordrecht, 1997.
70. S. Baglio, R. Cristaudo, L. Fortuna, G. Manganaro, "Complexity in an industrial flyback converter," *International Journal of Circuits, Systems and Computers*, vol. 5, no. 4, pp. 627–633, 1995.
71. Salvatore Baglio, *comportamenti non lineari e fenomeni caotici nei circuiti e nei sistemi dinamici*, Ph.D. thesis, University of Catania - Italy, 1994, ing. elettrotecnica VI ciclo, in italian.
72. E. Ott, C. Grebogi, J.A. Yorke, "Controlling chaos," *Physical Review Letters*, vol. 64, no. 11, pp. 1196–1199, 1990.
73. Fan Zou, Josef A. Nossek, "A chaotic attractor with Cellular Neural Networks," *IEEE Transactions on Circuits and Systems–Part I: Fundamental Theory and Applications*, vol. 38, no. 7, pp. 811–812, 1991.

74. Fan Zou, Josef A. Nossek, "Bifurcation and chaos in Cellular Neural Networks," *IEEE Transactions on Circuits and Systems–Part I: Fundamental Theory and Applications*, vol. 40, no. 3, pp. 166–173, 1993.
75. Fan Zou, Axel Katérle, Josef A. Nossek, "Homoclinic and heteroclinic orbits of the three-cell Cellular Neural Network," *IEEE Transactions on Circuits and Systems–Part I: Fundamental Theory and Applications*, vol. 40, no. 11, pp. 843–848, 1993.
76. Edward Ott, *Chaos in dynamical systems*, Cambridge University Press, Cambridge, 1993.
77. M.W. Hirsch, S. Smale, *Differential equations, dynamical systems and linear algebra*, Academic Press, New York, NJ, 1974.
78. Gabriele Manganaro, Mario Lavorgna, Matteo Lo Presti, Luigi Fortuna, "Cellular neural network to obtain the so-called unfolded Chua's Circuit," *European Patent*, , no. 96830137.4-2201, 1996.
79. P. Arena, S. Baglio, L. Fortuna, G. Manganaro, "State controlled CNN: A new strategy for generating high complex dynamics," *IEICE Transactions on Fundamentals of Electronics, Communications and Computer Sciences*, vol. E79-A, no. 10, pp. 1647–1657, 1996.
80. P. Arena, S. Baglio, L. Fortuna, G. Manganaro, "A simplified scheme for the realisation of the chua's oscillator by using SC-CNN cells," *IEE Electronics Letters*, vol. 31, no. 21, pp. 1794–1795, 1995.
81. Leon O. Chua, "Global unfolding of chua's circuit," *IEICE Transactions on Fundamentals of Electronics, Communications and Computer Sciences*, vol. E76-A, pp. 704–734, 1993.
82. Vincente Pérez-Muñuzuri, Vincente Pérez-Villar, Leon O. Chua, "Autowaves for image processing on a two-dimensional CNN array of excitable nonlinear circuits: flat and wrinkled labyrinths," *IEEE Transactions on Circuits and Systems–Part I: Fundamental Theory and Applications*, vol. 40, no. 3, pp. 174–181, 1993.
83. Alberto Pérez-Muñuzuri, Vincente Pérez-Muñuzuri, Vincente Pérez-Villar, Leon O. Chua, "Spiral waves on a 2-D array nonlinear circuits," *IEEE Transactions on Circuits and Systems–Part I: Fundamental Theory and Applications*, vol. 40, no. 11, pp. 872–877, 1993.
84. M.P. Kennedy, "Chaos in the Colpitts oscillator," *IEEE Transactions on Circuits and Systems–Part I: Fundamental Theory and Applications*, vol. 41, pp. 771–774, 1994.
85. M.P. Kennedy, "On the relationship between the chaotic Colpitts oscillator and Chua's Oscillator," *IEEE Transactions on Circuits and Systems–Part I: Fundamental Theory and Applications*, vol. 42, pp. 376–379, 1995.
86. B.Z. Kaplan G. Sarafian, "Is the Colpitts oscillator a relative of Chua's Circuit?," *IEEE Transactions on Circuits and Systems–Part I: Fundamental Theory and Applications*, vol. 42, pp. 373–376, 1995.
87. P. Arena, S. Baglio, L. Fortuna, G. Manganaro, "A CNN to generate the chaos of the colpitts oscillator," in *International Symp.on Nonlinear Theory and its Applications (NOLTA'95)*, 1995, vol. 2, pp. 689–694.
88. P. Arena, S. Baglio, L. Fortuna, G. Manganaro, "How state controlled CNN cells generate the dynamics of the colpitts-like oscillator," *IEEE Transactions on Circuits and Systems–Part I: Fundamental Theory and Applications*, vol. 43, no. 7, pp. 602–605, 1996.
89. T.Saito, "An approach toward higher dimensional hysteresis chaos generators," *IEEE Transactions on Circuits and Systems*, vol. 37, pp. 399–409, 1990.

90. L.O. Chua, C. A. Desoer, E.S. Kuh, *Linear and nonlinear circuits*, McGraw-Hill, New York, NJ, 1987.
91. P. Arena, S. Baglio, L. Fortuna, G. Manganaro, "Hyperchaos from cellular neural networks," *IEE Electronics Letters*, vol. 31, no. 4, pp. 250–251, 1995.
92. L.O. Chua, M. Komuro, T. Matsumoto, "The double scroll family," *IEEE Transactions on Circuits and Systems*, vol. 33, no. 11, pp. 1072–1118, 1986.
93. A.I. Mees, P.B. Chapman, "Homoclinic and heteroclinic orbits in the double scroll attractor," *IEEE Transactions on Circuits and Systems*, vol. 34, no. 9, pp. 1115–1120, 1987.
94. J.A.K. Suykens, J. Vandewalle, "Generation of n-double scrolls ($n = 1, 2, 3, 4, \ldots$)," *IEEE Transactions on Circuits and Systems–Part I: Fundamental Theory and Applications*, vol. 40, pp. 861–867, 1993.
95. P. Arena, S. Baglio, L. Fortuna, G. Manganaro, "Generation of n-double scrolls via cellular neural networks," *International Journal on Circuit Theory and Applications*, vol. 24, no. 3, pp. 241–252, 1996.
96. T. Katagiri, T. Saito, M. Komuro, "Lost solution and chaos," in *IEEE International Symposium on Circuits and Systems*, 1993, pp. 2616–2619.
97. K. Hat'Ta, T. Saito, "Chaos and bifurcation from a serial resonant circuit including a saturated inductor," in *IEEE International Symposium on Circuits and Systems*, 1993, pp. 2612–2615.
98. M. Itoh, R. Tomiyasu, "Canards and irregular oscillations in a nonlinear circuit," in *IEEE International Symposium on Circuits and Systems*, 1991, pp. 850–853.
99. M. Itoh, L.O. Chua, "Canards and chaos in nonlinear systems," in *IEEE International Symposium on Circuits and Systems*, 1992, pp. 2789–2792.
100. Y. Nishio, A. Ushida, "Multimode chaos in two coupled chaotic oscillators with hard nonlinearities," in *IEEE International Symposium on Circuits and Systems*, 1994, vol. 6, pp. 109–112.
101. Adel S.Sedra, Kenneth C. Smith, *Microelectronic Circuits*, Electrical Engineering Series. Oxford University Press, Oxford, 3rd edition, 1991.
102. Randall L. Geiger, Phillip E. Allen, Noel R. Strader, *VLSI: design techniques for analog and digital circuits*, Electronic Engineering Series. McGraw-Hill, New York, NJ, 1990.
103. Kenneth R. Laker, Willy M.C. Sansen, *Design of Analog Integrated Circuits and Systems*, Electrical and Computer Engineering. McGraw-Hill, New York, NJ, 1994.
104. L.M. Pecora, T.L. Carroll, "Synchronization in chaotic systems," *Physical Review Letters*, vol. 64, pp. 821–824, 1990.
105. Martin Hasler, *Synchronization principles and applications*, pp. 314–327, In *Circuits and Systems Series* [109], 1996.
106. T.L. Carroll, "Communicating with use of filtered, synchronized, chaotic signal," *IEEE Transactions on Circuits and Systems–Part I: Fundamental Theory and Applications*, vol. 42, pp. 105–110, 1995.
107. M.P. Kennedy, M.J. Ogorzałek, Ed.s, "Special issue on chaos synchronization and control: theory and applications," *IEEE Transactions on Circuits and Systems–Part I: Fundamental Theory and Applications*, vol. 44, no. 10, 1997.
108. P. Arena, S. Baglio, L. Fortuna, G. Manganaro, "Dynamics of state controlled CNN's," in *IEEE International Symposium on Circuits and Systems*, 1996, vol. 3, pp. 56–59.
109. Chris Toumazou, Nick Battersby, Sonia Porta, *Circuits and Systems Tutorials '94*, Circuits and Systems Series. IEEE Press, Piscataway, NJ, 1996.

110. K.M. Cuomo, A.V. Oppenheim, S.H. Strogatz, "Synchronization of lorenz-based chaotic circuits with applications to communications," *IEEE Transactions on Circuits and Systems–Part II: Analog and Digital Signal Processing*, vol. 40, no. 10, pp. 626–633, 1993.
111. P. Arena, S. Baglio, L. Fortuna, G. Manganaro, "Experimental signal transmission by using synchronized state controlled cellular neural networks," *IEE Electronics Letters*, vol. 32, no. 4, pp. 362–363, 1996.
112. R. Caponetto, L. Fortuna, M. Lavorgna, G. Manganaro, L. Occhipinti, "Experimental study on chaotic synchronization with non-ideal transmission channel," in *IEEE Global Telecommunications Conference (GLOBECOM'96)*, 1996, vol. 3, pp. 2083–2087.
113. R. Caponetto, L. Fortuna, G. Manganaro, M.G. Xibilia, "Chaotic system identification via genetic algorithm," in *First IEE/IEEE Int.Conference on Genetic Algorithms in Engineering Systems: Innovations and Applications (GALESIA'95)*, 1995, pp. 170–174.
114. R. Caponetto, L. Fortuna, G. Manganaro, M.G. Xibilia, "Synchronization-based nonlinear chaotic circuit identification," in *SPIE's International Symposium on Information, Communications and Computer Technology, Applications and Systems, "Chaotic Circuits for Communication"*, 1995, number 2612, pp. 48–56.
115. R. Caponetto, L. Fortuna, G. Manganaro, M.G. Xibilia, *Chaotic systems identification*, vol. 55 of *IEE Control Engineering*, chapter 6, pp. 118–133, Peter Peregrinus, London, 1997.
116. D.Goldberg, *Genetic Algorithms in Search, optimization and machine learning*, Addison-Wesley, Reading, MA, 1989.
117. Mitsuo Gen, Runwei Cheng, *Genetic algorithms and engineering design*, Engineering Design and Automation. John Wiley & Sons, New York, NJ, 1997.
118. Fan Zou, Josef A. Nossek, "Stability of Cellular Neural Networks with opposite-sign templates," *IEEE Transactions on Circuits and Systems–Part I: Fundamental Theory and Applications*, vol. 38, no. 6, pp. 675–677, 1991.
119. V.I. Krinsky, Ed., *Self-organization: autowaves and structures far from equilibrium*, Springer-Verlag, Berlin, Heidelberg, 1984.
120. P. Arena, S. Baglio, L. Fortuna, G. Manganaro, "Complexity in a two-layer CNN," In A. Rodríguez-Vázquez [18], pp. 127–132.
121. P. Arena, S. Baglio, L. Fortuna, G. Manganaro, "Self-organization in a two-layer CNN," *IEEE Transactions on Circuits and Systems–Part I: Fundamental Theory and Applications*, vol. 45, no. 2, pp. 157–163, 1998.
122. J.D. Murray, Ed., *Mathematical biology*, Springer-Verlag, Berlin, Heidelberg, 1989.
123. G. Nicolis, I. Prigogine, Ed., *Exploring complexity: an introduction*, W.H. Freeman Publishing, 1989.
124. Ladislav Pivka, "Autowaves and spatio-temporal chaos in CNN's - part I: A tutorial," *IEEE Transactions on Circuits and Systems–Part I: Fundamental Theory and Applications*, vol. 42, no. 10, pp. 638–649, 1995.
125. Ladislav Pivka, "Autowaves and spatio-temporal chaos in CNN's - part II: A tutorial," *IEEE Transactions on Circuits and Systems–Part I: Fundamental Theory and Applications*, vol. 42, no. 10, pp. 650–664, 1995.
126. Liviu Goraş, Leon O. Chua, Domine W. Leenaerts, "Turing patterns in CNN's – part I: once over lightly," *IEEE Transactions on Circuits and Systems–Part I: Fundamental Theory and Applications*, vol. 42, no. 10, pp. 602–611, 1995.
127. Liviu Goraş, Leon O. Chua, "Turing patterns in CNN's – part II: equations and behaviors," *IEEE Transactions on Circuits and Systems–Part I: Fundamental Theory and Applications*, vol. 42, no. 10, pp. 612–626, 1995.

128. Liviu Goraş, Leon O. Chua, Ladislav Pivka, "Turing patterns in CNN's - part III: computer simulation results," *IEEE Transactions on Circuits and Systems-Part I: Fundamental Theory and Applications*, vol. 42, no. 10, pp. 627–637, 1995.
129. P. Arena, L. Fortuna, G. Manganaro, "A CNN cell for pattern formation and active wave propagation," in *European Conference on Circuit Theory and Design*, 1997, vol. 1, pp. 371–376.
130. P. Arena, R. Caponetto, L. Fortuna, G. Manganaro, "Cellular Neural Networks to explore complexity," *Soft Computing*, vol. 1, no. 3, pp. 120–136, 1997.
131. S. Kondo, R. Asai, "A reaction-diffusion wave on the skip of the marine Angelfish Pomacanthus," *Nature*, vol. 376, pp. 765–768, 1995.
132. P.S.G. Stein, "Motor systems, with specific reference to the control of locomotion," *Annals Review of Neuroscience*, , no. 1, pp. 61–81, 1978.
133. P. Arena, M. Branciforte, L. Fortuna, "A cnn based experimental frame for patterns and autowaves," *International Journal on Circuit Theory and Applications*, , no. in press, 1998.
134. Arena Paolo, Caponetto Riccardo, Fortuna Luigi, Occhipinti Luigi, "Circuit for motion control realized by analog cells locally connected," *European Patent*, , no. pending, 1998.
135. G.K. Pearson, "Common principles of motor control in vertebrates and invertebrates," *Annals Review of Neuroscience*, , no. 16, pp. 265–297, 1993.
136. R.L. Calabrese, "Oscillation in motor pattern-generating networks," *Current Opinion in Neurobiology*, vol. 5, pp. 816–823, 1995.
137. W.J. Davis, D. Kennedy, "Command interneurons controlling swimmeret movements in the lobster. i. types of effects on motoneurons," *Journal of Neurophysiology*, vol. 35, pp. 1–12, 1972.
138. R.M.Harris-Warrik, E. Marder, "Modulation of neural networks for behavior," *Annals Review of Neuroscience*, , no. 14, pp. 39–57, 1991.
139. E. Niebur, P. Erdos, "Theory of the locomotion of nematodes: control of the somatic motor neurons by interneurons," *Mathematical Bioscience*, vol. 118, pp. 51–82, 1993.
140. A.H. Cohen, GB.Ermentrout, T. Kiemel, N. Kopell, K. Sigvardt, T. L. Williams, "Modelling of intersegmental coordination in the lamprey central pattern generator for locomotion," *Trends in Neuroscience*, vol. 15, pp. 434–438, 1992.
141. G.K. Pearson, "The control of locomotion," *Scientific American*, vol. 235, no. 6, pp. 72–86, 1976.
142. Sunjung Park, Joonho Lim, Soo-Ik Chae, "Discrete-time cellular neural networks using distributed arithmetic," *IEE Electronics Letters*, vol. 31, no. 21, pp. 1851–1852, 1995.
143. K. Halonen, V. Porra, T. Roska, L.O. Chua, *Programmable analogue VLSI CNN chip with local digital logic*, pp. 135–144, In J. Vandewalle [36], 1993.
144. G.C. Cardarilli, F. Sargeni, "Very efficient VLSI implementation of CNN with discrete templates," *IEE Electronics Letters*, vol. 29, no. 14, pp. 1286–1287, 1993.
145. Fausto Sargeni, Vincenzo Bonaiuto, "A fully digitally programmable CNN chip," *IEEE Transactions on Circuits and Systems-Part I: Fundamental Theory and Applications*, vol. 42, no. 11, pp. 741–745, 1995.
146. Mario Salerno, Fausto Sargeni, Vincenzo Bonaiuto, "6 × 6DPCNN: a programmable mixed analogue-digital chip for cellular neural networks," In A. Rodríguez-Vázquez [18], pp. 451–456.

147. Angel Rodríguez-Vázquez, Servando Espejo, Rafael Domínguez-Castro, Jose L. Huertas, Edgar Sánchez-Sinencio, "Current-mode techniques for the implementation of continuous- and discrete-time cellular neural networks," In *IEEE Transactions on Circuits and Systems–Part II: Analog and Digital Signal Processing* [13], pp. 132–146.
148. Joseph E. Varrientos, Edgar Sánchez-Sinencio, Jaime Ramírez-Angulo, "A current-mode cellular neural network implementation," In *IEEE Transactions on Circuits and Systems–Part II: Analog and Digital Signal Processing* [13], pp. 147–155.
149. Ari Paasio, Adam Dawidziuk, Kari Halonen, Veikko Porra, "Fast and compact 16 by 16 cellular neural network implementation," *Analog Integrated Circuits and Signal Processing*, vol. 12, pp. 59–70, 1997.
150. G.F. Dalla Betta, S. Graffi, Zs.M. Kovács, G. Masetti, "CMOS implementation of an analogically programmable cellular neural network," In *IEEE Transactions on Circuits and Systems–Part II: Analog and Digital Signal Processing* [13], pp. 206–215.
151. José M. Cruz, Leon O. Chua, *Design of High-speed, High-density CNN's in CMOS technology*, pp. 117–134, In J. Vandewalle [36], 1993.
152. Josef A. Nossek, Gerhard Seiler, *Cellular Neural Networks: theory and circuit design*, pp. 95–115, In J. Vandewalle [36], 1993.
153. Sa H. Bang, Bing J. Sheu, "A neural network for detection of signals in communication," *IEEE Transactions on Circuits and Systems–Part I: Fundamental Theory and Applications*, vol. 43, no. 8, pp. 644–655, 1996.
154. Peter Kinget, Michiel Steyaert, "Analogue CMOS VLSI implementation of cellular neural networks with continuously programmable templates," in *IEEE International Symposium on Circuits and Systems*, 1994, vol. 6, pp. 367–370.
155. Mancia Anguita, Francesco J. Pelayo, Alberto Prieto, Julio Ortega, "Analog CMOS implementation of a discrete time CNN with programmable cloning templates," In *IEEE Transactions on Circuits and Systems–Part II: Analog and Digital Signal Processing* [13], pp. 215–218.
156. G.C. Cardarilli, R. Lojacono, M. Salerno, F. Sargeni, "VLSI implementation of a cellular neural network with programmable control operator," in *IEEE Midwest Symposium on Circuits and Systems*, 1993, pp. 1089–1092.
157. Gunhee Han, Jose Pineda de Gyvez, Edgar Sánchez-Sinencio, "Optimal manufacturable CNN array size for time multiplexing schemes," In A. Rodríguez-Vázquez [18], pp. 387–392.
158. Ari Paasio, Asko Kananen, Kari Halonen, Veikko Porra, "TOPS information processing on as single chip," *IEEE Circuits and Devices*, vol. 14, no. 3, pp. 13–15, 1998.
159. Rudy J. van de Plassche, Willy M.C. Sansen, Johan H. Huijsing, Ed.s, *Analog Circuit Design*, Kluwer, Dordrecht, 1995.
160. Johan H. Huijsing, Rudy J. van de Plassche, Willy M.C. Sansen, Ed.s, *Analog Circuit Design*, Kluwer, Dordrecht, 1996.
161. J.M. Cruz, L.O. Chua, "A 16x16 cellular neural network universal chip," *Analog Integrated Circuits and Signal Processing*, vol. 15, pp. 227–237, 1998.
162. M. Salerno, F. Sargeni, V. Bonaiuto, "A 6x6 cells interconnection-oriented programmable chip for CNN," *Analog Integrated Circuits and Signal Processing*, vol. 15, pp. 239–250, 1998.
163. Lei Wang, Jose Pineda de Gyvez, Edgar Sánchez-Sinencio, "Time multiplexed color image processing based on a CNN with cell-state outputs," *IEEE Transactions on Very Large Scale Integration (VLSI) Systems*, vol. 6, no. 2, pp. 314–322, 1998.

164. N.C. Battersby C. Toumazou, J.B. Hughes, *Switched-currents an analogue technique for digital technology*, vol. 5 of *Circuits and Systems series*, Peter Peregrinus, London, 1993.
165. Rolf Unbehauen, Andrzej Cichocki, *MOS switched-capacitor and continuous-time integrated circuits and systems*, vol. 13 of *Communications and Control Engineering*, Springer-Verlag, Berlin, Heidelberg, 1989.
166. D.M.W. Leenaerts, G.H.M. Joordens, J.A. Hegt, "A 3.3v 625khz switched-current multiplier," *IEEE International Journal of Solid State Circuits*, vol. 31, no. 9, pp. 1340–1343, 1996.
167. Gunhee Han, Edgar Sánchez-Sinencio, "*CMOS* continuous multipliers: A tutorial," submitted for publication.
168. Gabriele Manganaro, Jose Pineda de Gyvez, "Design and implementation of an algorithmic S^2I switched current multiplier," in *IEEE International Symposium on Circuits and Systems*, 1998.
169. Gabriele Manganaro, Jose Pineda de Gyvez, "A four quadrant S^2I switched-current multiplier," *IEEE Transactions on Circuits and Systems–Part II: Analog and Digital Signal Processing*, vol. 45, no. 7, pp. 791–799, 1998.
170. John B. Hughes, Kenneth W. Moulding, "S^2I: a switched-current technique for high performance," *IEE Electronics Letters*, vol. 29, no. 16, pp. 1400–1401, 1993.
171. Geir E. Sæter, Chris Toumazou, Gaynor Taylor, Kevin Eckersall, Ian M. Bell, "Concurrent self test of switched current circuits based on the S^2I technique," in *IEEE International Symposium on Circuits and Systems*, 1995, vol. 2, pp. 841–844.
172. John B. Hughes, Kenneth W. Moulding, "Enhanced S^2I switched-current cells," in *IEEE International Symposium on Circuits and Systems*, 1996.
173. M. Bracey, W. Redman-White, J. Richardson, J.B. Hughes, "A full Nyquist 15 MS/s 8-b differential switched-current A/D converter," *IEEE International Journal of Solid State Circuits*, vol. 31, no. 7, pp. 945–951, 1996.
174. Terri S. Fiez, Guojin Liang, David J. Allstot, "Switched-current circuit design issues," *IEEE International Journal of Solid State Circuits*, vol. 26, no. 3, pp. 192–201, 1991.
175. Terri S. Fiez, David J. Allstot, "CMOS switched-current ladder filters," *IEEE International Journal of Solid State Circuits*, vol. 25, no. 6, pp. 1360–1367, 1990.
176. Minkyu Song, Yongman Lee, Wonchan Kim, "A clock feedthrough reduction circuit for switched-current systems," *IEEE International Journal of Solid State Circuits*, vol. 28, no. 2, pp. 133–137, 1993.
177. B. Jonsson, S. Eriksson, "New clock feedthrough compensation scheme for switched-current circuits," *IEE Electronics Letters*, vol. 29, no. 16, pp. 1446–1447, 1993.
178. Markus Helfenstein, George S. Moschytz, "Clock feedthrough compensation technique for switched-current circuits," *IEEE Transactions on Circuits and Systems–Part II: Analog and Digital Signal Processing*, vol. 42, no. 3, pp. 229–231, 1995.
179. H.-K. Yang, E.I. El-Masry, "Clock feedthrough analysis and cancellation in current sample/hold circuits," *IEE Proc.-Circuits Devices and Systems*, vol. 141, no. 6, pp. 510–516, 1994.
180. Klaas Bult, Hans Wallinga, "A class of analog CMOS circuits based on the square-law characteristic of an MOS transistor in saturation," *IEEE International Journal of Solid State Circuits*, vol. 22, no. 3, pp. 357–365, 1987.
181. Daniel H. Sheingold, *Nonlinear circuits handbook*, Analog Devices Inc., Norwood, MA, 1976.

182. Fikret Dülger, "A programmable fuzzy controller emulator chip," M.S. thesis, Istanbul Technical Univerisity, 1996.
183. N. Weste, K. Eshraghian, *Principles of CMOS VLSI design*, Addison-Wesley, Reading, MA, 2nd edition, 1993.
184. Sergio Franco, *Design with operational amplifiers and analog integrated circuits*, McGraw-Hill, New York, NJ, 1988.
185. Oscar Moreira-Tamayo, *Analog Systems for Spectral Analysis and Signal Processing*, Ph.D. thesis, Texas A&M University, 1996.
186. O. Moreira-Tamayo, J.Pineda de Gyvez, "Filtering and spectral processing of 1-d signals using cellular neural networks," in *IEEE International Symposium on Circuits and Systems*, 1996, vol. 3, pp. 76–79.
187. Bruno Andó, Salvatore Baglio, Salvatore Graziani, Nicola Pitrone, "A novel analog modular sensor fusion architecture for developing smart structures," in *Proc. Of IEEE IMTC97*, 1997.
188. C. Sidney Burrus, Ramesh A. Gopinath, Haitao Guo, *Wavelets and Wavelet transforms: A primer*, Prentice Hall, New York, NJ, 1998.
189. Gabriele Manganaro, Jose Pineda de Gyvez, "1-D discrete time CNN with multiplexed template hardware," In V.Tavsanoglu [19].
190. C. Toumazou, J.B. Hughes, D.M. Pattullo, "Regulated cascode switched-current memory cell," *IEE Electronics Letters*, vol. 26, no. 5, pp. 303–305, 1990.
191. J. Alvin Connelly, Pyung Choi, *Macromodeling with SPICE*, Prentice Hall, New York, NJ, 1992.
192. Uwe Helmke, John B.Moore, *Optimization and dynamical systems*, Communications and Control Engineering Series. Springer-Verlag, Berlin, Heidelberg, 1994.
193. A. Cichocki, R.Unbehauen, *Neural Networks for optimization and signal processing*, John Wiley & Sons, New York, NJ, 1993.
194. A.A. Andronov, A.A. Vitt, S.E. Khaikin, *Theory of oscillators*, Pergamon Press, Oxford, 1966.
195. T. Roska, L. Kek, Ed., *Analogic CNN Prognram Library Version 6.1*, DNS-5-94 Budapest, 1994.
196. P.L. Venatier, F. Werblin, T. Roska, L.O. Chua, "Analogic cnn algorithms for some image compressions and restoration tasks," *IEEE Transactions on Circuits and Systems–Part I: Fundamental Theory and Applications*, vol. 42, no. 5, pp. 279–284, 1995.
197. L.O. Chua, P. Thiran, "An analytic method for designing simple cellular neural networks," *IEEE Transactions on Circuits and Systems–Part I: Fundamental Theory and Applications*, vol. 38, no. 11, pp. 1332–1341, 1991.
198. A. Zarandy, A. Stoffels, T. Roska, L.O.Chua, "Implementation of binary and grey-scale mathematical morphology on the cnn universal machine," *IEEE Transactions on Circuits and Systems–Part I: Fundamental Theory and Applications*, vol. 45, no. 2, pp. 163–168, 1998.

Index

activator–inhibitor 120, 121, 150
advection 36, 38, 39
air quality modeling 35–39
analog multiplier 164, 173–180
– bandwidth 185–186
– input feedthrough 185
– internal trim error 182, 184
analog processor 10, 29
artificial locomotion 149–160
– control 154
aspect ratio 181
autowaves 116–120, 135–136, 140–142, 149, 152, 154, 159–160

Bendixon's criterion 109, 222
bifurcation 92, 108–110, 221
– Hopf 109, 221, 223
boundary conditions 7, 22–23, 36
– fixed (Dirichlet) 23
– periodic 23
– zero-flux (Neuman) 23, 115, 123, 124, 138
bulk effect 174

canards 70
cascode 165, 175, 190, 191
cell 4–5
– boundary 7
– dynamics 21
– implementation 192–199
– inner 7
– missing 23
chaos synchronization 79–104
– cascade 81
– cryptography 104
– definition 79
– inverse system 81
– master-slave 80, 81, 93
– non-ideal channel effects 86–87
– Pecora–Carroll 80–81, 93
– response system 80
– transceiver 101

chaos theory 43
chaotic system identification 93–101
charge injection 168, 194
Chua's diode 47
clock feedthrough 168–171
CNN
– Chua and Yang model 3–15, 105, 132, 165
– definition 20–23
– delay 15–17, 22
– discrete-time 17–19, 189
– dynamic rules 7
– generalized 15–20
– graph 12
– multilayer 9, 39, 112
– multiple neighborhood size 17
– nonlinear 15–17, 22
– nonuniform processor 17
– stability 11
– state controlled 44
– universal machine 19–20, 151, 160
coaxial cable 86
common gate amplifier 194
communication channel
– adapted 86
– characteristic impedance 86
– disturbed 87
– noise 87
– non-ideal 86
– not adapted 86
complexity 132, 133
convolution operator 7, 9
cross-correlation 91
current mirror 164
– adjustable 174, 179, 181
– cascode 190, 192
current offset 171, 175, 176, 179–180, 184, 191

diffusion 36, 37, 121
disturbances 91
double scroll 48, 49

dual algorithms   10, 20, 30, 32–35
dynamics   21
– propagation effect   6, 116
– slow–fast   70, 110–112, 114
– structural stability   101, 220

eigenfunction   123–124, 143
eigenspace   60
eigenvalue   60, 123, 143, 213
Einstein summation   18, 21, 206

fixed point   60, 107–109, 121, 213–217
– characteristic equation   108
– definition   210
– hyperbolic   213, 216
– non-hyperbolic   108, 218
– real   107, 108
– virtual   107–109, 111

genetic algorithms   93, 95

Hartman–Großman theorem   106, 108, 213, 216
heteroclinic orbit   110, 221
– definition   215
homoclinic orbit   60, 219, 221
– definition   214
hyperchaos   55, 220

image processing   26
– gray scale image   10
– noise removal   26, 28, 29
initial condition constraint   5
input constraint   5

Jacobian matrix   80, 108, 121, 123, 213

level shifter   194
limit cycle   109–110, 136, 217–219
– definition   211
linear $I - V$ converter   194, 199
low voltage circuits   165
Lyapunov exponent   55
– conditional   80
– definition   219–220

mechatronic   149
Monte Carlo analysis   192
MOS switches   168, 182, 191
– CMOS   182
– current steering   176
multi–chip   165

n-double scroll   58–66

neighborhood set   5–6, 21, 190
– radius   5
NMR   28–35
– TOCSY   30
noise   26, 87, 129, 140, 142
– generator   87
– pink   91
– white   91, 201
nonlinear distortion   180
nonlinear inductor   68

operational transconductance amplifier   164
optimization   95
oscillator
– Bonhoeffer–Van der Pol   70
– Chua   47, 49, 77, 93, 95, 105, 114, 132
– Colpitts   51
– coupled   72, 74, 152
– hysteresis hyperchaotic   55
– n-double scroll   58
– non-autonomous   66, 68

parametric uncertainties   127–130
partial differential equations   35, 112, 126, 133, 149
patterns   112, 120–127
– checkerboard   145, 158
– Turing   121–124, 130, 136, 142–148, 154
– Turing space   122, 123
phase plane
– linear region   106
– partial saturation region   106, 110
– saturation region   106, 110
Poincaré–Bendixon theorem   110, 222
programmable chaos   78
PWL function   58, 108, 206

reaction–diffusion   112, 115, 120, 126, 132, 133, 149, 150, 154
replica circuits   168, 192
reverberator   117, 118
ring counter   195
robot
– READIBELT   159–160
– REXABOT   154–159
– WORMBOT   152–154

S^2I
– coarse memory   169
– fine memory   169
– memory   168–173, 190–191
– multiplier   173–180, 193–194

self–organization   112, 132
signal to noise ratio   186, 201
Šilnikov theorem   60, 221–222
spatial mode   122
spiral waves   117, 127–129
squarer circuit   173–174
stability   212–220
– complete   8, 10, 110, 215
– linear CNN   11
state controlled CNN   44–47, 79, 81–92
– definition   45
– discrete component realization   45
switched-current   167, 189
synaptic law   21
– implementation   193

tapped delay line   189–192
template   7–8, 10, 225–258
– cell–linking   12
– circulant diffusion   114
– cloning   7
– control   6, 21, 190, 199, 203
– design   19
– feedback   6, 21, 114, 190, 203
– Laplacian   114, 122, 134, 138–139
– learning   19
– negative cell–linking   13

– non–reciprocal   11
– nonlinear   22
– opposite–sign   12–14
– positive   11, 14
– positive cell–linking   12, 13
– space invariant   5, 7
– strictly sign–symmetric   12
– symmetric   8, 11, 19
– value–asymmetric   13
time multiplexing   165, 189
topologic conjugacy   48, 70
– definition   220
topology   205–206
total harmonic distortion   185
traveling wavefronts   115–120, 132, 149

variational equation   80, 218
VLSI implementation
– analog and mixed–signal   164
– digital   164
– fixed template   164
– programmable template   164, 189

wavelet   201

yield and reliability   165

Springer and the environment

At Springer we firmly believe that an international science publisher has a special obligation to the environment, and our corporate policies consistently reflect this conviction.

We also expect our business partners – paper mills, printers, packaging manufacturers, etc. – to commit themselves to using materials and production processes that do not harm the environment. The paper in this book is made from low- or no-chlorine pulp and is acid free, in conformance with international standards for paper permanency.